l'actualité chimique
LIVRES

COLLECTION DIRECTED BY **Paul RIGNY**

RADIATION CHEMISTRY

FROM BASICS TO APPLICATIONS IN MATERIAL AND LIFE SCIENCES

EDITED BY :

Mélanie SPOTHEIM-MAURIZOT,
Mehran MOSTAFAVI,
Thierry DOUKI,
Jacqueline BELLONI

EDP SCIENCES

17 avenue du Hoggar
Parc d'activité de Courtaboeuf, BP 112
91944 Les Ulis Cedex A, France

Couverture, maquette intérieure et mise en page : Thierry Gourdin

Imprimé en France

ISBN : 978-2-7598-0024-7

Contents

Foreword

"L'Actualité Chimique" is a monthly scientific journal meant to convey information on progress in the chemical sciences to a public endowed with a certain ability to master scientific matters. The articles were written by scientists who took time out of their laboratories to explain their studies and their knowledge with a pedagogy and an appeal suitable for non-specialists. Mostly written in French, it creates a bond in the chemical community in French-speaking countries where it is very much appreciated by scientists, teachers and engineers.

However, the scope of the journal implies some limits that we want to erase with this new collection "L'Actualité Chimique – Livres", which will be complementary in two directions: the first one is illustrated by the present book, as it addresses readers more specialized than the journal usually does, and being written in English, it has the ambition of attracting attention worldwide on a field of chemistry where recent progress is noted. The second direction that will be found in the new collection is, in contrast, that of disseminating the progress of chemistry for the benefit of a large, French-speaking, not necessarily professional public. The first trend will produce books that we will find in many laboratories; books produced according to the second trend will instead be largely found in public libraries, in schools or even in the homes of scientifically curious people.

This first volume of "L'Actualité Chimique – Livres" is of the first kind and devotes itself to **Radiation Chemistry – From basic science to applications in biology and material science**.

This field of research is undergoing a true and fruitful rejuvenation. Already active in the mid - 20th century, the development of this scientific field had been somewhat slowed down by the high cost of short-pulse particle accelerators and specialized construction. Recent progress in instrumentation *e.g.*: the shaping of picosecond radiation pulses, faster time-resolved detection techniques, and powerful molecular structure determination techniques, has coincided to enhance the capacity of radiation chemistry sufficiently to warrant new investments and the start of new laboratories. Radiation chemistry today is responsible for major progress in the understanding of the elementary chemical event and powerful enough to unravel the mechanisms of the damage induced by radiation to living matter (a question of great concern in the public) or the transformations induced in irradiated materials.

These aspects are developed in the book by international-level specialists and will be of interest to scientists who are starting in the field, to more experienced ones, and also to students and teachers; it will also be very useful to many professionals who apply or deal with radiation in their activities to improve materials or to avoid radiation-induced damage to them.

Paul RIGNY
Chief Editor of "L'Actualité Chimique"
March 2008

Preface

Radiation chemistry deals with the chemical reactions resulting from the interaction of high-energy photons or particles with matter. Such radiation possesses energy high enough to induce ionisation of the components of the material and the breaking and building of chemical bonds.

In the present volume, our purpose is to familiarise the larger communities of students and chemists in other specialities with this relatively little-known but essential domain of chemistry. The covered topics range from the basics (primary phenomena and mechanisms) to the broad fields of their application. Understanding radiation-induced chemical and biochemical reactions is essential for improving existing processes and developing new ones.

Therefore we have called upon internationally recognized experts who kindly agreed to contribute to this volume with clear, instructive and pedagogically presented chapters abundantly illustrated with attractive colour figures.

The first chapters of Part I deal with primary radiolytic phenomena and describe recent developments at the facilities used to create radiation-induced species, as well as the most advanced methods for their detection and study. The mechanisms of radiation-matter interactions and their consequences for the physical chemistry of liquids and solutions are discussed.

Part II describes specific mechanisms and key processes in space and nuclear chemistry, as well as in material sciences and pharmaceutical and food chemistry. The high energy of ionizing radiation offers the specific advantage of easy and homogeneous sample penetration. Therefore, by targeting chemical bonds at room temperature via cost-competitive, chemical additive-free processes, ionizing radiation can be used for many interesting purposes. For example, thanks to the understanding of radiation-induced nucleation/growth processes, the final size and properties of metal nanoclusters can be controlled for applications in catalysis, electronics, and photography. High-performance polymeric materials, obtained using the cleavage or the formation of chemical bonds by irradiation, have a multitude of uses in everyday life. Remediation of waste-water requires the destruction of toxic chemicals, which is efficiently accomplished by irradiation. The use of ionizing radiation for food treatment and the sterilization of pharmaceuticals and medical devices operate *via* the efficient destruction of micro-organisms, but they require systematic confirmation of the absence of any toxic molecules that could be produced during irradiation.

The search for new means of improving the success of cancer radiotherapy motivates an increasing interest in the chemical mechanisms underlying radiobiology. Part III of the volume is devoted to this very active research domain, and in particular, to studies of the damage induced by ionizing radiation in biomolecules (DNA, proteins, lipids). Answers are given as to what are the mechanisms of the reactions in DNA and other biomolecules following the initial ionization and excitation, how they can be simulated by computational models, how radiation-induced lesions are repaired or prevented, and finally how this improved knowledge is used to specifically eradicate tumours (cancer radiotherapy).

With no pretence of exhaustively covering in detail all the topics of radiation chemistry, this volume will hopefully fulfil the expectation of the reader to learn about a domain that we consider a most exciting and promising area of chemistry.

We cannot end this preface without addressing our thanks to Yann Gauduel and Paul Rigny, respectively former and present Chief Editors of " L'Actualité Chimique", who solicited and accompanied us in the realisation of this work. All the other members of the editorial board of the journal and of EDP Sciences are equally warmly thanked.

Mélanie SPOTHEIM-MAURIZOT,
Mehran MOSTAFAVI,
Thierry DOUKI,
Jacqueline BELLONI
March 2008

List of Authors

AMOURETTE Christine
Service de Santé des Armées / Centre de Recherches
24, Av. des Maquis du Grésivaudan - 38702 La Tronche / FRANCE *camourette@crssa.net*

BALDACCHINO Gérard
Commissariat à l'Énergie Atomique / Laboratoire de Radiolyse
Bât. 546 CEA/Saclay - 91191 Gif-sur-Yvette / FRANCE *gerard.baldacchino@cea.fr*

BALOSSO Jacques
CHU A. Michallon / Service de Cancérologie-Radiothérapie
BP 217 - 38043 Grenoble cedex 9 / FRANCE *jbalosso@chu-grenoble.fr*

BELLONI Jacqueline
CNRS-Université Paris-Sud / Laboratoire de Chimie Physique-ELYSE
Bât. 349 Université Paris-Sud - 91405 Orsay / FRANCE *jacqueline.belloni@lcp.u-psud.fr*

BERLIN Yuri
Northwestern University / Department of Chemistry
2145 Sheridan Road - Evanston, IL 60208-3113 / USA *berlin@chem.northwestern.edu*

BERNHARD William
University of Rochester / Department of Biochemistry and Biophysics
575 Elmwood avenue, Box 712
Rochester, NY 14642 / USA *william_bernhard@urmc.rochester.edu*

BOBROWSKI Krysztof
Institute of Nuclear Chemistry and Technology
Dept of Radiation Chemistry and Technology
Dorodna 16, 03-195 Warsaw / POLAND *kris@ichtj.waw.pl*

BOUNIOL Pascal
Commissariat à l'Énergie Atomique Saclay / Laboratoire des Bétons
Bât. 158 - 91191 Gif-sur-Yvette / FRANCE *pascal.bouniol@cea.fr*

BUXTON Georges
1A Hollin Crescent - Leeds LS16 5ND / UNITED KINGDOM *george.buxton@talktalk.net*

CADET Jean
CEA Grenoble / DRFMC/SCIB
17 rue des Martyrs - 38054 Grenoble cedex 9 / FRANCE *jcadet@cea.fr*

CASTAING Bertrand
CNRS / Centre de Biophysique Moléculaire
Rue Charles Sadron - 45071 Orléans cedex 2 / FRANCE *castaing@cnrs-orleans.fr*

COQUERET Xavier
Université de Reims Champagne-Ardenne / Réactions Sélectives et Applications
Europol'Agro - 51687 Reims cedex 2 / FRANCE *xavier.coqueret@univ-reims.fr*

DAVIDKOVA Maria
Nuclear Physics Intitute / Dept of Radiation Dosimetry
Na Truhlarce 39/64 - 18086 Praha 8 / CZECH REPUBLIC *davidkova@ujf.cas.cz*

DAWES Anita
The Open University / Department of Physics and Astronomy
Walton Hall - Milton Keynes MK7 6AA / UNITED KINGDOM *a.dawes@open.ac.uk*

DOUKI Thierry
CEA Grenoble / DRFMC/SCIB
17 rue des Martyrs - 38054 Grenoble cedex 9 / FRANCE *tdouki@cea.fr*

EMMI Salvatore Silvano
CNR / ISOF
Via P. Gobetti, 101 - 40129 Bologna / ITALY *emmi@isof.cnr.it*

FORAY Nicolas
INSERM / European Synchrotron Radiation Facility
BP 220-38043 Grenoble cedex / FRANCE *foray@esrf.fr*

GARDES-ALBERT Monique
Université René Descartes Paris V / Laboratoire de Chimie Physique
45 Rue des Saints-Pères - 75270 Paris cedex 06 / FRANCE *Monique.Gardes@univ-paris5.fr*

HICKEL Bernard
Commissariat à l'Énergie Atomique / Laboratoire de Radiolyse
Bât. 546 CEA/Saclay - 91191 Gif-sur-Yvette / FRANCE *b.hickel@wanadoo.fr*

HOLTOM Philip
The Open University / Department of Physics and Astronomy
Walton Hall - Milton Keynes MK7 6AA / UNITED KINGDOM *p.holtom@open.ac.uk*

HOUEE-LEVIN Chantal
Université Paris-Sud / Laboratoire de Chimie Physique
bât. 350 - 91405 Orsay / FRANCE *chantal.houee-levin@lcp.u-psud.fr*

KISTER Jacky
Université Paul Cezanne Aix-Marseille / Systèmes chimiques complexes
CNRS-UMR Faculté de St-Jérôme 6171
13397 Marseille cedex 20 / FRANCE *Jacky.kister@univ-cezanne.fr*

LAMPRE Isabelle
Université Paris-Sud / Laboratoire de Chimie Physique
Bât. 349 - 91405 Orsay / FRANCE *isabelle.lampre@lcp.u-psud.fr*

MASON Nigel J
The Open University / Department of Physics and Astronomy
Walton Hall - Milton Keynes MK7 6AA / UNITED KINGDOM *n.j.mason@open.ac.uk*

MOSTAFAVI Mehran
Physical Chemistry Institute, Centre ELYSE-CLIO
CNRS / University Paris-Sud, Orsay / FRANCE *mehran.mostafavi@lcp.u-psud.fr*

PROUILLAC Caroline
Université Paul Sabatier / Laboratoire Hétérochimie Fondamentale et Appliquée
118 Route de Narbonne - 31062 Toulouse / FRANCE *prouilla@chimie.ups--tlse.fr*

RAFFI Jacques
CEA Université Paul Cezanne Aix-Marseille III
Laboratoire de Radiolyse de la Matière organique
Avenue Escadrille Normandie-Niémen
13397 Marseille cedex 20 / FRANCE *j.raffi@univ.u-3mrs.fr*

REMITA Hynd
Université Paris-Sud / Laboratoire de Chimie Physique
Bât. 349 - 91405 Orsay / FRANCE *hynd.remita@lcp.u-psud.fr*

RIMA Ghassoub
Université Paul Sabatier / Laboratoire Hétérochimie Fondamentale et Appliquée
118 Route de Narbonne - 31062 Toulouse / FRANCE *rima@chimie.ups--tlse.fr*

SAGE Evelyne
CNRS / Institut Curie
Centre Universitaire - 91405 Orsay / FRANCE *evelyne.Sage@curie.u-psud.fr*

SEVILLA Michael D.
Oakland University / Department of Chemistry
Rochester - 48309 Michigan / USA *sevilla@oakland.edu*

SIEBBELES Laurens
Delft University of Technology
Mekelweg 15 - Delft 2629 JB / THE NETHERLANDS *L.D.A.Siebbeles@tnw.tudelft.nl*

SPOTHEIM-MAURIZOT Mélanie
INSERM senior scientist, Molecular Biophysics
Centre – CNRS, Orléans / FRANCE *spotheim@cnrs-orleans.fr*

TAKACS Erzsebet
Hungarian Academy of Sciences / Institute of isotopes and surface chemistry
PO Box 77 - 1525 Budapest / HUNGARY *takacs@iki.kfki.hu*

TILQUIN Bernard
Université Catholique Louvain 72-30
Unité d'Analyse Chimique et Physico-Chimique des Médicaments
72 Avenue E. Mounier - B1200 Bruxelles / BELGIQUE *tilquin@cham.ucl.ac.be*

WISHART James
Brookhaven National Laboratory / Chemistry Department
Upton, NY 11973 / USA *wishart@bnl.gov*

Part I

Primary radiation-induced phenomena

Chapter 1

An overview of the radiation chemistry of liquids

George V. BUXTON

Introduction

Radiation chemistry is the chemistry initiated by the interaction of high-energy photons and atomic particles with matter, so-called ionising radiation. As a method of generating free radicals for applications in general chemistry the most commonly used sources of ionising radiation are ^{60}Co γ-rays, which are photons having energies of 1.17 and 1.33 MeV (1 eV = 1.6×10^{-19} J), or fast electrons from an accelerator with energies typically in the range 2-20 MeV. The dose absorbed by the material is expressed in grays (1 Gy = 1 J kg^{-1}) and the dose rate in Gy s^{-1}. In each case the result of the interaction of high energy particles with molecules is the ejection of a single electron, called a secondary electron which itself may have sufficient energy to cause further ionisations, but which rapidly (< 10^{-12} s) reaches thermal equilibrium with the liquid and becomes trapped as a so-called solvated electron (e_s^-). In this way, stable molecules (M) are converted into solvated electrons and highly reactive free radicals ($M^{\cdot+}$):

$$M \rightsquigarrow M^{\cdot+} + e_s^- \quad (1)$$

Pulse radiolysis experiments have provided clear evidence for solvated electrons in both polar (water, alcohols, etc.) and non-polar (alkanes) liquids through their optical absorption spectra.

An important characteristic of ionising radiation is that it is absorbed non-selectively so that molecules are ionised according to their relative abundance in the medium of interest.

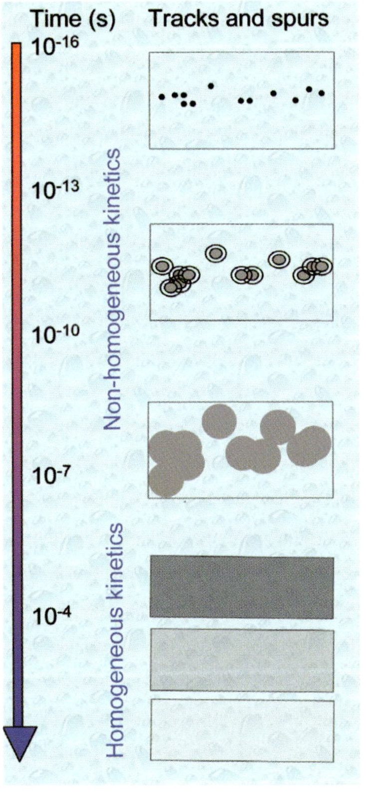

Figure 1: *Initial non-homogeneous spatial distribution of ionisation events in spurs along the track, and evolution with time by diffusion and reaction up to homogeneous distribution at ≈10^{-7} s. A qualitative, albeit rather crude, way of visualising the distribution and evolution of spurs along the track is to recall the pattern of splashes produced when you throw a small flat stone and make it skip across the smooth surface of a pond. Initially the splashes are well-separated (low LET) but as the stone loses its kinetic energy they get closer together and finally merge (high LET) at the end of its trajectory, which is generally not a straight line. With the elapse of time the circular ripples produced by each splash expand and eventually overlap with those of their neighbours, which is analogous to homogeneous distribution along the track of an ionising particle in the radiolysis of water.*

For example, in dilute solution ([solute] ≤ 0.1 mol dm^{-3}) the ionised molecules (M) are essentially those of the solvent so that knowledge of the radiation chemistry of the solvent is of paramount importance for chemistry initiated by ionising radiation. However, the rate of energy loss of an energetic particle passing through a liquid is non-uniform; it increases as the kinetic energy of the particle decreases. The rate of energy loss per unit length of track of the particle is known as the stopping power or linear energy transfer (LET). When the ionising particle is a high energy electron, the ionisation events occur in small clusters called spurs that are widely separated along the track and the radiation is classed as low LET (**Fig. 1**). For energetic nuclear ions, *e.g.* H^+, He^{2+}, Li^+, etc., the rate of energy loss is much higher so that the spurs are much closer together and can overlap to form cylindrical columns of ionised molecules in the liquid. In this case the radiation is classed as high LET. Actually, low LET radiation has a high LET component due to low-energy secondary electrons, and high LET radiation has a low LET component due to high-energy secondary electrons, so it is common practice to classify types of radiation by their average LET over the length of the particle track. In the case of ^{60}Co γ-rays it is the secondary electrons that cause most of the ionisations but the γ-photons, being uncharged, are very much more penetrative than electrons of the same energy.

Radiation chemistry of water

It was in early studies of water radiolysis that the concept of spurs was developed [1]. The current state of knowledge can be summarised by the following reactions (the time by which each event is estimated to be complete is indicated) (**Fig. 2**) [2]:

$$H_2O \rightsquigarrow H_2O^{\cdot+} + e^- \quad \text{and} \quad H_2O^* \qquad 10^{-16}\,s \qquad (2)$$
$$H_2O^{\cdot+} + H_2O \longrightarrow {}^{\cdot}OH + H_3O^+ \qquad 10^{-14}\,s \qquad (3)$$
$$H_2O^* \longrightarrow H^{\cdot} + {}^{\cdot}OH, \quad H_2 + O^{\cdot} \qquad 10^{-13}\,s \qquad (4)$$
$$e^- + nH_2O \longrightarrow e_{aq}^- \qquad 10^{-12}\,s \qquad (5)$$

Reaction (2) represents ionisation and electronic excitation of water molecules; this occurs on the timescale of an electronic transition. The positive radical ion $H_2O^{\cdot+}$ is believed to undergo the ion molecule reaction (3) in ~ 10^{-14} s. The electronically excited states H_2O^* are known to dissociate in the vapour phase in reaction (4), and the electron released in reaction (1) is known to be solvated by ~ 10^{-12} s. At this time, for low LET radiation such as ^{60}Co γ-rays and fast electrons from an accelerator, the products of reactions (3) – (5) are clustered together in small widely separated spurs, which on average contain 2 to 3 ion pairs. Next, these products begin to diffuse randomly, with the result that a fraction of them encounter one another and react to form molecular and secondary radical products, while the remainder escape into the bulk liquid and effectively become homogeneously distributed with respect to

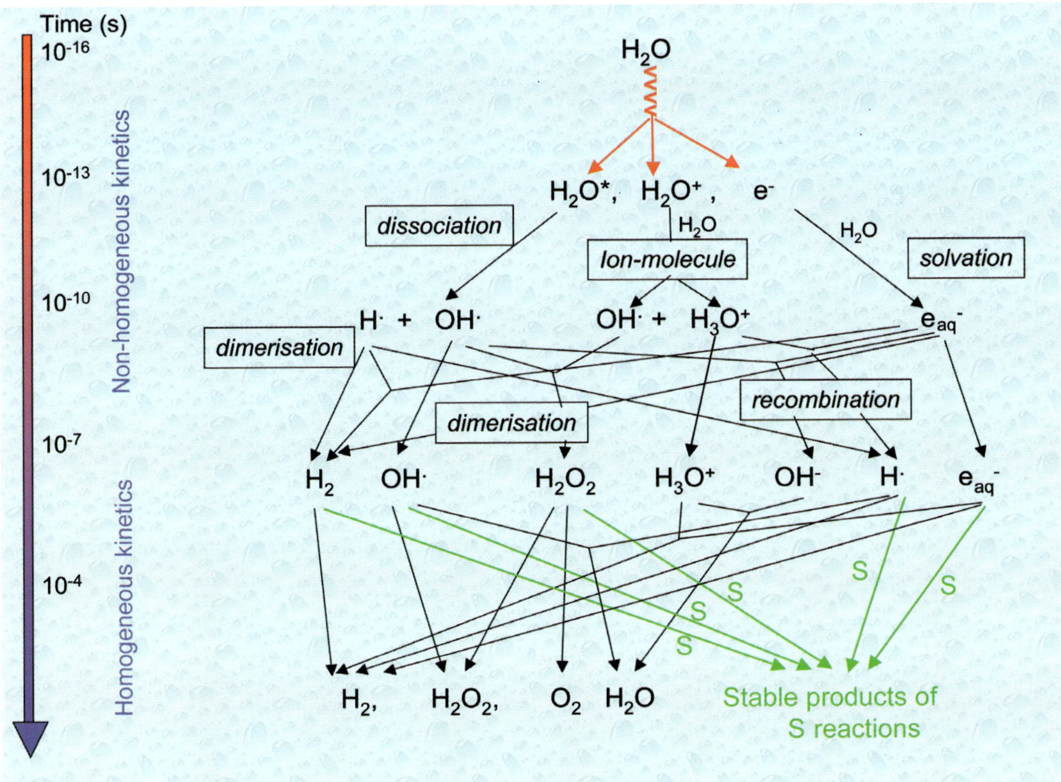

Figure 2 : *Scheme of reactions of transient species produced by irradiation in water without or with a diluted solute S acting as a radical scavenger.* The processes in reaction (2) are shown in red.

reaction with solutes acting as radical scavengers (**Fig. 1**). These spur processes of reaction and diffusive-escape are complete within ~ 10^{-7} s, at which time the radiolysis of water for low LET radiation (*e.g.* 0.23 eV nm^{-1}) can be represented as reaction (6) [2] (**Fig. 2**):

$$H_2O \rightsquigarrow 0.28\ e_{aq}^-,\ 0.062\ H^\bullet,\ 0.28\ {}^\bullet OH,\ 0.047\ H_2,\ 0.073\ H_2O_2,\ 0.28\ H_3O^+ \quad (6)$$

Here the numbers are the radiation chemical yields (known as *G* values) in units of μmol J^{-1}. In the early literature (and sometimes still today) *G* values are quoted as molecules/100 eV, or just as a number in which case the units molecules/100 eV are implied. The conversion factor is:

$$1\ \text{molecule}/100\ \text{eV} = 1.036 \times 10^{-7}\ \text{mol J}^{-1}\ (\text{or } 0.1036\ \mu\text{mol J}^{-1})$$

For historical reasons the *G* values in reaction (6) are known as *primary* yields and it is estimated that about 40% of the *initial* yields (G^o) produced in reactions (3) – (5) are consumed by the spur reactions, *i.e.* $G^o(e_{aq}^-)$ ~ 0.5 μmol J^{-1}. The spur reactions are listed in **Table 1**.

Table 1. Spur Reactions in Water [2].

Reaction	k (10^{10} dm^3 mol^{-1} s^{-1})
$e_{aq}^- + e_{aq}^- \longrightarrow H_2 + 2OH^-$	0.54
$e_{aq}^- + {}^\bullet OH \longrightarrow OH^-$	3.0
$e_{aq}^- + H_3O^+ \longrightarrow H^\bullet + H_2O$	2.3
$e_{aq}^- + H^\bullet \longrightarrow H_2 + OH^-$	2.5
$H^\bullet + H^\bullet \longrightarrow H_2$	1.3
${}^\bullet OH + {}^\bullet OH \longrightarrow H_2O_2$	0.53
${}^\bullet OH + H^\bullet \longrightarrow H_2O$	3.2
$H_3O^+ + OH^- \longrightarrow 2H_2O$	14.3

The extent of these reactions increases with increasing LET due to overlapping spurs; for example, for LET = 108 eV nm^{-1} the primary yields (*G* values) become those in reaction (6'):

$$H_2O \rightsquigarrow 0.044\ e_{aq}^-,\ 0.028\ H^\bullet,\ 0.056\ {}^\bullet OH,\ 0.11\ H_2,\ 0.11\ H_2O_2,\ 0.044\ H_3O^+ \quad (6')$$

Radiation chemistry of organic liquids

The complexity of organic molecules compared to H_2O means that there is a greater variety of products arising from reactions following the ionisation event in reaction (1).

Furthermore, the relative permittivity ε_r (or dielectric constant) of the liquid is an important parameter. The probability P of an electron (e_s^-) which is thermalised at a distance r metres from its geminate positive ion ($M^{\bullet+}$) escaping recombination with it is $\exp(-r_c/r)$ where r_c is the distance at which the Coulomb potential between e_s^- and $M^{\bullet+}$ is equal to thermal energy (kT) and is given by the Onsager expression $r_c = e^2/4\pi\varepsilon_0\varepsilon_r k_B T$, where e is the elementary charge ($e = 1.6 \times 10^{-19}$ C), ε_0 is the vacuum permittivity ($\varepsilon_0 = 8.854 \times 10^{-12}$ J^{-1} C^2 m^{-1}), k_B is the Boltzmann constant ($k_B = 1.38 \times 10^{-23}$ J K^{-1}) and T is temperature in Kelvin. In water at 298 K, where $\varepsilon_r = 78.5$, $r_c \sim 0.7$ nm so that the probability of recombination of an electron with its positive ion is small since r is a few nanometres. In contrast to this in a hydrocarbon such as cyclohexane or benzene, for which $\varepsilon_r \sim 2$, $r_c \sim 30$ nm, most of the electrons and ions produced in reaction (1) recombine. Those ions that escape geminate recombination are called free ions. The yields of free ions (G_{fi}) in these hydrocarbons are 0.005 µmol J^{-1} (benzene) and 0.015 µmol J^{-1} (cyclohexane). More polar organic molecules such as alcohols show larger values of G_{fi}. For example for ethanol with $\varepsilon_r = 24.3$, $G_{fi} = 0.18$ µmol J^{-1}. In simple terms, one should be able to estimate thermalisation distances r from G_{gi}, but other factors have to be considered. Thus the values of G_{fi} for nonpolar liquids tend to increase with the sphericity of the molecules, ranging from 0.005 µmol J^{-1} for benzene to 0.11 µmol J^{-1} for neopentane [3].

Aqueous solutions

The largest contribution of radiation chemical techniques to general free-radical chemistry has been made in aqueous solution because they provide a very convenient way of generating an enormous variety of highly reactive species which cannot readily be generated by thermal or photochemical methods. In particular, the technique of pulse radiolysis has provided a wealth of kinetic and mechanistic information in inorganic, organic, and biochemistry [4,5].

Properties of the primary radicals from water radiolysis

As shown in reaction (6), the principal primary radicals are the hydrated electron (e_{aq}^-) which is a powerful reductant (standard reduction potential $E^\circ = -2.87$ V [6]) and the hydroxyl radical ($^\bullet OH$) which is a powerful oxidant ($E^\circ(^\bullet OH/OH^-) = 1.90$ V in neutral solution, and $E^\circ(H^+, {}^\bullet OH/H_2O) = 2.72$ V in acidic solution). The hydrogen atom ($H^\bullet$) is not an important species in neutral or alkaline solution, but it becomes the major reductant ($E^\circ(H^+/H^\bullet) = -2.31$ V) in acidic solution through reaction (7):

$$e_{aq}^- + H^+ \longrightarrow H^\bullet \qquad (7)$$

The hydrated electron may be visualised as an excess electron surrounded by oriented water molecules (Chapter 3), although the precise details of its structure are not yet settled. It reacts by transferring into a vacant orbital of an acceptor molecule or ion:

$$e_{aq}^{-} + S^{n} \longrightarrow S^{n-1} \qquad (8)$$

Here n is the charge on the acceptor S. Rate constants for reaction (8) range from 16 l mol^{-1} s^{-1} ($S = H_2O$) up to the diffusion-controlled limit, although the measured activation energy is invariably small (6-30 kJ mol^{-1}), which suggests that the entropy of activation is the dominant kinetic parameter. This can be rationalised in terms of the accessibility of a vacant orbital on S for the electron to enter and explains why solvated electrons are sufficiently long-lived to be observed by pulse radiolysis in liquids such as water, simple alcohols, ethers, amines and alkanes whose molecules have no low-energy vacant orbitals. Organic molecules with low-energy vacant orbitals (*e.g.* most aromatics, halides, thiols, disulfides, carbonyl- and nitro-compounds) all react rapidly, e_{aq}^{-} acting as a nucleophile. Thus its reactivity is greatly enhanced by electron-withdrawing substituents adjacent to double bonds or attached to aromatic rings. Although the first step is electron addition to a vacant orbital, in some cases, notably with organic halides (RX), bond breakage occurs very rapidly and the overall reaction essentially appears as a dissociative electron capture process with elimination of the halide ion X^{-}:

$$e_{aq}^{-} + RX \longrightarrow (RX^{\bullet -}) \longrightarrow R^{\bullet} + X^{-} \qquad (9)$$

Inorganic ions react rapidly with e_{aq}^{-} when their reduction potentials are more positive than – 2.9 V. In general, cations react faster than anions, reflecting the effect of the negative charge of e_{aq}^{-}.

The hydroxyl radical readily oxidises inorganic ions and the reaction is often represented as a simple one-electron transfer process:

$$^{\bullet}OH + S^{n} \longrightarrow S^{n+1} + OH^{-} \qquad (10)$$

However, there is abundant evidence from pulse radiolysis studies to indicate that $^{\bullet}OH$ forms an adduct with the inorganic ion in the case of halides (X^{-}) and simple metal cations (M_{aq}^{n+}) and it is suggested that this is the reaction path for most oxidisable inorganic ions. A good example for halides is the reaction with chloride ion which only occurs in acidic solution because $E^{o}(Cl^{\bullet}/Cl^{-}) = 2.41$ V. The mechanism comprises reactions (11) and (12) and the intermediate $HOCl^{\bullet -}$ has been identified by its absorption spectrum [7]:

$$^{\bullet}OH + Cl^{-} \rightleftharpoons HOCl^{\bullet -} \qquad (11)$$

$$HOCl^{\bullet -} + H^+ \longrightarrow Cl^\bullet + H_2O \qquad (12)$$

Similarly, pulse radiolysis studies of the oxidation of simple aquo-complexes of metal ions by $^\bullet OH$ reveal that the reaction can be represented by the following steps:

$$^\bullet OH + M^{n+}_{aq} \longrightarrow [M^{n+}_{aq}, {}^\bullet OH] \longrightarrow M^{(n+1)+}_{aq} + OH^- \qquad (13)$$

The hydroxyl radical reacts with many inorganic anions at near-diffusion controlled rates ($k \sim 10^{10}$ dm^3 mol^{-1} s^{-1} [5]), but with aquated metal ions ($M^{n+} = Tl^+$, Ag^+, Cu^{2+}, Sn^{2+}, Fe^{2+}, Mn^{2+}) the rate constants (k) have an upper limit of $\sim 3 \times 10^8$ dm^3 mol^{-1} s^{-1}. There is no correlation between k and the rates of exchange of water molecules coordinated to M^{n+}, which rules out ligand substitution as a general mechanism. Other possibilities are abstraction of H from a coordinated water molecule or $^\bullet OH$ entering and expanding the coordination shell of the metal ion.

In alkaline solution, $^\bullet OH$ is rapidly converted to its conjugate base $O^{\bullet -}$:

$$^\bullet OH + OH^- \rightleftharpoons O^{\bullet -} + H_2O \qquad (14)$$

$O^{\bullet -}$ is generally less reactive than $^\bullet OH$ with inorganic anions and with Br^-, CO_3^{2-} and $Fe(CN)_6^{4-}$ the reaction is immeasurably slow, a property which was used to measure the equilibrium constant for reaction (14) and hence obtain $pK_a(^\bullet OH) = 11.9$ [8].

In its reactions with organic molecules, $^\bullet OH$ is an electrophile whereas $O^{\bullet -}$ is a nucleophile. For example, in their reaction with aromatic compounds containing an aliphatic side chain, $^\bullet OH$ adds preferentially to the aromatic ring but $O^{\bullet -}$ abstracts H from the side chain:

$$^\bullet OH + C_6H_5CH_3 \longrightarrow (HOC_6H_5CH_3)^\bullet \qquad (15)$$
$$O^{\bullet -} + C_6H_5CH_3 \longrightarrow C_6H_5CH_2^\bullet + OH^- \qquad (16)$$

This can result in a change of reaction mechanism when the pH is raised so that $O^{\bullet -}$ replaces $^\bullet OH$ as the oxidant. $^\bullet OH$ is rather unselective when it abstracts H from C–H bonds because of the significantly larger energy of the H–OH bond being formed, *e.g.* with 1-propanol three different radicals are produced:

$$^\bullet OH + CH_3CH_2CH_2OH \longrightarrow CH_3CH_2{}^\bullet CHOH \quad (53.4\%)$$
$$\longrightarrow (CH_3{}^\bullet CHCH_2OH + {}^\bullet CH_2CH_2CH_2OH) \quad (46\%) \qquad (17)$$

Although it is a reducing radical, the reactions of the hydrogen atom with organic molecules are similar to those of $^\bullet OH$. Thus it adds to C=C double bonds and aromatic rings and abstracts H from C–H bonds in saturated compounds; in the latter case it is generally less reactive and more selective than $^{\bullet}OH$. In its reactions with metal ions $H^{\bullet}$ is less reactive than e_{aq}^{-} and in some cases it reacts via the formation of a hydride complex and effectively becomes an oxidant:

$$H^{\bullet} + Fe^{2+} \longrightarrow [Fe^{3+}, H^{-}] \xrightarrow{[H^{+}]} Fe^{3+} + H_2 \quad (18)$$

Generation of secondary radicals

The radiolysis of water produces approximately equal yields of reducing ($e_{aq}^{-} + H^{\bullet}$) and oxidising ($^{\bullet}OH$) radicals, but in its applications in general chemistry it is desirable to have either totally reducing or totally oxidising conditions. These conditions can readily be obtained in the following ways:

(a) by interconversion of the primary radicals into a single kind;
(b) by converting all the primary radicals into a single kind of secondary radical;
(c) by removing the unwanted primary radical by reaction to form a relatively inert secondary radical.

In this way the radiolysis of water provides a ready source of one-electron redox agents that can be finely tuned in terms of reduction potential and electric charge, and this has been widely used in obtaining kinetic and mechanistic data for free radical chemistry in aqueous solution [5].

Oxidising conditions

A very convenient and widely used method of converting e_{aq}^{-} into $^{\bullet}OH$ is to saturate the aqueous solution with N_2O so that reaction (19) followed by (14) occur:

$$e_{aq}^{-} + N_2O \longrightarrow N_2 + O^{\bullet -} \quad (19)$$

Under these conditions, the conversion e_{aq}^{-} to $O^{\bullet -}$ is essentially complete in ~ 15 ns but the half-life of its protonation in reaction (14) is ~ 8 ns so complete conversion to $^{\bullet}OH$ takes ~ 40 ns. In pulse radiolysis experiments designed to measure reactions of $^{\bullet}OH$ one should take care, therefore, to choose conditions that allow this radical to be fully formed before the reaction of interest begins. The hydrogen atom reacts only slowly with N_2O ($k = 2.1 \times 10^6$ dm^3 mol^{-1} s^{-1} [5]) so that approximately 90% of the radicals available in N_2O-saturated neutral solution are $^{\bullet}OH$ and the remaining 10% are $H^{\bullet}$.

Oxidising radicals that are more selective than $^{\bullet}OH$ in their reactions can be obtained by converting $^{\bullet}OH$ into another inorganic radical such as $Br_2^{\bullet-}$ ($E^o = 1.66$ V), $I_2^{\bullet-}$ ($E^o = 1.05$ V), $(SCN)_2^{\bullet-}$ ($E^o = 1.33$ V), $CO_3^{\bullet-}$ ($E^o = 1.5$ V), $N_3^{\bullet}$ ($E^o = 1.33$ V). These radicals are particularly useful for studying redox changes in metalloproteins and organometallic complexes because they are more likely to react at the metal centre whereas $^{\bullet}OH$ will also attack the organic moiety to generate a reducing radical there, either by abstraction of H or by addition to an aromatic ring or C=C bond.

A very useful way of generating oxidising radicals more powerful than $^{\bullet}OH$ in neutral solution is via reaction (20):

$$e_{aq}^- + S_2O_8^{2-} \longrightarrow SO_4^{\bullet-} + SO_4^{2-} \qquad (20)$$

$SO_4^{\bullet-}$ has $E^o = 2.43$ V and so can be used to generate other oxidants such as $Cl^{\bullet}$ ($E^o = 2.4$ V) and $NO_3^{\bullet}$ ($E^o = 2.5$ V) through reaction with Cl^- and NO_3^-, respectively.

Reducing conditions

A good method of obtaining totally reducing conditions is to convert the primary radicals e_{aq}^-, $H^{\bullet}$ and $^{\bullet}OH$ into a single kind of secondary radical; for example in N_2O-saturated solution containing formate ion (HCO_2^-), $CO_2^{\bullet-}$ ($E^o = -1.9$ V) is produced in reactions (21):

$$^{\bullet}OH/H^{\bullet} + HCO_2^- \longrightarrow H_2O/H_2 + CO_2^{\bullet-} \qquad (21)$$

A similar strategy can be used to generate a host of organic radicals as exemplified in reactions (22) – (24):

$$^{\bullet}OH + RH \longrightarrow R^{\bullet} + H_2O \qquad (22)$$
$$^{\bullet}OH + \varphi H \longrightarrow (HO\varphi)^{\bullet} \qquad (23)$$
$$^{\bullet}OH + HR{=}R'H \longrightarrow H(OH)R{-}^{\bullet}R'H \qquad (24)$$

Here R is an alkyl group and φ is an aryl group.

When reactions of e_{aq}^- are to be studied the usual method is to add 2-methyl-2-propanol ($(CH_3)_3COH$) to convert $^{\bullet}OH$ into the relatively unreactive radical $^{\bullet}CH_2C(CH_3)_2OH$. Reaction of e_{aq}^- with organic halides, RX, is also a useful source of organic radicals:

$$e_{aq}^- + RX \longrightarrow R^{\bullet} + X^- \qquad (25)$$

The hydrated electron reacts rapidly with simple aquated metal ions, M^{n+}_{aq}, to produce hyper-reduced oxidation states such as Ag^0, Au^0, Cd^+, Co^+, Ni^+, Zn^+, etc. that are not easily accessible by other methods:

$$e_{aq}^{-} + M^{n+} \longrightarrow M^{(n-1)+} \qquad (26)$$

These hyper-reduced metal ions are themselves powerful reductants and can be used to reduce metalloproteins at negatively charged sites which are less accessible to e_{aq}^-.

One of the important fields of study opened up by reaction (26) is the production and characterisation of noble and non-noble, mono- and multi-metallic clusters. Nano-particles of many metals have been produced and their catalytic properties investigated in this way [9] (Chapter 7).

One-electron reduction and oxidation in organic solvents

The advantage of organic liquids over water is the range of solvent polarities available so that species that are insoluble in water can be studied in a suitable solvent or mixture of solvents. A selection of illustrative examples is given in the next sections [10].

Alcohols

Methanol

The generation of free radicals in the radiolysis of methanol can be represented by reaction (27), followed rapidly by (28):

$$CH_3OH \rightsquigarrow e_s^-, H^\bullet, {}^\bullet CH_2OH, CH_3O^\bullet \qquad (27)$$
$$H^\bullet/CH_3O^\bullet + CH_3OH \longrightarrow {}^\bullet CH_2OH + H_2/CH_3OH \qquad (28)$$

Thus this system generates the reducing radical $^\bullet CH_2OH$ so that when the solubility of a compound is higher in methanol than in water it is advantageous to effect its one-electron reduction by the radiolysis of its methanolic solution. This technique has been utilised, for example, for one-electron reduction of some organometallic complexes and quinones. As in water, e_s^- can be converted to $^\bullet CH_2OH$ *via* reactions (19) (with N_2O) and (29):

$$O^{\bullet -} + CH_3OH \longrightarrow {}^\bullet CH_2OH + OH^- \qquad (29)$$

In alkaline solution $^{\bullet}CH_2OH$ deprotonates to form $^{\bullet}CH_2O^-$ which is a stronger reductant than $^{\bullet}CH_2OH$:

$$^{\bullet}CH_2OH + CH_3O^- \rightleftharpoons {}^{\bullet}CH_2O^- + CH_3OH \quad (K = 2.1 \times 10^3 \text{ [11]}) \qquad (30)$$

2 - Propanol

The radiation chemistry of 2-propanol is analogous to that of methanol; the major reducing radicals are e_s^- and $(CH_3)_2{}^{\bullet}COH$, and $(CH_3)_2{}^{\bullet}CO^-$ in alkaline solution. This solvent has been used to investigate the one-electron reduction of fullerenes. For example, reduction of C_{60}, which is insoluble in water, by e_s^- and $(CH_3)_2{}^{\bullet}COH$ generates $C_{60}{}^{\bullet -}$ which is stable for hours in the absence of oxygen.

Chlorinated hydrocarbons

Radiolysis of solvents such as dichloromethane (CH_2Cl_2), tetrachloromethane (CCl_4), and 1,2-dicholoroethane ($ClCH_2CH_2Cl$) generates radical cations or radicals that are strong oxidants. Thus, $CH_2Cl_2{}^{\bullet +}$ oxidises the fullerenes $C_{60,}$ C_{76} and C_{78} to the corresponding fullerene radical cations $C_n{}^{\bullet +}$ ($n = 60, 76, 78$). The oxidising radical in CCl_4 is $CCl_3{}^{\bullet}$ formed in reaction (30) by elimination of chloride ion:

$$e_s^- + CCl_4 \longrightarrow CCl_3{}^{\bullet} + Cl^- \qquad (31)$$

The one-electron oxidation of a wide range of compounds has been investigated in this way, including phenols and metallotetraphenylporphyrins. One-electron oxidation of metallotetraphenylporphyrins is also achieved in 1,2-dichloroethane, but in this case pyridine is added as a base to prevent demetallation of the complexes by HCl which is a radiolysis product.

Acetonitrile

This is a versatile solvent in general chemistry because it is aprotic and has a high dielectric constant. The main radical is the reductant $CH_3CN^{\bullet -}$ and this makes acetonitrile a convenient solvent in which to study the one-electron reduction of transition metal complexes that are not stable in hydroxylic media such as water and alcohols.

Hydrocarbons

In these nonpolar liquids, geminate ion recombination is the dominant process as described earlier. The initial result of this recombination of electron and parent positive ion is the formation of electronically excited states of the solvent molecule that are liable to have quite high energies. Unless fragmentation of the molecule takes place through bond breakage, the excited molecules fall to their lowest excited state through internal conversion within $\sim 10^{-12}$ s [12]. After this, the various processes known from photochemistry take place, *i.e.* intersystem crossing, energy transfer, luminescence, bond rupture, etc. [12]. Aromatic compounds are much more resistant to radiation damage than alkanes and alkenes, because the excitation energy is shared by all the π-electrons so that this energy is not localised sufficiently in any bond to cause it to break. It is informative to examine the radiation chemistry of two simple, symmetrical hydrocarbons, cyclohexane (C_6H_{12}) and benzene (C_6H_6). In each case all C–H and C–C bonds are equivalent.

Cyclohexane

The major products and their *G*-values are H_2 ($G = 0.58$ μmol J^{-1}), cyclohexene (C_6H_{10}, $G = 0.33$ μmol J^{-1}) and bicyclohexyl ($(C_6H_{11})_2$, $G = 0.18$ μmol J^{-1}), which can be understood in terms of the decomposition of excited states of cyclohexane ($C_6H_{12}^*$), formed by ion recombination or directly by the ionising radiation, in the following ways:

$$C_6H_{12}^* \longrightarrow {}^{\cdot}C_6H_{11} + H^{\cdot} \qquad (32)$$

$$C_6H_{12}^* \longrightarrow C_6H_{10} + H_2 \qquad (33)$$

$$H^{\cdot} + C_6H_{12} \longrightarrow {}^{\cdot}C_6H_{11} + H_2 \qquad (34)$$

$$2{}^{\cdot}C_6H_{11} \longrightarrow (C_6H_{11})_2 \qquad (35)$$

$$2{}^{\cdot}C_6H_{11} \longrightarrow C_6H_{12} + C_6H_{10} \qquad (36)$$

Reaction (32) in which a C–H bond is broken is the major fate of $C_6H_{12}^*$. There are only minor yields of products with less than six carbon atoms, indicating that rupture of C–C bonds in cyclohexane is relatively unimportant.

Benzene

The resistance of benzene to radiation damage is reflected in $G(H_2) \sim 0.004$ μmol J^{-1}; it is more than 100-fold less than the yield in cyclohexane. Other decomposition products are formed in similarly low yields and the radiation chemistry of benzene is dominated by its lowest singlet ($^1B^*$) and triplet ($^3B^*$) excited-state molecules. By adding suitable solutes the yields of these states have been measured as $G(^1B^*) = 0.17\ \mu mol\ J^{-1}$ and $G(^3B^*) = 0.44\ \mu mol\ J^{-1}$. Since $^3B^*$ is a major primary species in the radiolysis of liquid benzene, pulse radiolysis provides a ready method of producing and studying triplet excited states of other (solute) molecules whose triplet energy lies below that of $^3B^*$ (3.56 eV) by triplet-triplet energy transfer [12]. These triplet states are not always accessible by photo-excitation which predominantly generates excited singlet states if intersystem crossing from singlet to triplet excited sate is inefficient [12].

Concluding remarks

This brief description of the radiation chemistry of liquids shows how versatile is the use of high energy ionising radiation in chemistry. Despite the large energy involved in the radiolytic step, the radiolysis of most liquids produces solvated electrons and relatively simple free radicals, and the technique can be tuned to solve specific problems. Two of the more important applications are the following. The first is the identification of transient intermediates, and the measurement of their reaction rates [13], which enables one to unravel complex mechanisms initiated by free radicals in inorganic, organic and bio-organic chemistry. The second is the synthesis of stable and mono-disperse nanoparticles of metals that are very efficient catalysts (Chapter 7).

References

[1] Allen A.O., "The Radiation Chemistry of Water and Aqueous Solutions", Van Nostrand, Princeton, N.J., 1961.

[2] Buxton G.V., Radiation Chemistry of the Liquid State: (1) Water and Homogeneous Aqueous Solutions *in* "Radiation Chemistry: Principles and Applications", Rodgers M.A.J., Farhataziz (eds), VCH, Weinheim, 1987, 321-349.

[3] Holroyd R.A., The Electron: Its Properties and Reactions *in* "Radiation Chemistry: Principles and Applications", Rodgers M.A.J., Farhataziz (eds), VCH, Weinheim, 1987, 201-235.

[4] Rodgers M.A.J., Farhataziz (eds), "Radiation Chemistry: Principles and Applications", VCH, Weinheim, 1987.

[5] Ross A.B., Mallard W.G., Helman W.P., Buxton G.V., Huie R.E., Neta P., NDRL-NIST Solution Kinetics Database:- Ver. 3.0, Notre Dame Radiation Laboratory, Notre Dame, IN and National Institute of Standards and Technology, Gaithersberg, MD, 1998.

[6] Stanbury D.M., Reduction Potentials Involving Inorganic Free Radicals in Aqueous Solution, Adv. Inorg. Chem., 1989, **33**, 6-138. (All values of E^o quoted in the text are from this source).

[7] Hug G.L., Optical Spectra of Nonmetallic Inorganic Transient Species in Aqueous solution, Nat. stand. ref. dat. ser.: NSRDS-NBS 69, 1981.

[8] Buxton G.V., Greenstock C.L., Helman W.P., Ross A.B., Critical Review of Rate Constants for Reactions of Hydrated Electrons, Hydrogen Atoms and Hydroxyl Radicals ($^{\bullet}OH/^{\bullet}O^{-}$) in Aqueous Solution, J. Phys. Chem. Ref. Data, 1988, **17**, 513-886.

[9] Belloni J., Mostafavi M., Remita H., Marignier J.-L., Delcourt M.O., Radiation-induced synthesis of mono- and multimetallic clusters and nanocolloids, New J. Chem., 1998, **22**, 1239-1255.

[10] Buxton G.V., Mulazzani Q.G., Radiation-Chemical Techniques *in* "Electron Transfer in Chemistry. Volume 1: Principles, Theories, Methods and Techniques", Balzani V. (ed), Wiley-VCH, Weinheim, 2001, p. 503-557.

[11] Johnson D.W., Salmon G.A., Pulse Radiolysis of Methanol and Ethanol: Acid-Base Behaviour of Hydroxymethyl and Hydroxyethyl Radicals, J. Chem. Soc., Faraday Trans., I, 1975, **71**, 583-591.

[12] Calvert J.G., Pitts J.N., "Photochemistry", Wiley, New York, 1966.

[13] Buxton G.V., Measurements of Rate Constants for Radical Reactions in the Liquid Phase *in* "General Aspects of the Chemistry of Free Radicals", Z.B. Alfassi (ed), Wiley, Chichester, 1999, 51-78.

Chapter 2

Tools for radiolysis studies

James F. WISHART

Since the first report of the chemical effects of radiation by Pierre and Marie Curie, researchers have needed tools to deliver ionizing radiation for their scientific studies in increasingly precise ways. In the earliest stages, this was accomplished by the development of radioactive sources of increasing refinement and activity, and by the construction of X-ray tubes of increasing power. In the middle of the 20th Century, particle (primarily electron) accelerators took over as the primary tools of radiation chemists. At first, these accelerators were employed as continuous radiation sources like their predecessors. However, the development of pulse radiolysis techniques in the 1960s vastly increased the ability to study radiation-induced chemical kinetics. Before long, time resolution was extended into the picosecond regime [1-5]. In recent years, a new generation of radiolysis facilities has been developed to extend temporal resolution to even shorter times, at the same time providing a range of new transient detection capabilities [6-13].

Detailed accounts of the development of radiation chemistry and its tools can be found elsewhere. The purpose of this chapter is to describe the basic characteristics of continuous and pulsed sources of ionizing radiation for radiolysis studies, and to provide a broad overview of the present and near-future status of radiolysis instrumentation worldwide, for the benefit of readers who would like to use these powerful techniques to advance their own research. It is inevitable under the circumstances that some facilities may be missed and that future developments will soon render this overview out-of-date, however the substantial progress that has been made in the years since the previous reviews appeared [14-16] merits description here.

Types of ionizing radiation and their methods of generation

Ionizing radiation for chemical studies comes in many types with a vast range of properties that allow adaptation of experimental design to the chemistry of interest. The fundamental point to remember is that radiolytic energy deposition within the sample is inhomogeneous on short time (nanosecond) and length (nanometer) scales, and that the spatial pattern of reactive species produced from this energy deposition depends on the type of incident radiation and its energy. Photonic radiations (X- and gamma rays) deposit energy in well-separated interactions within the sample, while highly charged nucleons produce much denser deposition patterns with a higher probability of overlapping regions of ionized species. This effect is quantified as "Linear Energy Transfer" or LET. Photons and electrons are considered to be low LET radiations, whereas protons and heavier nucleons are high LET. There is also a dependence of LET on particle kinetic energy. Reactions of (and between) primary radiation-produced species convert the spatial inhomogeneity into variations in yield of radiolysis products on longer time scales (microseconds and beyond) depending on the type and energy of radiation. Chapter 1 of this book and other publications [14-17] discuss these effects in greater detail. The experimental choice of radiation type and energy depends on the application. Electron beams and X-rays (gammas) are well suited for general kinetics studies, while high LET radiations provide important details about the physical mechanisms of radiolysis and its chemical effects, and they are also becoming quite important in the study of clustered damage in nanostructured materials, including biological systems (DNA and cells) and synthetic polymers and resists (*e.g.*, ion beam lithography).

X- and gamma rays

Ionizing photonic radiation (X- and gamma rays) can be produced by the decay of certain radioactive isotopes or generated by stopping or deflecting a particle beam (*bremsstrahlung* and synchrotron radiation, respectively). X- and gamma rays, by virtue of their low LET, can penetrate sample vessels of moderate thickness such as cryostats or pressure vessels, however by the same token the deposited radiolytic dose (energy per unit sample mass) is low compared to particle radiations of the same fluence. Radioactive gamma sources based on ^{60}Co or ^{137}Cs isotopes are used in continuous radiation mode for radiolysis product studies and competition kinetics measurements. There are two basic types of source-based irradiators for chemical research. In the first, samples are placed within a shielded vault containing a radioactive source in a shielded container. After the experimenter exits the vault, the source is removed from the shielded container to expose the samples. In a modification of this design, samples may be transported into and out of the irradiation vault by a conveyor system. The second type of irradiator consists of a hollow cylindrical radioactive source permanently fixed in shielding. A mechanism is used to transport samples into the center of the cylindrical source, where they receive uniform radiation by virtue of the geometry. Due to the configuration, sample dimensions are limited to what will fit inside the source cylinder. Some gamma irradiators can be fitted with flow systems that transport fluids through the irradiation zone for controlled exposure times.

Particle accelerators can be used as continuous or pulsed X-ray sources by stopping the particle beam (typically electrons) in a high-atomic-number material such as gold or tungsten. The process of *bremsstrahlung* ("braking radiation" in German) produces a broad continuum of X-radiation that peaks at one third of the incident particle energy. This method can be used to produce nanosecond or picosecond X-ray pulses for time-resolved kinetics studies. For example, accelerator-produced X-rays were used to initiate experiments to measure the mobilities of excess electrons in non-polar liquids as functions of temperature and pressure, using transient conductivity measurement techniques. The experiments depend on the ability of X-rays to penetrate pressure vessel walls that would stop particle beams.

Other types of accelerators can be used as X-ray sources for specialized purposes. Synchrotron facilities can provide very intense radiation over a wide but selectable range of energies. Depending on the operating mode of the synchrotron it is possible to do time-resolved studies. Extremely short, sub-picosecond pulses of X-rays can be generated by laser wake-field accelerators (described below), which use terawatt electromagnetic fields from femtosecond lasers to accelerate electrons but also produce intense pulses of X-rays. When a second femtosecond laser pulse is used to interrogate the sample at various delay intervals with respect to the X-ray pulse, it would be possible to follow the very earliest steps of the radiolytic process.

Because of their high intensity, X-ray tubes were commonly used as laboratory radiation sources for radiation chemistry experiments until they were superceded by particle accelerators during the middle part of the 20th Century. They still retain specialized uses in research applications such as being used as the radiation source for MARY (MAgnetic field effect on Reaction Yield) spectroscopy studies of radical cation lifetimes and reactivity in alkane solvents [14,18]. MARY spectroscopy uses fluorescence to detect variations in singlet-triplet dynamics in radical ion pairs as a function of magnetic field. It is particularly useful for short-lived transients that are difficult to study by ESR.

Particle accelerators for radiolysis

There are several methods used to accelerate charged particle beams for pulse radiolysis. Acceleration requires a force applied by an electric field. The field may be a continuous gradient (or "DC" as in direct current) as produced by an electrostatic potential (**Fig. 1a**), or oscillating in time and space as produced by radio frequency (RF), microwave, or optical laser radiation (**Fig. 1b**).

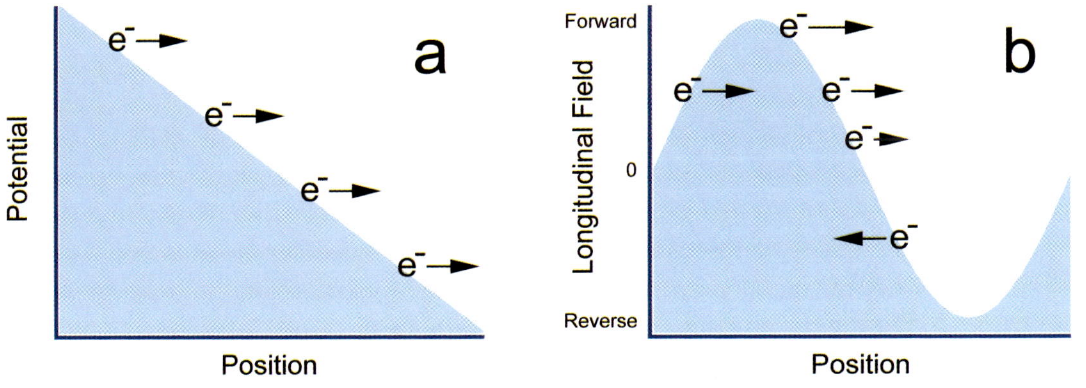

Figure 1: *Schematic depictions of electrostatic (a) and oscillating electromagnetic (b) fields for charged particle acceleration.*

Examples of the electrostatic type of accelerator are the Van de Graaff generator and capacitor-driven DC impulse generators. The characteristics of the two are quite different. The Van de Graaff generator (**Fig. 2**) develops and maintains a continuous electrostatic field by transporting charge (typically electrons) on a belt of non-conducting material that passes between two terminals at ground and high potential, respectively. The terminals are separated by stacks of insulating material alternating with metal plates. The metal plates are each connected to a resistor chain that slowly drains the charge off the high potential terminal so that the accelerating potential can be controlled by the belt charging rate. Typical operating potentials of such accelerators are 2-5 MV. Standing electrostatic potentials higher than 5 MV become progressively more difficult to sustain without extraordinary measures, and beam energies of 2-5 MeV are adequate for many pulse radiolysis applications. An evacuated beam tube runs from the high-potential terminal to ground. The insulating glass spacers along the tube are interrupted at intervals by the metal plates connected to the resistor chain, so that a uniform accelerating gradient is applied to the particle beam.

Electron beams are by far the most common usage of Van de Graaff generators for pulse radiolysis, although ion sources may also be used. Typically, grid-gated cathodes are used to produce electron pulses varying in length from several nanoseconds to several microseconds with beam currents on the order of one Ampere. On some systems, continuous emission is possible at significantly lower currents (~ 1 mA). The minimum pulse width is determined by the response of the pulsing circuit. Specialized circuitry providing a

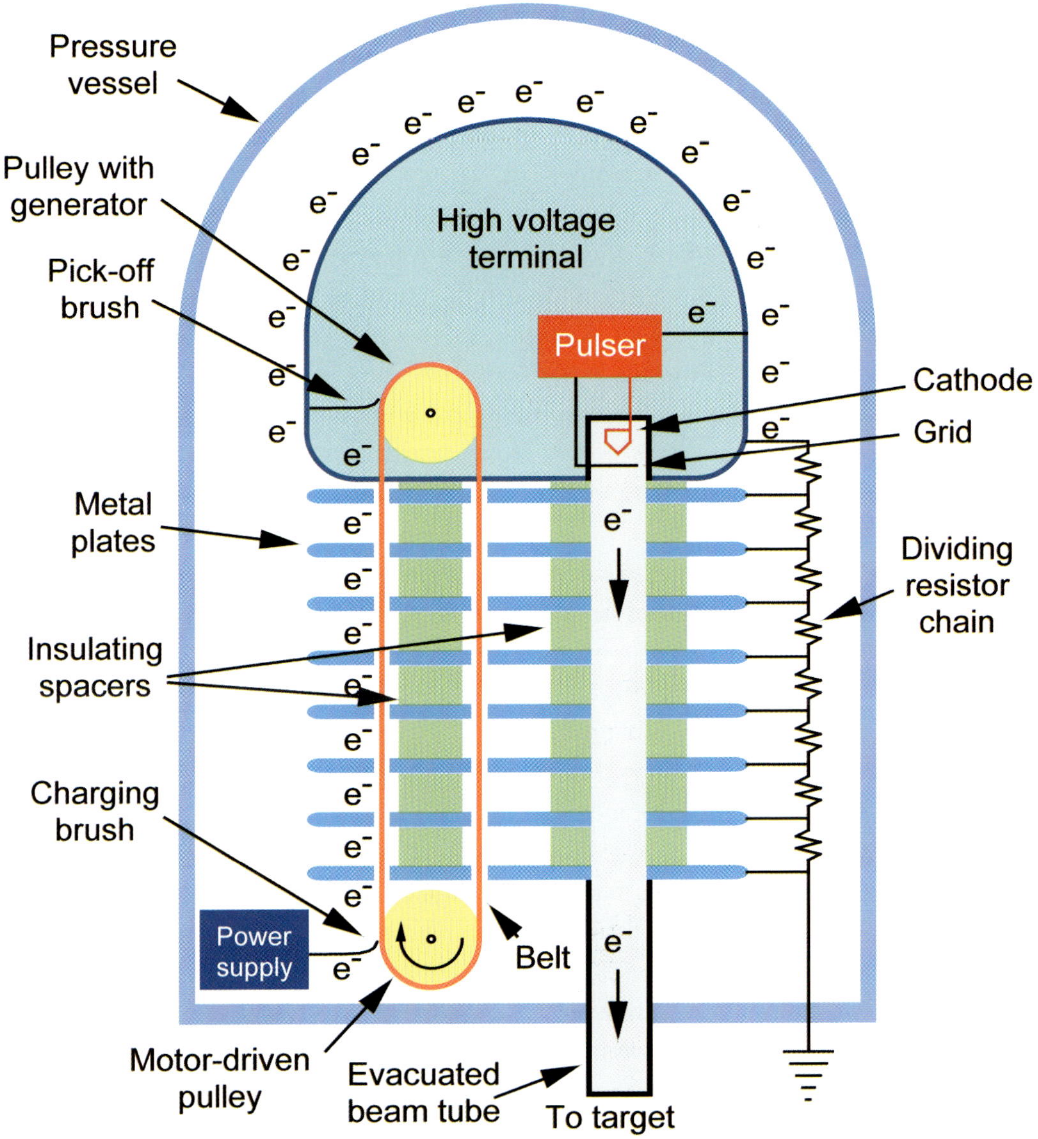

Figure 2 : *A Van de Graaff accelerator.*

sub-nanosecond electron pulse was developed on the 3 MV Van de Graaff accelerator at the Technical University of Delft [19]. Van de Graaff accelerators emit no electromagnetic noise apart from the impulse generated from the electron beam itself, since they are electrostatic devices. This is an advantage for the implementation of real-time detection systems such as time-resolved microwave conductivity (TRMC) and transient electron paramagnetic resonance (EPR). The Delft Van de Graaff has been the site of significant development of the TMRC technique, which measures the migration and recombination processes of charge carriers in materials [20]. Transient EPR detection is useful for identifying and observing the reactions of radical species that are difficult to characterize by optical absorption spectroscopy [21]. A 3 MeV Van de Graaff accelerator with EPR detection system is located at the Notre Dame Radiation Laboratory (NDRL) in the United States. NDRL also hosts a time-resolved resonance Raman spectroscopy (TRRR) system on a 2 MeV Van de Graaff. The high specificity of vibrational spectroscopy provides information about the structure of radical species [22]. The 4 MeV ELBENA Van de Graaff at the Hahn-Meitner Institute in Berlin supports advanced transient absorption detection capabilities as well as AC and DC conductivity detection methods with nanosecond time resolution (DC) [23] The 2 MeV Van de Graaff at Brookhaven National Laboratory incorporates a sensitive transient absorption detection system and time-of-flight charge carrier mobility measurements in non-polar media by DC conductivity.

The second series of electrostatic accelerators are based on direct-current impulse generators. The best-known examples of these are the Febetron units produced by the Field Emission Corporation. These machines are based on a Marx-bank impulse generator, whereby a high (0.6 to 2.3 MV) potential is generated by triggering a series of spark gaps to switch a bank of capacitors from charging in parallel to discharging in series through a large cathode tube. Very large electron currents (~ 7000 A) are passed in pulses lasting several to tens of nanoseconds, delivering radiolytic doses up to 20 kGy. The large doses are well suited for the study of radical-radical reactions and the radiation chemistry of gases. The electrical noise generated by the spark gaps and the discharge of the Febetron is a major problem for detection systems, requiring much signal averaging and background correction, however the Febetron is unmatched in its ability to deliver large radiolytic doses in short pulse durations. Although Febetron-based pulse radiolysis facilities were once located in many laboratories, today only a few remain, notably a 600 keV unit at the Laboratoire de Chimie Physique of the University of Paris XI, in Orsay, France.

Another kind of impulse generator-based accelerator based on a pulse transformer was developed in Novosibirsk, USSR, in the 1970s. One such unit (ELIT-1, 1 MeV) has been operating at the Time-Resolved Spectroscopy Laboratory of the University of Leipzig. It is equipped with a beam deflection unit that can reduce the electron pulse width within the sample to less than a nanosecond and it has been coupled to a Fourier transform EPR detection system.

Accelerators that use electromagnetic radiation to produce the accelerating field can accelerate particles to much higher energies than electrostatic accelerators. Higher beam energies provide more penetrating power for elaborate experimental setups such as pressure vessels and cryostats, and the ability to better focus more charge into a smaller target to produce a higher radiolytic dose. As a practical consideration however, beam energies above 40 MeV (for electrons) are less desirable for pulse radiolysis because an increasing fraction of the beam energy is converted into *bremsstrahlung* instead of being deposited in the sample in radiolytic events. In addition, the *bremsstrahlung* produced by high-energy beams can induce radioactivity in typical substances found in scientific equipment and building materials, creating an additional safety hazard.

Linear accelerators (or linacs) are the most common type of oscillating-field accelerator for pulse radiolysis. The accelerating sections consist of a series of resonant cavities for radio-frequency or microwave radiation (100 MHz to 10 GHz, with 1-3 GHz being most common). Charged particles are accelerated by the oscillating electric field to different degrees depending on their position (phase) with respect to the RF cycle. The differential acceleration collects the particles into bunches that are clustered around the optimal acceleration phase in each RF cycle. This fact underscores a significant difference in beam temporal profile between electrostatic and oscillating field accelerators. Beams emitted from the former type are continuous for the duration of the emission, whereas for the latter the beam is structured into a pulse train or "macropulse" consisting of a series of bunches separated in time by the period of the accelerating radiation. Each bunch is inherently short in time (~ 30 picoseconds); but significant technical effort is required to exploit the time resolution offered by such a short pulse, as explained below.

The majority of linear accelerators used for pulse radiolysis work exclusively in the pulse train mode and are typically referred to as "nanosecond" linacs, in contrast to the "picosecond" linacs that can operate in the single bunch mode. Nanosecond linacs for pulse radiolysis are distributed throughout the world, including the Notre Dame Radiation Laboratory in the U.S., the Commissariat à l'Énergie Atomique (CEA) in Saclay, France, the Institute of Applied Radiation Chemistry in Łódź, Poland, the Institute for Nuclear Chemistry and Technology in Warsaw, Poland, the ISOF-CNR in Bologna, Italy, the Institute of Isotope and Surface Chemistry in Budapest, Hungary, the Bhabha Atomic Research Center (BARC) in Mumbai, India, the National Centre for Free Radical Research in Pune, India, the Australian Radiation Protection and Nuclear Safety Agency in Melbourne, the University of Auckland, New Zealand, and at the Shanghai Institute of Nuclear Research in China. A linac for pulse radiolysis will be installed at the University of Manchester, UK, as part of the establishment of a new research group in Radiation Chemistry. Transient optical absorption spectroscopy is the standard technique for kinetics measurements at each of these facilities. In addition, transient mid-infrared detection methods are under development at the linac in Saclay [24].

Picosecond electron linacs require special techniques to generate single electron bunches for ultrafast kinetic studies. The problem stems from the fact that it is not practical to electrically gate a cathode electron source on and off in less than a nanosecond, thus cathode gating is too slow to fill only one period (350 to 770 ps) of the RF cycle at the accelerating frequencies of typical accelerators. In a clever development, a sweeping beam deflection device was used on the nanosecond Novosibirsk linac to select a single picosecond bunch from the pulse train [5]. Kinetics of charge recombination were followed by using a streak camera to monitor emission from excited states produced *via* electron-hole recombination. Aside from that special case, historically the standard method of generating picosecond single pulses (**Fig. 3**) is to inject the electrons into a lower frequency RF field operating at a sub-harmonic of the accelerating frequency (for example, 476 MHz for an S-band, 2856 MHz linac), allow the electron packet to evolve into a compact single bunch in the lower-frequency section, then to inject the compressed bunch into the higher-frequency accelerating section. The system used for injecting the single bunches is called a "sub-harmonic pre-buncher".

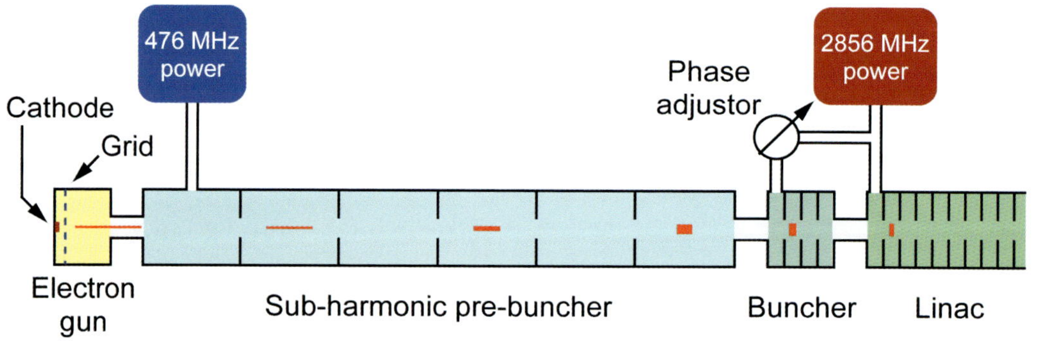

Figure 3 : *Schematic representation of an S-band pre-bunched picosecond linear accelerator.*

Because it involves more accelerating structures with multiple RF frequencies and phase relationships, and a high-performance electron gun source to inject the electrons in about one nanosecond, a sub-harmonic pre-buncher adds a lot of cost and complexity of operation of the linac facility. Consequently, only a few pre-bunched picosecond linacs were built for pulse radiolysis studies. The first one was installed at Argonne National Laboratory in the 1970s [2]; subsequently other picosecond installations were built at the University of Tokyo Nuclear Energy Research Laboratory (NERL) in Tokai-Mura [3,4] and Osaka University [4]. Very recently, a new pre-bunched linac for pulse radiolysis has been built at the Shanghai Institute of Applied Physics.

During the late 1980s and into the 1990s, a new technology arose for generating picosecond electron beams [14,15]. Instead of using thermionic emission from a hot cathode

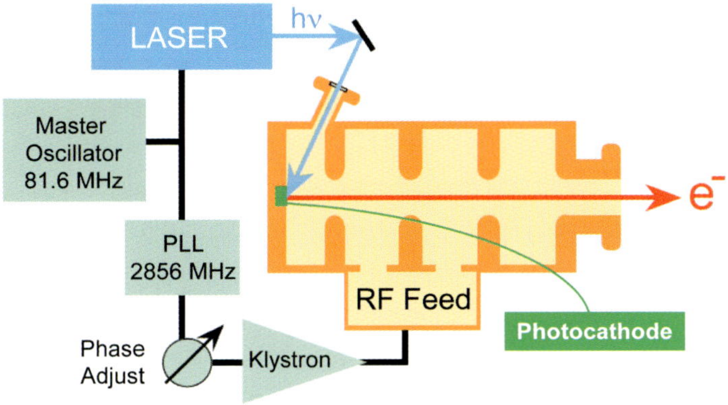

Figure 4 : *Schematic representation of the BNL LEAF photocathode electron gun accelerator showing the laser-microwave synchronization relationship.*

to produce the electron beam, a metal (Cu or Mg) or semiconductor (*e.g.*, Cs_2Te) cathode is struck with an ultraviolet laser pulse to cause it to emit electrons via the photoelectric effect. The electron pulse emitted from the photocathode is as short as the incident laser pulse (although it tends to expand during beam transport). Thus it is possible to inject single bunches of electrons directly into the high-frequency accelerating structure, if the laser pulse is synchronized to the correct RF phase for proper acceleration. Commercial lasers with the ability to synchronize themselves to external frequency sources became available in the same time frame. The electron source and accelerator functions can be combined into a single unit called a photocathode electron gun. A schematic of the photocathode electron gun of the Brookhaven National Laboratory Laser-Electron Accelerator Facility (LEAF) [6,14] is shown in **Figure 4** and a picture of the accelerator with the photocathode back plate removed is shown in **Figure 5**. While the LEAF accelerator consists of an integral structure of 3.5 microwave cavities that accelerates electron bunches to a final energy of 9 MeV when

Figure 5 : *The LEAF accelerator opened to show the 6 mm diameter magnesium photocathode in the center of the back plate.*

driven with 15 MW of microwave power, 1.5 cavity photocathode electron guns that produce beam energies of 4-5 MeV are often used as injectors for systems with further accelerating stages, including most of the photocathode radiolysis facilities listed further below.

Photocathode-based picosecond electron accelerators are conceptually simpler than pre-bunched thermionic systems, although they require reasonably powerful, multicomponent femtosecond or picosecond laser systems to drive the photocathode. In addition, the availability of synchronized laser pulses allows the development of advanced detection capabilities with unprecedented time resolution. The combination of ease of use and powerful detection methods has stimulated strong interest in photocathode gun systems. Since the installation of the first photocathode electron gun pulse radiolysis system at BNL [5,13], four additional photocathode-based facilities have become operational and two more are in progress. The operational centers include the ELYSE facility at the Université de Paris-Sud XI in Orsay, France [7,8], NERL in Tokai-Mura, Japan [9,10], Osaka University [11,12], and Waseda University in Tokyo [13]. Facilities under development are located at the Technical University of Delft, the Netherlands, and the BARC in Mumbai, India.

The latest development in ultrafast electron sources for pulse radiolysis does away completely with accelerating structures. Instead, femtosecond laser pulses at terawatt power levels are focused tightly onto jets of helium atoms. The extremely strong electromagnetic field of the laser radiation strips the electrons off the helium atoms and accelerates them to very high energies, in a process called laser wake-field acceleration [25,26] but sometimes referred to as "Table-Top Terawatt" or T^3. Efforts are being made to characterize and control the energy spectrum of the accelerated electrons [26]. Experiments have shown that it is possible to use beams from these systems to do pulse radiolysis measurements [27], although energy spread of the electron beam and geometrical constraints place the practical time resolution in the few picosecond range despite the use of femtosecond laser pulses. Laser wake-field acceleration is studied in several laboratories around the world, but work on its development as a radiolysis method has occurred at the Terawatt Ultrafast High Field Facility at Argonne National Laboratory (**Fig. 6**) [28], the Laboratoire d'Optique Appliquée, École Polytechnique-ENSTA, Palaiseau, France [25-27], NERL in Tokai-Mura, Japan, and the Rutherford Appleton Laboratory in the UK. T^3 systems also hold promise as ultrafast X-ray sources.

Experimental Detection Techniques for Ultrafast Radiolysis

Transient absorption optical detection methods for picosecond and faster resolution are subject to a number of considerations. First, the velocities of high-energy electrons are approximately that of the speed of light in vacuum, $v_{elect} \approx c$, whereas light itself is slowed by the refractive index n of the sample $v_{light} = c/n$. Therefore, visible light falls about one picosecond behind the electron beam for each millimeter of transit through water when the laser and electron beams are co-linear, thus the time resolution is constrained by the path length through the sample and must be traded off against signal magnitude. An arrangement that uses beams that cross at an angle can be used to compensate for the difference in velocities [31]. Another factor is that real-time, digitizer-based optical detection systems have response (bandwidth) limitations that limit time resolution, although technology has improved over the years. Still today, extremely fast biplanar phototube detectors and high-bandwidth oscilloscopes (≥ 6 GHz) offer resolution in the visible region down to approximately 80 picoseconds, but that is still slower than the capabilities of the accelerator to produce short electron pulses. Detectors in other wavelength regions are slower and have complex responses that must be deconvoluted from the kinetics.

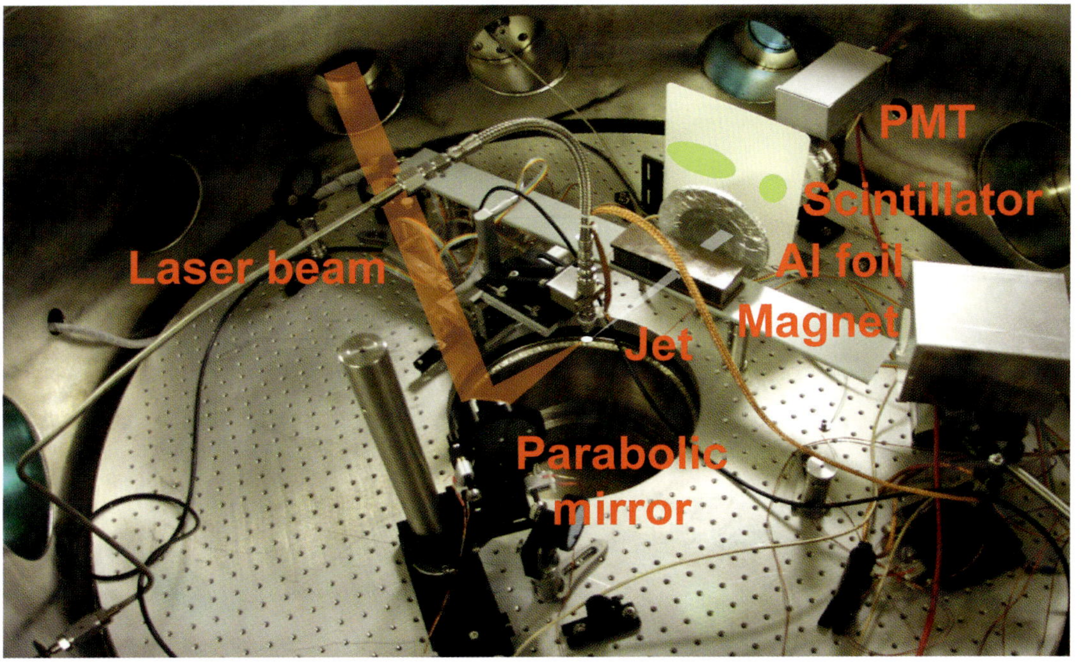

Figure 6 : *Vacuum chamber of the Terawatt Ultrafast High Field Facility (TUHFF) at ANL, configured for a measurement of electron beam energy dispersion (courtesy of R. Crowell).*

Another method of continuous detection is the use of a streak camera in absorption mode. Measuring radiation-induced emission with a streak camera is relatively straightforward, but sensitive transient absorption measurements are difficult to make because of limited dynamic range of the streak camera. The first work in this area was done in the 1970s, but little was done to follow up until the development of high-dynamic range streak cameras in recent years. One such streak camera has been used to great effect to measure transient absorption spectra at the ELYSE facility in Orsay, France [8].

To overcome these limitations, stroboscopic detection methods have been used since the earliest picosecond radiolysis measurements (**Fig. 7**). Analogous to laser pump-probe experiments, in pulse-probe transient absorption spectroscopy a short pulse of light is used to measure the absorbance of a sample as a function of time delay between the electron pulse and the probe beam. Time resolution using this method is typically on the order of the electron pulse width. Originally, Čerenkov light generated from the electron beam itself was used as a variable-delay probe beam [1,2]. Later, a free-running Ti-Sapphire oscillator was used with a time-amplitude converter to measure spur decay of the solvated electron. With the advent of laser photocathode electron gun systems that provide picosecond-synchronized laser pulses, high-resolution pulse-probe kinetics measurements have nearly become routine. Timing improvements [12] and streak-camera jitter detection have made such measurements possible on thermionic, pre-bunched picosecond linacs as well. For transient spectroscopy, two methods of probe pulse generation are used to measure kinetics at wavelengths other than the fundamental of the gun-driving laser system (~ 800 nm). At some facilities, a white light continuum is generated by focusing intense laser pulses into a substrate such as a sapphire plate, fused silica or D_2O [10,11,13], while at others a specified probe wavelength is produced by an optical parametric amplifier pumped by the gun-driving laser system [6,12]. Broadband detection of the white-light probe is accomplished with a spectrograph/CCD or diode array combination [11], while single-wavelength detection is done with a pair of diodes [6,8,11,13]. Broadband detection works within the sensitivity range of silicon detectors (up to 950 nm), while NIR-sensitive photodiodes (Ge and InGaAs) extend the range of single-wavelength detection to 1700 nm. Pulse-probe transient spectroscopy has been used to follow the formation of Xe_2^* excimers in the radiolysis of supercritical xenon [29], measure dissociation rates of aryl halide radical anions, to re-evaluate the yield of hydrated electron at picosecond times [30], and to observe the solvation of excess electrons in ionic liquids.

While affording high time-resolution, pulse-probe absorption measurements typically require large numbers of shots to assemble a profile of the reaction by time and wavelength. Since cumulative radiation damage of the sample can be a problem, sample solutions are often flowed once-through or recirculated during experiments. Many interesting studies are thus not practical to perform because samples are not available in the necessary quantities or they cannot be flowed. To address this problem, techniques to measure complete time profiles or spectra, in one shot or just a few shots, have recently been developed. A group at ANL employed the Frequency-Domain Single Shot (FDSS) spectroscopy technique where a high-bandwidth femtosecond pulse is stretched to several hundred picoseconds, passed through the sample, then dispersed by a monochromator onto a photodiode array, where each wavelength corresponds to a different time. The Osaka group has obtained spectra with high signal-to-noise in a single-shot, using a CCD to detect broadband absorption by reducing the time interval between sample and reference shots to 1 ms [11]. At BNL's LEAF facility, an optical fiber bundle is being used to create imaged probe beams containing 100 different time delays in a single shot.

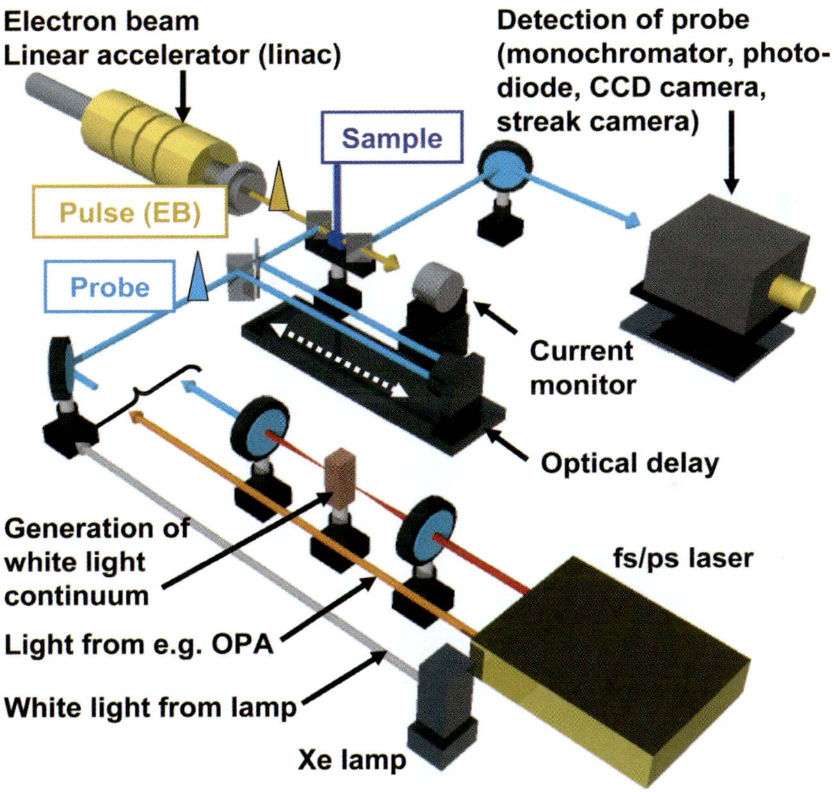

Figure 7 : *Schematic of a general detection system for picosecond pulse radiolysis (courtesy of Dr. A. Saeki, Osaka University).*

Heavy ion sources

Radiolysis experiments using heavy ion beams (protons and heavier atomic nuclei stripped of all electrons) occupy an important role, despite the relatively small number of investigators in this area. Because heavy ions have much higher LET values, the pattern of energy deposition, or track structure, is much denser than with "lighter" radiations. The variation of yields of radiolytic products as a function of LET, and the effects of variable concentrations of scavengers on those yields, have provided critical information for the understanding of the radiolytic process [14,17]. The high local density of ionizations produced by heavy ions creates localized areas of damage that can be useful, as in the ion beam modification of polymers [14]. Ion beams have been used to fragment polymers to form membranes, and conversely, to polymerize precursor substrates to form forests of polymer rods on surfaces. The localized damage caused by heavy ions has important implications for radiation biology and radiation medicine. The American space agency NASA has established a heavy ion radiolysis facility at BNL (the NASA Space Radiation Laboratory) for *in vitro* and *in vivo* studies to prolonged exposure to heavy ion radiation in interplanetary space. Proton radiotherapy is a precise way of delivering large radiation doses to kill tumors. Heavy ion radiolysis of water produces the hydroperoxyl radical $HO_2^{\cdot}$, which is not formed by low-LET radiolysis in the absence of molecular oxygen. It is an important means of causing oxidative damage to hypoxic tumors.

Generally, heavy ion radiolysis experiments are performed at large, multipurpose accelerator facilities using cyclotrons or tandem Van de Graaff accelerators. The *Grand accélérateur national d'ions lourds* (GANIL) in Caen is the primary facility in France for heavy ion experiments. Major facilities in Japan include TIARA at the Japan Atomic Energy Agency, the Heavy Ion Medical Accelerator (HIMAC) in Chiba, and the High-fluence Irradiation Facility of the University of Tokyo. Heavy ion sources in the U.S. include ATLAS at ANL, the National Superconducting Cyclotron Laboratory at Michigan State University, and the Triple Ion Irradiation Laboratory at Oak Ridge National Laboratory. Because heavy ions are stopped in very short distances within samples, special techniques and equipment configurations are used to optimize transient signals.

Future trends

The past decade has been an encouraging period in the development of radiolysis capabilities that has reversed an earlier trend of decline in number and accessibility. Two new technologies, photocathode electron guns and laser wake-field accelerators, have emerged and spawned a large new generation of ultrafast accelerator facilities. These installations are developing advanced experimental techniques and making sophisticated experiments available to a larger community of researchers than ever before. Earlier-generation picosecond accelerators have been upgraded to high levels of performance. New nanosecond linacs were installed at Notre Dame, Saclay and Pune, and the University of Manchester has founded a program and Chair in Radiation Chemistry that will reinforce a field that was in danger of disappearing from the U.K. These developments attest to the fact that major research support establishments in several countries recognize that radiation chemistry and radiolysis experimentation have important roles to play in solving future energy needs in many areas (including nuclear, solar and high-performance materials) and in developing science to protect human health and the environment.

It is not taking a risk to predict that performance and capabilities of the new radiolysis installations will markedly improve as these young facilities mature. But what other developments can we look forward to? Certainly there is strong interest in bringing additional spectroscopic tools into the pulse radiolysis laboratory. Efforts are underway to adapt transient mid-infrared detection techniques to pulse radiolysis, to take advantage of the specificity of vibrational spectroscopy [24]. Strong interest in nanoscience and the mechanisms of reactions in heterogeneous systems will push the development of interface-specific spectroscopies in radiation chemistry, for example surface-enhanced Raman spectroscopy and second-harmonic or sum-frequency generation.

Acknowledgments

This work was carried out at Brookhaven National Laboratory under contract DE-AC02-98CH10886 with the U.S. Department of Energy and supported by its Division of Chemical Sciences, Office of Basic Energy Sciences. The author wishes to thank R. Crowell and A. Saeki for Figures 6 and 7.

References

[1] Bronskill M.J., Taylor W.B., Wolff R.K., Hunt J.W., Design and performance of a pulse radiolysis system capable of picosecond time resolution, Rev. Sci. Instr., 1970, **41**, 333-340.

[2] Jonah C.D., A wide-time range pulse radiolysis system of picosecond time resolution, Rev. Sci. Instr., 1975, **46,** 62-66.

[3] Kobayashi H., Ueda T., Kobayashi T., Washio M., Tabata Y., Tagawa S., Picosecond single electron pulse for pulse radiolysis studies, Radiat. Phys. Chem., 1983, **21**, 13-19.

[4] Tabata Y., Kobayashi H., Washio M., Tagawa S., Yoshida Y., Pulse radiolysis with picosecond time resolution, Radiat. Phys. Chem., 1985, **26**, 473-479.

[5] Grigoryants V.M., Lozovoy V.V., Chernousov Yu.D., Shebolaev I.V., Arutyunov V.A., Anisimov O.A., Molin Yu.N., Pulse radiolysis system with picosecond time resolution referred to Cherenkov radiation, Radiat. Phys. Chem., 1989, **34**, 349-352.

[6] Wishart J.F., Cook A.R., Miller J.R., The LEAF picosecond pulse radiolysis facility at Brookhaven National Laboratory, Rev. Sci. Instr., 2004, **75**, 4359–4366.

[7] Belloni J., Monard H., Gobert F., Larbre J.-P., Demarque A., De Waele V., Lampre I., Marignier J.-L., Mostafavi M., Bourdon J.C., Bernard M., Borie H., Garvey T., Jacquemard B., Leblond B., Lepercq P., Omeich M., Roch M., Rodier J., Roux R., ELYSE - A picosecond electron accelerator for pulse radiolysis research, Nucl. Instr. Meth. A, 2005, **539**, 527–539.

[8] Marignier J.-L., de Waele V., Monard H., Gobert F., Larbre J.-P., Demarque A., Mostafavi M., Belloni J., Time-resolved spectroscopy at the picosecond laser-triggered electron accelerator ELYSE, Radiat. Phys. Chem., 2006, **75**, 1024-1033.

[9] Muroya Y., Lin M., Watanabe T., Wu G., Kobayashi T., Yoshii K., Ueda T., Uesaka M., Katsumura Y., Ultra-fast pulse radiolysis system combined with a laser photocathode RF gun and a femtosecond laser, Nucl. Instrum. Meth. A, 2002, **489**, 554–562.

[10] Muroya Y., Lin M., Iijima H., Ueda T., Katsumura Y., Current status of the ultra-fast pulse radiolysis system at NERL, the University of Tokyo, Res. Chem. Intermed., 2005, **31**, 261-272.

[11] Saeki A., Kozawa T., Tagawa S., Picosecond pulse radiolysis using femtosecond white light with a high S/N spectrum acquisition system in one beam shot, Nucl. Instr. and Meth. A, 2006, **556**, 391-396.

[12] Saeki A., Kozawa T., Kashiwagi S., Okamoto K., Isoyama G., Yoshida Y., Tagawa S., Synchronization of femtosecond UV-IR laser with electron beam for pulse radiolysis studies, Nucl. Instr. and Meth. A, 2005, **546**, 627-633.

[13] Nagai H., Kawaguchi M., Sakaue K., Komiya K., Nomoto T., Kamiya Y., Hama Y., Washio M., Ushida K., Kashiwagi S., Kuroda R., Improvements in time resolution and signal-to-noise ratio in a compact pico-second pulse radiolysis system, Nucl. Instr. Meth. B, 2007, **265**, 82-86.

[14] "Radiation Chemistry: Present Status and Future Trends", Jonah C. D., Rao B.S.M. (ed.), Studies in Physical and Theoretical Chemistry, Vol. 87, Elsevier Science, 2001.

[15] "Photochemistry and Radiation Chemistry: Complementary Methods for the Study of Electron Transfer", Wishart J.F., Nocera D.G., (eds), Adv. Chem. Ser. 254, American Chemical Society, Washington, DC, 1998.

[16] "Radiation Chemistry: Principles and Applications", Rodgers M.A.J., Farahtaziz (eds), VCH, Weinheim, 1987.

[17] Mozumder A., "Fundamentals of Radiation Chemistry", Academic Press, San Diego, 1999.

[18] Stass D.V., Sviridenko F.B., Molin Y.N., Magnetic field effect study of solvent hole deprotonation in X-irradiated liquid n-alkanes, Radiat. Phys. Chem., 2003, **67**, 207-210.

[19] Luthjens L.H., Vermeulen M.J.W., Hom M.L., de Loos M.J., van der Geer S.B., Revision of (sub) nanosecond pulser for IRI Van de Graaff electron accelerator aided by field propagation calculations, Rev. Sci. Instr., 2005, **76**, 024702.

[20] de Haas M.P., Choffat F., Caseri W., Smith P., Warman J.M., Charge Mobility in the Room-Temperature Liquid-Crystalline Semiconductor Poly(di-n-butylstannane), Advanced Materials, 2006, **18**, 44-47.

[21] Werst D.W., Trifunac A.D., Observation of Radical Cations by Swiftness or by Stealth, Chem. Res., 1998, **31**, 651-657.

[22] Tripathi G.N.R., Time resolved resonance Raman observation of the extreme protonation forms of a radical zwitterion in water, J. Chem. Phys., 2005, **122**, 071102.

[23] Janata E., Lilie J., Martin M., Instrumentation of kinetic spectroscopy-11. An apparatus for a.c.-conductivity measurements in laser flash photolysis and pulse radiolysis experiments, Radiat. Phys. Chem., 1994, **43**, 353-356.

[24] Le Caër S., Vigneron G., Renault J.P., Pommeret S., First coupling between a LINAC and FT-IR spectroscopy: The aqueous ferrocyanide system, Chem. Phys. Lett., 2006, **426**, 71-76.

[25] Malka V. et al., Electron acceleration by a wake field forced by an intense ultrashort laser pulse, Science, 2002, **298**, 1596–1600.

[26] Malka V., Faure J., Glinec Y., Lifschitz A.F., Laser-plasma accelerator: status and perspectives, Philosophical Transactions of the Royal Society A, 2006, **364**, 601-610 and references therein.

[27] Brozek-Pluska B., Gliger D., Hallou A., Malka V., Gauduel Y., Direct observation of elementary radical events: low and high energy radiation femtochemistry in solutions, Radiat. Phys. Chem., 2005, **72**, 149–157.

[28] Crowell R.A., Shkrob I.A., Oulianov D.A., Korovyanko O., Gosztola D.J., Li Y., Rey-de-Castro R., Motivation and development of ultrafast laser-based accelerator techniques for chemical physics research, Nucl. Inst. Meth. Phys. Res. B, 2005, **241**, 9-13.

[29] Holroyd R.A., Wishart J.F., Nishikawa M., Itoh K., Reactions of Charged Species in Supercritical Xenon as Studied by Pulse Radiolysis, J. Phys. Chem. B, 2003, **107**, 7281-7287.

[30] Muroya Y., Lin M.Z., Wu G., Iijima H., Yoshii K., Ueda T., Kudo H., Katsumura Y., A re-evaluation of the initial yield of the hydrated electron in the picosecond time range, Radiat. Phys. Chem., 2005, **72**, 169-172.

[31] Yang J., Kondoh T., Kozawa T., Yoshida Y., Tagawa S., Pulse radiolysis based on a femtosecond electron beam and a femtosecond laser light with double-pulse injection technique, Radiat. Phys. Chem., 2006, **75**, 1034-1040.

Chapter 3

The solvated electron: a singular chemical species

Mehran MOSTAFAVI and Isabelle LAMPRE

Introduction

In this chapter, we pay close attention to the solvated electron, this so peculiar "chemical species" already introduced in the first chapter as one situated at the origin of most phenomena characterizing radiation chemistry. Its description, the analysis of its true nature and of its main properties thus appears as essential in chemistry. Denoted by e_s^-, it can be visualized as a negative elementary charge surrounded by solvent molecules. The peculiarity comes from the fact that the charge is delocalised but without a nucleus, and that solvent molecules rally around it (**Fig. 1**). The solvated electron is a thermodynamically stable radical but like most free radicals it has a short lifetime due to its great chemical reactivity. As early as the nineteenth century, the solvated electron was observed but not identified (**Tab. 1**). In 1808, Sir H. David and later, in 1864, W. Weyl reported the intense blue colouring obtained by dissolving alkali metals in ammonia (NH_3). Weyl also found that the resulting solutions had reducing properties when used in organic synthesis, but he did not discover the nature of these blue solutions. Nowadays it is known that the blue colour is due to the solvated electron, as alkali metals dissolved in ammonia give the metal cation and a solvated electron according to, for example in the case of sodium:

$$Na_{metal} \xrightarrow{NH_{3\,liq}} Na^+_{NH_3} + e^-_{NH_3} \quad (1)$$

Nearly one century later, in 1904, thanks to conductivity measurements, C. Kraus showed that, in liquid ammonia containing dissolved alkali metals, there was a negatively

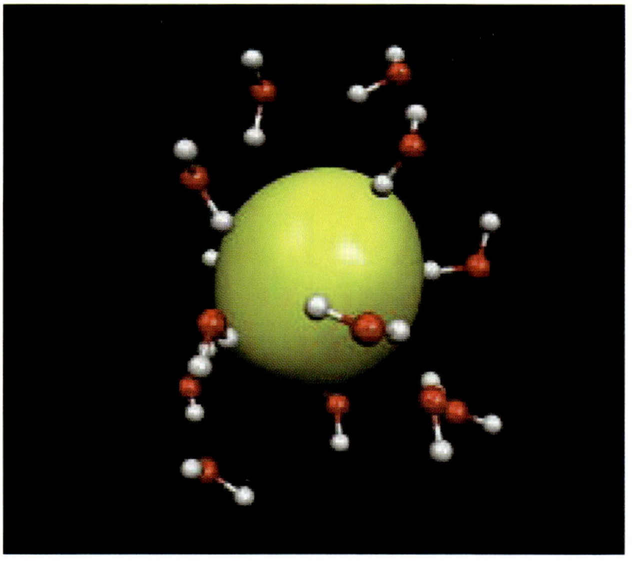

Figure 1 : *Schematic representation of the hydrated electron obtained by molecular simulations* [10]. *The delocalised negative charge is surrounded by water molecules creating a cavity with a radius of about 2.5 Å.*

charged species with a conductivity larger than that of anions, but independent of the counter-ion (Li^+, Na^+, K^+). He suggested that this species was an electron surrounded by ammonia molecules behaving like an anion. In 1952, to explain the bleaching of aqueous solutions containing methyl blue upon irradiation in the presence of carbon dioxide (CO_2), G. Stein also proposed the transient formation of a hydrated electron, similar to the solvated electron in ammonia. The direct spectroscopic observations in 1962 by E.J. Hart and J.W. Boag of transient solvated electrons produced by pulse radiolysis of aqueous solutions [1] opened the way to an extensive investigation on the properties and theoretical models of the solvated electron. The solvated electron in water is known as the hydrated electron, denoted by e_{aq}^-. After these first observations, the solvated electron was soon detected in various solvents through its intense optical absorption band in the visible or near infrared domain [2]. Owing to this property, the reactivity of the solvated electron has been studied by transient absorption measurements in many solvents using pulse techniques. In addition to pulse radiolysis, other methods can be used to produce a solvated electron and allow its study in different environments (**Inset**). Due to the development of ultrashort laser pulses, great strides have been made towards the understanding of the solvation and short-time reactivity of the electron, mainly in water but also in other polar solvents. So, despite its short life-time, the solvated electron is a unique chemical moiety whose properties may be compared in many solvents. In this chapter, we consider the main properties of the solvated electron, its reactivity and recent results concerning the solvation dynamics of the electron.

Some physical properties of the solvated electron

The identification and the understanding of the chemical properties of the solvated electron can be made through the knowledge of its physical properties. The properties of the solvated electron depend on several factors such as solvent, temperature and pressure.

Table 1. *Main dates for the discovery and study of the solvated electron.*

Years	Strides	Authors
1808	Report on blue colouring of ammonia at the touch of alkali metals.	H. David [a]
1864	Blue colouring observed for solutions of alkali metals in liquid ammonia, methylamine and ethylamine; reducing properties of these solutions.	W. Weyl [a,b]
1908	Identification of a species with a negative elementary charge in ammonia solutions of alkali metals, by conductivity measurements.	C. Kraus [a]
1952	Suggestion of transient formation of solvated electrons to explain the bleaching of aqueous solutions containing methyl blue under irradiation in the presence of CO_2.	G. Stein, R.L. Platzman [b]
1962	First pulse-radiolysis experiments of aqueous solutions revealing the formation of solvated electrons.	E.J. Hart, J.W. Boag [b] [1]
1971	Solvation dynamics of the electron in propan-1-ol at −120 °C observed by pulse-radiolysis.	J.H. Baxendale [23]
1987	Solvation dynamics of the electron in water at room temperature observed by ultrashort pulse photolysis.	A. Migus *et al.* [26]
2007	Probing of the electron in water at attosecond time scale	D. Nordlund *et al.* [34]

See reviews: **a**) *Edwards P.P., The electronic properties of metal solutions in liquid ammonia and related solvents, Adv. Inorg. Chem. Radiochem., 1982,* **25***, 135-185.* **b**) *Boag J.W., Pulse radiolysis: a historical account of the discovery of the optical absorption spectrum of the hydrated electron, in "Early developments in radiation chemistry", Kroh J. (Ed.), Royal Society of Chemistry, Cambridge, 1989, 7-20.*

Volume

In 1921, by dissolving an alkali metal in liquid ammonia, C. Kraus and W. Lucasse observed a volume expansion of the solution greater than that obtained for the dissolution of ordinary salts [3]. They attributed this volume expansion to the formation of the solvated electron with a cavity, regarded as a particle since the electron itself has a negligible volume. For example, the dissolution of 3 moles of sodium in one litre of liquid ammonia induces an increase in volume of 43 cm^3 compared to the pure liquid. Assuming that all the metal is dissociated, it may be deduced that in ammonia the electron occupies a spherical volume with a radius of 0.18 nm. In fact, the cavity radius of the solvated electron in ammonia is greater than that value and is about 0.3 nm.

Inset : Ways of producing the solvated electron

The solvated electron may be generated by different methods, more or less appropriate according to the media and the detection systems.

Chemistry. Historically, the first method was the dissolution of alkali metals in amine solvents; nevertheless, this is useful only in media in which the lifetime of the solvated electron is long enough (at least a few hours in the pure solvent). For example, in liquid ammonia, the solvated electron may be formed from sodium metal:

$$Na_{metal} + NH_{3\ liq} \rightarrow Na_{NH_3}^{+} + e_{NH_3}^{-} \qquad (1)$$

Electrochemistry. Electrolysis reactions may also be used to produce the solvated electron in similar media; for instance, in liquid ammonia the reactions are:
At the anode:

$$4\,NH_3 - 2\,e^- \rightarrow 2\,NH_2^{\bullet} + 2\,NH_4^+ \rightarrow N_2H_4 + 2\,NH_4^+ \qquad (2)$$

At the cathode:

$$NH_3 + e^- \rightarrow e_{NH_3}^{-} \qquad (3)$$

Radiolysis. Under high-energy radiations (γ- or X-rays, beams of accelerated electrons or positive ions), electrons may be ejected from the most abundant (solvent) molecules in the medium. These ejected electrons have excess kinetic energy that is lost in collision with solvent molecules, which may be electronically excited, or ionised to produce more electrons in a cascade scheme. When their kinetic energy falls below the ionisation/excitation threshold of the solvent, the electrons are "thermalised" and become "solvated" as solvent molecules get reorganised around them.

Photolysis. Absorption of UV or visible light from a flash lamp or a laser is used to produce the solvated electron. Two processes may occur: (1) the photon energy is sufficient to ionise a solute and produce the electron, for instance, photo-detachment from an anion (Cl^-, $Fe(CN)_6^{4-}$, Na^-...), (2) the laser intensity is high enough to induce ionisation of the solvent by multiphoton absorption (*e.g.* $H_2O + nh\nu \rightarrow H_2O^{\bullet+} + e^-$).

It is to be noted that the cavity models developed by theoreticians to describe the solvated electron originate from this observed volume expansion. In these models a cavity is occupied by the electron and the solvent molecules are organized around it (**Fig. 1**). The great volume of the cavity accounts for the volume expansion of the solution.

Charge

Conductivity measurements performed by C. Kraus in ammonia solutions of alkali metals gave evidence of a negatively charged species independent of the cation, the solvated electron. However, it was to be wondered whether the same species was formed in the radiolysis of liquids. By studying the rates of reaction of the reducing species produced by γ-radiolysis of aqueous solutions containing different neutral, cationic and anionic solutes, G. Czapski and H. A. Schwarz found that this reducing species has a unit negative charge and is a hydrated electron [4].

Mobility

Most methods for determining the electron mobility use pulse radiolysis techniques in which the concentration of electrons is followed during or after the ionizing pulse by measurement of the change in conductivity. However, due to the inherent conductance of polar liquids, direct conductivity measurements of solvated electrons are generally difficult in these media. Therefore, the diffusion coefficient and the mobility of the solvated electron in various solvents have been indirectly derived from the value of the rate constant of a fast reaction thought to be diffusion controlled [5]. In 1969 it was discovered that excess electrons in non polar solvents are from 100 to 100,000 times more mobile than ions in the same liquids. The electron mobility depends on the solvent and on the structure of the molecules. For example, the mobility in n-alkanes decreases with increasing chain length, but it increases with branching; at room temperature the value is 28.0, 0.15 and 70 $cm^2V^{-1}s^{-1}$ in ethane, n-pentane and 2,2-dimethylpropane, respectively [6]. In polar solvents the electron mobility is generally slightly greater than ionic mobilities, which are about $7 \times 10^{-4}\ cm^2V^{-1}s^{-1}$. In water the measured mobility of the hydrated electron is $1.9 \times 10^{-3}\ cm^2V^{-1}s^{-1}$. In a viscous solvent such as ethane-1,2-diol, the value falls down to $2.8 \times 10^{-4}\ cm^2V^{-1}s^{-1}$. The mobility increases with temperature as the viscosity decreases. The viscosity effect shows that the transport of the solvated electron is mainly due to diffusion and that transport *via* jump or tunnel effect is not predominant for this species. This strengthens the idea of the solvated electron as a "real" chemical entity and not as a free isolated charge.

Optical absorption

A major characteristic of the solvated electron is its optical absorption spectrum. The optical absorption spectrum of the hydrated electron was first determined in 1962 by Hart and Boag using transient absorption measurements in pulse radiolysis of pure water and aqueous solutions of carbonates; it appears as an intense broad structureless band with a maximum around 720 nm in pure water [1]. Then, optical absorption spectra were reported for the solvated electron in a large number of solvents. The position of the maximum and the width of the absorption band depend on the medium. **Figure 2** shows the optical absorption spectrum of the solvated electron in various solvents at room temperature. The solvents may be classified in three groups:

1/ Polar protic solvents with hydrogen bonds, like water and alcohols. The wavelength at the absorption maximum of the solvated electron spectrum lies in the visible domain, between 500 and 820 nm; for instance, it is around 525 nm for glycerol and 640 nm for methanol.

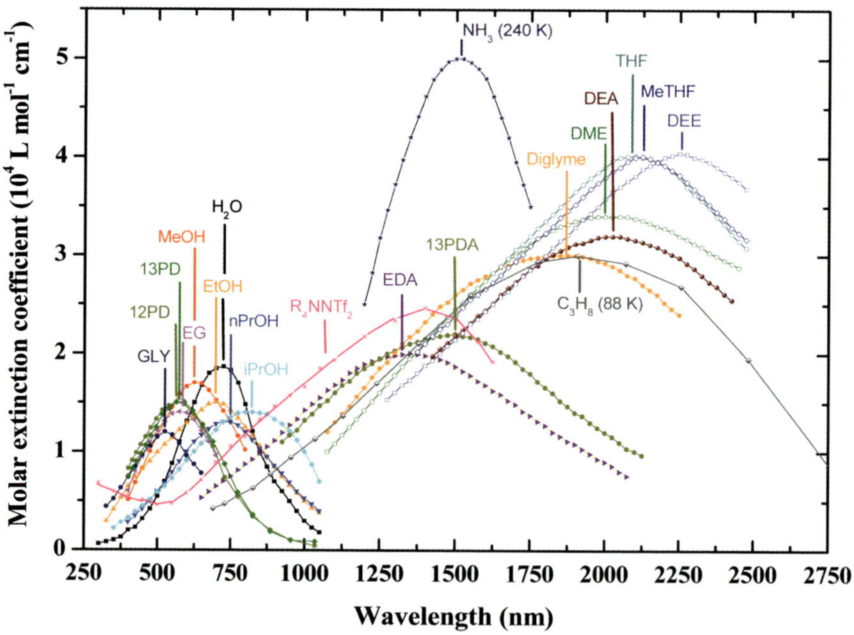

Figure 2 : *Optical absorption spectra of the solvated electron in various solvents (from [2] with addition of other data). Abbreviations: H_2O: water; MeOH: methanol; EtOH: ethanol; 1PrOH: propan-1-ol; 2PrOH: propan-2-ol; 12ED: ethane-1,2-diol or ethylene glycol; 12PD: propane-1,2-diol; 13PD: propane-1,3-diol; GLY: glycerol or propane-1,2,3-triol; R_4NNTf_2: methyl-tributyl-ammonium bis[trifluoromethyl-sulfonyl]imide; EDA: ethane-1,2 -diamine; 13PDA: propane-1,3-diamine; NH_3: ammonia; C_3H_8: propane; DME: dimethylether; DEA: diethanolamine; THF: tetrahydrofuran; MeTHF: methyltetrahydrofuran; DEE: diethylether.*

2/ Liquid ammonia, amines and ammonium ionic liquids. The absorption spectrum of the solvated electron is situated in the near infrared domain; in the case of liquid ammonia at room temperature the wavelength of the absorption maximum is 1830 nm.

3/ Slightly or non polar aprotic solvents like ethers and hydrocarbons. The absorption spectrum is observed in the infrared domain with a maximum at wavelengths higher than 2000 nm; for example, 2200 nm for tetrahydrofuran.

On the whole, the maximum of the absorption band of the solvated electron shifts to higher wavelengths as the dielectric constant of the solvent diminishes. However, the absence of a simple correlation between the maximum and the dielectric constant or the dipolar moment of the solvent suggests that the position of the absorption spectrum is mostly governed by the molecular structure of the liquids. Quantum simulations have indicated that the electronic ground state of the solvated electron is a 1s-like localized state and the excited states are three non-degenerate 2p-like states, also bound and localized, followed by a band of delocalised states. Hence, the broad absorption band of the electron corresponds mostly to an 1s $\rightarrow$ 2p transition with a contribution of the transition from the bound state to the continuum at high energies [7]. The latter contribution accounts for the asymmetry of the spectrum. As the energy levels of the exited states are generally close to the conduction band of the solvent, the position of the absorption maximum gives information on the depth of the potential energy well in which the solvated electron is localized. So, the shorter the wavelength of the absorption maximum is (the higher the transition energy is), the deeper is the well.

The absorption spectrum of the solvated electron depends on the nature of the solvent, but is also particularly sensitive to parameters such as pressure- and temperature-induced changes in solvent structure and properties (Chapter 4). The optical absorption band shifts to higher energies (shorter wavelengths) with increasing pressure up to 2000 bar; the shift is larger in primary alcohols than in water and it correlates with the increase in liquid density rather than with the rise in dielectric constant [8]. A rise in the temperature induces a red shift of the solvated electron absorption spectrum. Thus, the absorption maximum in water is located around 692 nm at 274 K and 810 nm at 380 K [9]. The asymmetric shape of the absorption band of the hydrated electron at temperatures below the critical region is well fitted by Gaussian and Lorentzian functions on the low- and high-energy sides, respectively [9]. Lately, quantum-classical molecular-dynamics simulations of an excess electron in water performed for a wide range of temperature and pressure suggest that the observed red shift of the optical absorption spectrum is a density effect rather than a temperature effect [10]. Indeed, by increasing the temperature the mean volume of the cavity of the solvated electron increases, the electron is less confined in the cavity, the well becomes less deep. The temperature dependence of the absorption spectrum of the solvated electron has been recorded not only in water but also in alcohols (**Figs. 3 and 4**). The studies in ethane-1,2-diol (12ED), propane-1,2-diol (12PD) and propane-1,3-diol (13PD) emphasize the significant influence of the length of the aliphatic chain, and the effect of the distance between the two OH groups on the behaviour of the solvated electron [11,12].

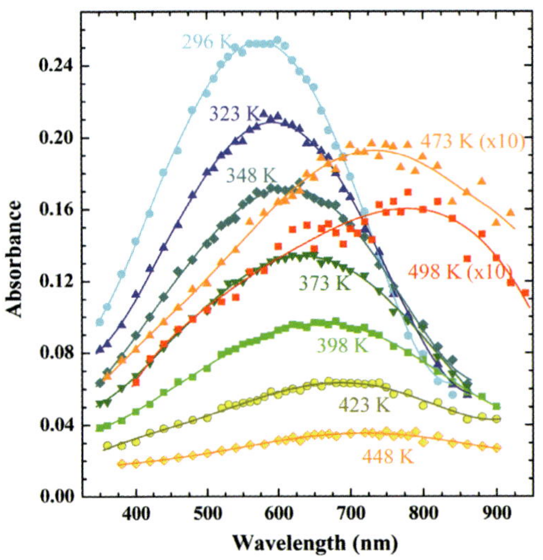

Figure 3 : *Optical absorption spectra of the solvated electron in propane-1,3-diol at different temperatures (from [12]). The absorption spectrum shifts to longer wavelength with increasing temperature.*

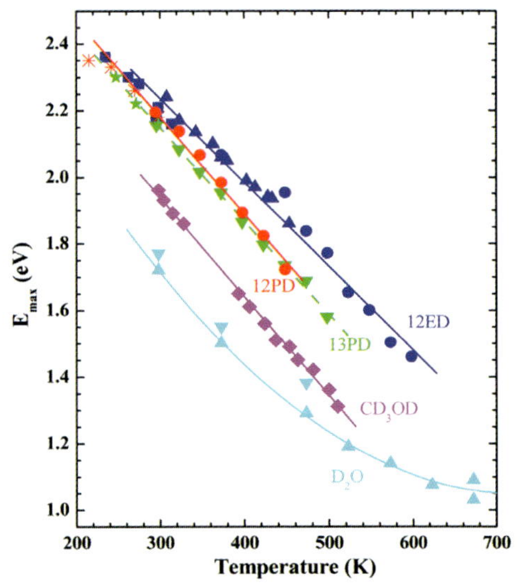

Figure 4 : *Energy of the maximum of the solvated electron absorption band (E_{max}) as a function of temperature in five different solvents (from [12]). By increasing the temperature, the absorption spectrum shifts to longer wavelength so the transition energy decreases. D_2O : deuterated water ; CD_3OD : deuterated methanol ; 13PD: propane-1,3-diol ; 12PD: propane-1,2-diol ; 12ED: ethane-1,2-diol.*

At room temperature, the wavelength of the absorption maximum is shorter in 12PD (565 nm) than in 13PD (575 nm), indicating that the two neighbouring OH create deeper electron traps in 12PD compared to 13PD. However, the traps in 12PD appear less deep than in 12ED since the energy of the absorption maximum of the solvated electron measured at a given temperature is lower in 12PD than in 12ED. This shows an influence of the additional methyl group on the solvent structure, in particular on the three-dimensional networks of hydrogen-bonded molecules. Moreover, an increase in temperature affects more greatly the absorption spectrum of the solvated electron in 12PD and 13PD than in 12ED. The effect is also greater in 12PD compared to 13PD. Indeed, the temperature coefficients are -2.9×10^{-3}, -2.8×10^{-3} and -2.5×10^{-3} eV K^{-1} for 12PD, 13PD and 12ED, respectively (**Fig. 4**). The decrease in viscosity versus temperature is also larger in 12PD compared to 13 PD and 12ED. The different behaviour for the three solvents results from larger modifications of the solvent structure and molecular interactions for 12PD than for 13PD and 12ED. So, the solvated electron interacting strongly with the solvent is very sensitive to solvent molecular structure.

Chemical reactivity of the solvated electron

Soon after the discovery of the absorption spectrum of the solvated electron in pulse radiolysis experiments (Chapter 2), the rates and mechanisms of its reaction with a wide variety of solutes was studied. Although it is a transient species, the solvated electron is a very important reducing agent. Indeed, its reduction potential is very negative; the value of $E°(H_2O/e_s^-)$ for the solvated electron in water is equal to -2.8 V with respect to the standard hydrogen electrode. The reaction rate constant and the probability of encounter with another species decide the so-called lifetime of the solvated electron, which therefore depends on the experimental conditions.

Geminate recombination and spur reaction

In radiation chemistry one of the most important reactions of solvated electrons is recombination with positive ions and radicals that are simultaneously produced in close proximity inside small volumes called spurs. These spurs are formed through further ionisation and excitation of the solvent molecules. So, in competition with diffusion into the bulk, leading to a homogeneous solution, the solvated electron may react within the spurs (Chapter 1).

Geminate recombinations and spur reactions have been widely studied in water, both experimentally and theoretically [13-16], and also in a few alcohols [17,18]. Typically, recombinations occur on a timescale of tens to hundreds of picoseconds. In general, the primary cation undergoes a fast proton transfer reaction with a solvent molecule to produce the stable solvated proton and the free radical. Consequently, the recombination processes are complex and depend on the solvent. The central problem in the theory of geminate ion recombination is to describe the relative motion and reaction between the two particles with opposite charges initially separated by a distance r_0. In water, the primary products of solvent radiolysis are the hydrated electron e_{aq}^-, the hydroxyl radical $^{\bullet}OH$ and the hydronium cation H_3O^+:

$$H_2O \xrightarrow{\gamma,\, nh\nu} H_2O^{\bullet+} + e^- \xrightarrow{H_2O} e_{aq}^- + {}^{\bullet}OH + H_3O_{aq}^+ \qquad (4)$$

Interestingly, it has been shown that the recombination of the hydrated electron is greatly dominated by reaction with •OH radical, because the reaction with H_3O^+ is not diffusion controlled despite the Coulombic attraction. Geminate ion recombination is usually considered to be negligible in water because the Onsager radius is small (0.7 nm) compared to the radius of the distribution of e_{aq}^- in the spur (~ 2.3 nm) (Chapter 1).

Figure 5 depicts the decay of the solvated electron due to spur reactions in two different solvents, water and tetrahydrofuran. In both liquids, the solvent relaxation is very fast (less than 1 ps), therefore, the absorption signals on the picosecond time scale are due to the fully solvated electron. As the dielectric constant of tetrahydrofuran is low ($\varepsilon = 7.6$ compared to 80 for water), the electrostatic attraction is not screened by the solvent and geminate recombination between the solvated electron and the cation can occur over long separation distances in contrast to water. Moreover, the mobility of e_s^- in THF is roughly three times higher than that in water. That explains why the decay of the solvated electron is more important in tetrahydrofuran compared to water [19].

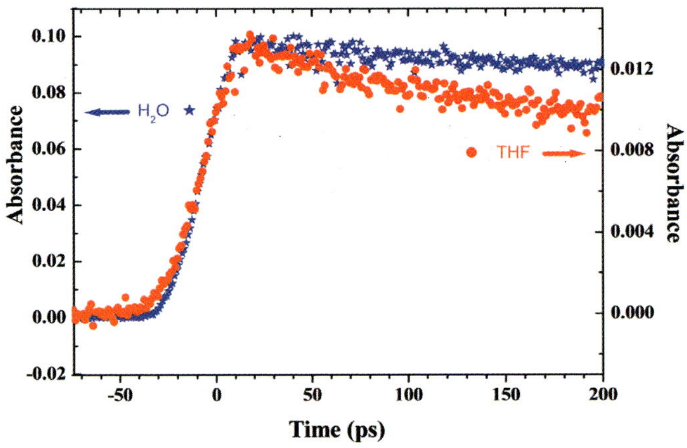

Figure 5 : *Decay at 790 nm of the solvated electron in water (H_2O) and in tetrahydofuran (THF) (from [19]). The decay of the solvated electron due to spur reactions is faster in a slightly polar solvent like THF than in water which is a very polar solvent.*

It is to be noted that, after geminate recombination, when diffusion takes place on the nanosecond and longer time scale, reactions between the radiolytic species still occur and, in the absence of any other solutes, those reactions are responsible for the disappearance of the solvated electron. Besides, the metastability of the "blue" solutions of alkali metals in liquid ammonia is due to the fact that the solvated electron does not react with another solvated electron and that it reacts extremely slowly with the protonated form (NH_4^+) which is at an extremely low concentration.

Reactions with solutes

During the last 35 years, the reactivity of the solvated electron has been widely investigated, mainly by pulse radiolysis (Chapter 1). Indeed, because of its intense optical absorption in the visible and near infrared spectral regions, the solvated electron is most conveniently and reliably observed directly using pulse radiolysis or photolysis. So, the majority of rate constants of reactions between the solvated electron and solutes or electron scavengers have been measured by the method of decay kinetics. The reaction is evidenced by the shortening of the lifetime of the solvated electron as the concentration of the solute is increased (**Fig. 6**). If the reaction between the solvated electron and the solute is dominant, the decay approaches a simple exponential form from which the reaction rate constant can be deduced. The scavenging factor is the product of the scavenger concentration by the rate constant ($[S] \times k_{S+e_s^-}$).

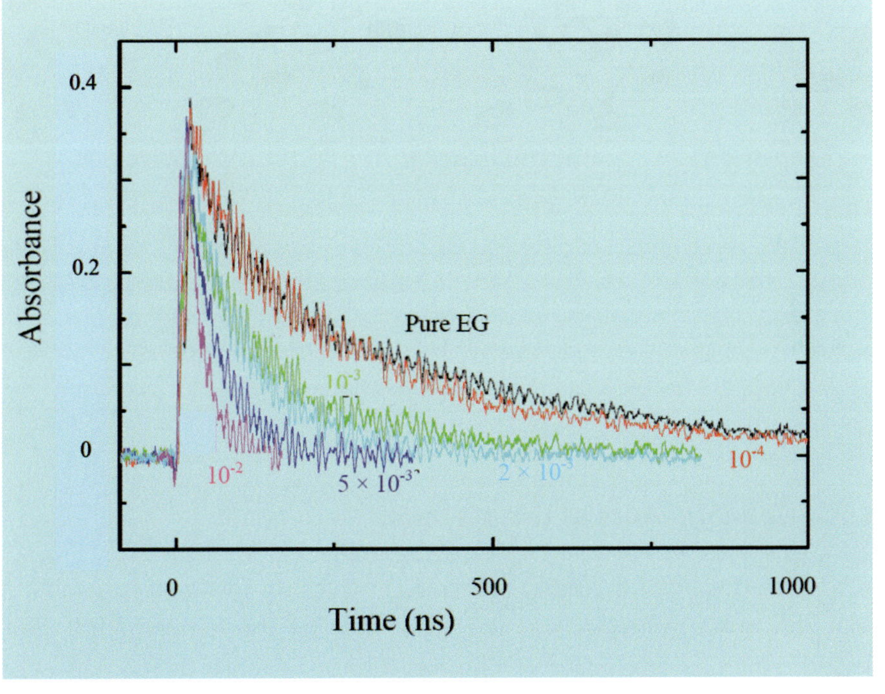

Figure 6 : *Decays of the solvated electron recorded at 575 nm in pulse-radiolysis of ethane-1,2-diol in the presence of silver cations, Ag^+, at various initial concentrations in mol l^{-1}. In pure ethane-1,2-diol and at the lowest concentration of Ag^+, the decay is mostly due to reactions of the solvated electron with other species produced by radiolysis. By increasing the concentration, the decay becomes faster as the solvated electron reacts predominantly with the Ag^+ cation.*

Many reactions of the solvated electron with different solutes, such as aliphatic, aromatic or heterocyclic compounds, and also anions and cations, have been studied.

A wealth of information on the reduction of metal ions in aqueous solutions has been obtained and a compilation was published in 1988 [20]. However, alkali or alkaline earth metal ions such as Li^+, Na^+, or Mg^{2+}, cannot be reduced by the hydrated electron in aqueous solution but can form an ion pair with the solvated electron in polar liquids. Among the various reactions of the solvated electron, the reduction of halogenated hydrocarbons is often used in radiation chemistry to produce well-defined radicals because of the selective cleavage of the carbon-halogen bond by the attack of the solvated electron. This reaction produces the halide ion and a carbon-centered radical, and is of great interest for environmental problems related to the destruction of halogenated organic contaminants in water and soil [21,22].

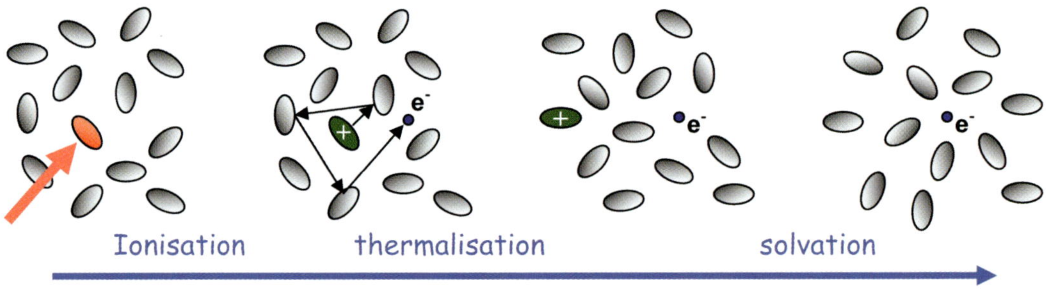

Figure 7 : *Schematic representation of the formation of the solvated electron following solvent ionisation. The fate of the electron starts with its ejection from a molecule upon ionisation by radiolysis or photolysis. The ejected electron progressively loses its excess kinetic energy by collisions with solvent molecules (thermalisation step). Then the electron is localised, trapped in a solvent site or cavity and becomes solvated when the solvent molecules have obtained their equilibrium configuration after relaxation.*

Solvation dynamics of the electron

A simplified view of the early processes in electron solvation is given in **Figure 7**. Initially, electron pulse radiolysis was the main tool for the experimental study of the formation and dynamics of electrons in liquids (Chapter 2), first in the nanosecond time range in viscous alcohols [23], later in the picosecond time range [24,25]. Subsequently, laser techniques have achieved better time resolution than pulse radiolysis and femtosecond pump-probe laser experiments have led to observations of the electron solvation on the sub-picosecond to picosecond time scales. The pioneering studies of Migus *et al.* [26] in water showed that the solvation process is complete in a few hundreds of femtoseconds and hinted at the existence of short-lived precursors of the solvated electron, absorbing in the infrared spectral domain (**Fig. 8**). The electron solvation process could thus be depicted by sequential stepwise relaxation cascades, each of the successive considered species or states of the electron having a fixed, individual spectrum [27,28]. However, subsequent studies on electron solvation following the photoionisation of water have favoured the so-called "continuous shift" model in which the existence of these IR-absorbing species is not strictly required [29,30]. In the latter model, only one localised electron is considered but its spectrum undergoes a continuous evolution, shifting to the blue (shorter wavelengths) during the solvation. In alcohols, the formation of the solvated electron is slower than in water [25,27]. The necessity to include both stepwise mechanism and continuous relaxation to successfully interpret experiments has also been suggested [31].

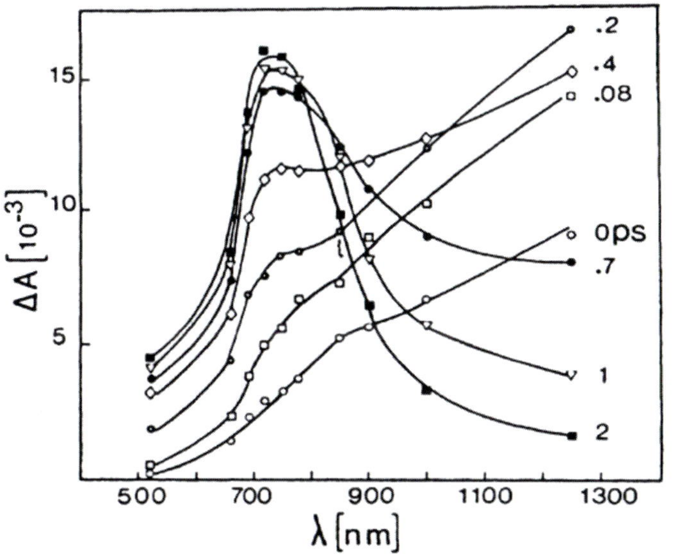

Figure 8 : *Absorption spectra of the electron at different delays after photo-ionisation of liquid water at 21 °C (from [26]). The solvation process is very fast as the hydrated electron (λ_{max} = 720 nm) is observed within 1 ps from precursors absorbing in the infrared domain (λ > 1200 nm).*

A continuous shift of the absorption spectrum of the solvated electron may be interpreted in two ways. First, this shift would be governed by solvent molecular reorientation around the electron (solvation), which has been suggested by a correlation between the relaxation time of the solvated electron and the dielectric relaxation time of the solvent [24]. Second, as suggested by Madsen *et al.* [30], the shift could be due to a "cooling" of the solvated electron *i.e.* a "cooling" of water around the solvation cavity since the spectrum of the solvated electron is known to shift to lower wavelengths at lower temperatures. This means that the absorption spectrum of the electron at any time during the relaxation process is identical to the spectrum of the solvated electron in the state of equilibrium with the solvent at some higher temperature. In this approach, the electron stabilisation is viewed as a succession of quasi-equilibrium states that are fully characterised by the time evolution of the local temperature. These two approaches of the "continuous shift" model are phenomenological: no explanation is given, on theoretical grounds, as to why such a picture of the electron solvation might be correct.

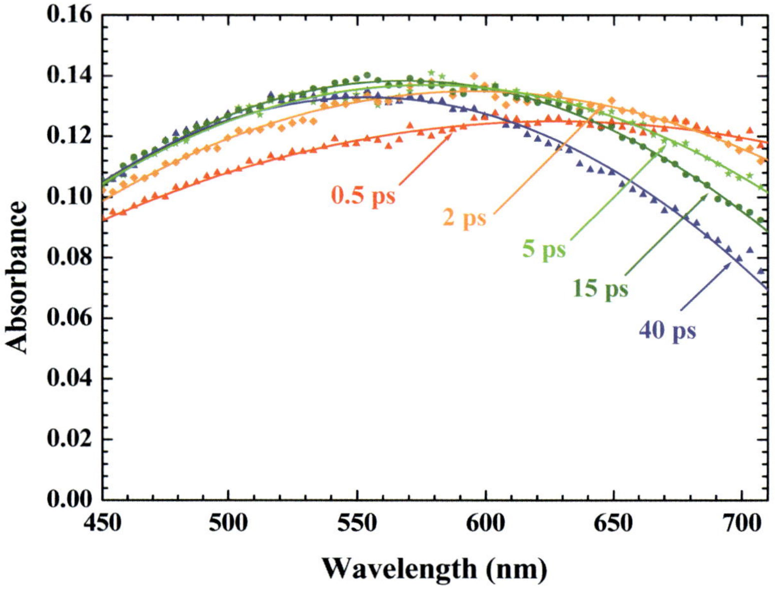

Figure 9 : *Transient absorption spectra of excess electrons in liquid ethane-1,2-diol for five time-delays after two-photon ionisation of the solvent by 200 fs pulses at 263 nm from [18]. The basic features of the solvated electron are already there after a few picoseconds.*

We recently studied the formation of the solvated electron in pure ethane-1,2-diol by photo-ionisation of the solvent [18,32]. The results showed that the excess electron presents a wide absorption band in the visible and near-IR domains at short delay times after the pump pulse, and that the red part of the absorption band drops rapidly in the first few picoseconds while the blue part increases slightly (**Fig. 9**). The time resolved spectra were fitted correctly by either one of two solvation models: a stepwise mechanism involving several distinct species and a continuous relaxation model. In **Figure 10** are reported, as an example, the kinetics and spectra of the three successive species (the weakly bound e_{wb}^-, the strongly bound e_{sb}^- and the solvated electron e_s^-) involved in the electron solvation dynamics according to the stepwise model.

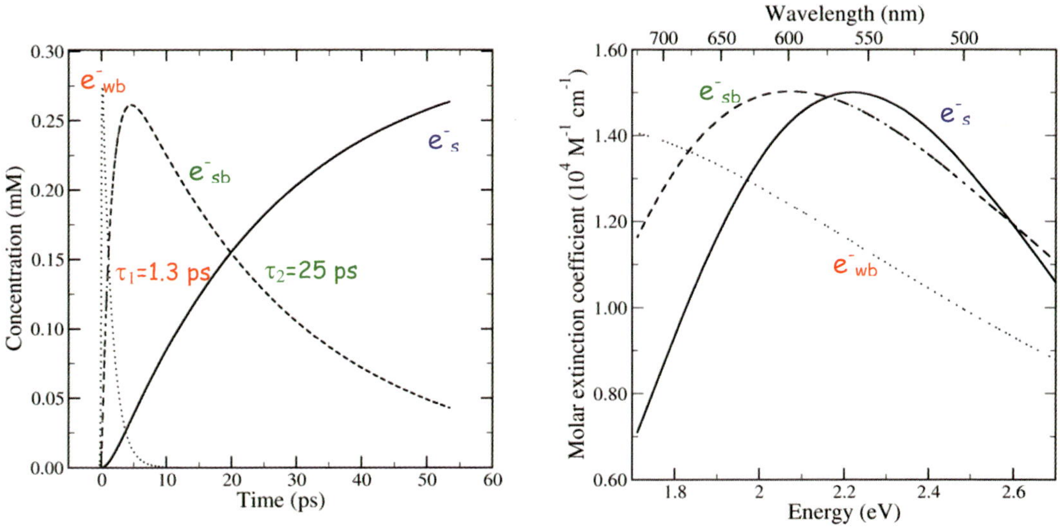

Figure 10 : *Time evolution of the concentrations and absorption spectra of the three electron species involved in the two-stepwise mechanism $e_{wb}^- \rightarrow e_{sb}^- \rightarrow e_s^-$ used to fit the experimental data obtained for the electron solvation in ethane-1,2-diol (Fig. 9) (from [32]).*

However, the fact that the time-evolution of the absorption spectrum of the solvated electron can be accurately described by the temperature-dependent absorption spectrum of the ground state solvated electron (**Fig. 11**) suggests that the spectral blue shift would be mostly caused by a continuous relaxation, or "cooling" of the electron trapped in a solvent cavity. To conclude, this analysis clearly indicates that it is not obvious to select a unique model to describe the solvation dynamics of electron in ethane-1,2-diol, and in other solvents.

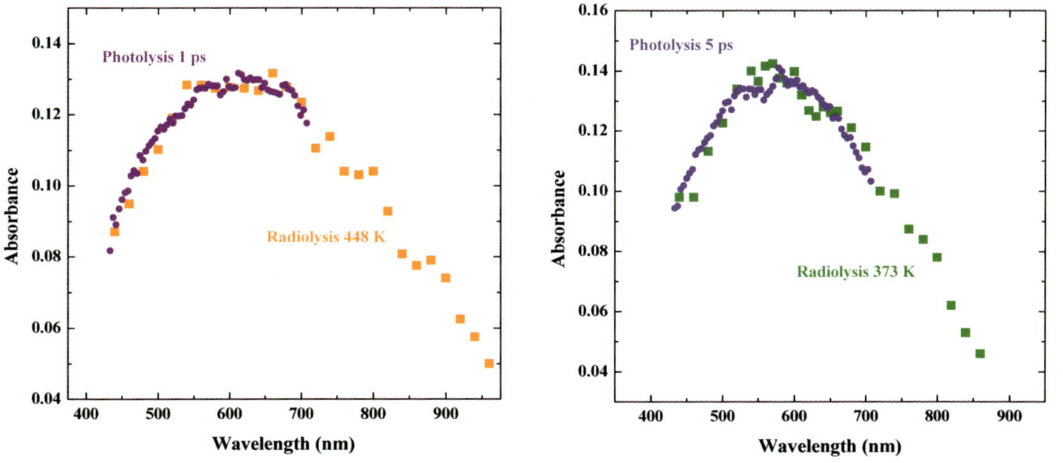

Figure 11 : *Similarity between the absorption spectra recorded in liquid ethane-1,2-diol at early time after two-photon ionisation (Fig. 9) and the absorption spectra of the solvated electron obtained at high temperature by pulse radiolysis (from [11] and [32]). These results suggest that the electron solvation is a continuous process similar to the continuous shift of the absorption spectrum of the solvated electron with temperature.*

Moreover, in a recent study on the of liquid water, Lian *et al.* [33] observed two distinct regimes of the spectral evolution of the electron. In the first picosecond after photoionisation, while the main absorption band of the electron progressively shifts to the blue (from 820 to 720 nm), a new peak at 1.15 µm and a shoulder at 1.4 µm are observed, which fully decayed in one picosecond. These features are attributed to the coupling between the excess electron and the modes of the water molecules lining the solvation cavity. At later delay times (t > 1 ps), the position of the band maximum is fixed, but the spectral profile continues to change by narrowing on the red side and broadening on the blue side. It appears that these results cannot be fitted by the presently available models and that more advanced models are needed. Very recently, by probing the electron delocalisation in water at attosecond time scale, D. Nordlund *et al.* [34] reported that an excess electron propagating along the H-bond network is trapped at a weakly H-bonded OH group. This trapping lasts long enough (> 20 fs) for electron solvation to occur, leading to the hydrated electron.

Conclusion

The solvated electron is a transient chemical species which exists in many solvents. The domain of existence of the solvated electron starts with the solvation time of the precursor and ends with the time required to complete reactions with other molecules or ions present in the medium. Due to the importance of water in physics, chemistry and biochemistry, the solvated electron in water has attracted much interest in order to determine its structure and excited states. The solvated electrons in other solvents are less quantitatively known, and much remains to be done, particularly with the theory. Likewise, although ultrafast dynamics of the excess electron in liquid water and in a few alcohols have been extensively studied over the past two decades, many questions concerning the mechanisms of localisation, solvation, and thermalisation of the electron still remain. Quantum and molecular dynamics simulations are necessary to unravel the structure of the solvated electron in many solvents and to better understand its properties.

References

[1] Hart E.J., Boag J.W., Absorption spectrum of the hydrated electron in water and in aqueous solutions, J. Am. Chem. Soc., 1962, **84**, 4090.

[2] Dorfman L.M., Jou F.-Y., Optical absorption spectrum of the solvated electron in ethers and in binary liquid systems, *in* "Electrons in fluids", Jortner J., Kestner N.R. (eds), Springer, New York, 1973,

447-459. Wishart J.F., Neta P, Spectrum and reactivity of the solvated electron in the ionic liquid methyltributylammonium bis(trifluoromethylsulfonyl)imide, J. Phys. Chem B, 2003, **107**, 7261-7267.

[3] Kraus C.A., Lucasse W.W., The conductance of concentrated solutions of sodium and potassium in liquid ammonia, J. Am. Chem. Soc., 1921, **43**, 2529-2539.

[4] Czapski G., Schwarz H.A., The nature of the reducing radical in water radiolysis, J. Phys. Chem., 1962, **66**, 471-474.

[5] Delaire, J.A., Delcourt, M.O., Belloni J., Mobilities of solvated electrons in polar solvents from scavenging rate constants, J. Phys. Chem., 1980, **84**, 1186-1189.

[6] Holroyd R.A., The electron: its properties and reactions, *in* "Radiation Chemistry, principles and applications", Farhaziz and M.A.J. Rodgers (eds), VCH, **1987**, 201.

[7] Rossky P.J., Schnitker J., The hydrated electron: quantum simulation of structure, spectroscopy, and dynamics, J. Phys. Chem., 1988, **92**, 4277-4285.

[8] Jou F.-Y., Freeman G.R., Shape of optical spectra of solvated electrons. Effect of pressure, J. Phys. Chem., 1977, **81**, 909-915.

[9] Jou F.-Y., Freeman G.R., Temperature and isotope effects on the shape of the optical absorption spectrum of solvated electrons in water, J. Phys. Chem., 1979, **83**, 2383-2387.

[10] Nicolas C., Boutin A., Levy B., Borgis D., Molecular simulation of a hydrated electron at different thermodynamic state points, J. Chem. Phys., 2003, **118**, 9689-9696.

[11] Mostafavi M., Lin M., He H., Muroya Y., Katsumura Y., Temperature-dependent absorption spectra of the solvated electron in ethylene glycol at 100 atm studied by pulse radiolysis from 296 to 598 K, Chem. Phys. Lett., 2004, **384**, 52-55.

[12] Lampre I., Lin M., He H., Zan Z., Mostafavi M., Katsumura Y., Temperature dependence of the solvated electron absorption spectra in propanediols, Chem. Phys. Lett., 2005, **402**, 192-196.

[13] Green N.J.B., Pilling M.J., Pimblott S.M., Clifford P., Stochastic modeling of fast kinetics in a radiation track, J. Phys. Chem., 1990, **94**, 251-258.

[14] Goulet T., Jay-Gerin J.-P., On the reactions of hydrated electrons with $OH^{\cdot}$ and H_3O^+. Analysis of photoionization experiments, J. Chem. Phys., 1992, **96**, 5076-5087.

[15] Crowell R.A., Bartels D.M., Multiphoton ionization of liquid water with 3.0-5.0 eV photons, J. Phys. Chem., 1996, **100,** 17940-17949.

[16] Thomsen C.L., Madsen D., Keiding S.R., Thögersen J., Christiansen O., Two-photon dissociation and ionization of liquid water studied by femtosecond transient absorption spectroscopy, J. Chem. Phys., 1999, **110**, 3453-3462.

[17] Kloepfer J.A., Vilchiz V.H., Lenchenkov V.A., Germaine A.C., Bradforth S.E., The ejection distribution of solvated electrons generated by the one-photodetachment of aqueous I^- and two-photon ionization of the solvent, J. Chem. Phys., 2000, **113**, 6288-6307.

[18] Soroushian B., Lampre I., Pernot P., De Waele V., Pommeret S., Mostafavi M., Formation and geminate recombination of solvated electron upon two-photon ionisation of ethylene glycol, Chem. Phys. Lett., 2004, **394**, 313-317.

[19] De Waele V., Sorgues S., Pernot P., Marignier J.-L., Monard H., Larbre J.-P., Mostafavi M., Geminate recombination measurements of solvated electron in THF using laser-synchronized picosecond electron pulse, Chem. Phys. Lett., 2006, **423**, 30-34.

[20] Buxton G.V., Greenstock C.L., Critical review of rate constants for reactions of hydrated electrons, hydrogen atoms and hydroxyl radicals in aqueous solution, J. Phys. Chem. Ref. Data, 1988, **17**, 513-886.

[21] Mackenzie K., Kopinke F.-D., Remmler M., Reductive destruction of halogenated hydro-carbons in liquids and solids with solvated electrons, Chemosphere, 1996, **33**, 1495-1513.

[22] Sun G.-R., He J.-B., Pittman Jr, C.U., Destruction of halogenated hydrocarbons with solvated electrons in the presence of water, Chemosphere, 2000, **41**, 907-916.

[23] Baxendale J.H., Wardman P., Direct observation of solvation of the electron in liquid alcohols by pulse radiolysis, Nature, 1971, **230**, 449-450.

[24] Chase W.J., Hunt J.W., Solvation time of the electron in polar liquids. Water and alcohols, J. Phys. Chem., 1975, **79**, 2835-2845.

[25] Kenney-Wallace G.A., Jonah C.D., Picosecond spectroscopy and solvation clusters. The dynamics of localizing electrons in polar fluids, J. Phys. Chem., 1982, **86**, 2572-2586.

[26] Migus, A., Gauduel Y., Martin J.L., Antonetti A., Excess electrons in liquid water: first evidence of a prehydrated state with femtosecond lifetime, Phys. Rev. Lett., 1987, **58**, 1559-1562.

[27] Shi X., Long F.H., Lu H., Eisenthal K.B., Electron solvation in neat alcohols, J. Phys. Chem., 1995, **99**, 6917-6922.

[28] Assel M., Laenen R., Laubereau A., Dynamics of excited solvated electrons in aqueous solution monitored with femtosecond-time and polarization resolution, J. Phys. Chem. A, 1998, **102**, 2256-2262.

[29] Hertwig A., Hippler H., Unterreiner A.-N., Transient spectra, formation and geminate recombination of solvated electrons in pure water UV-photolysis: an alternative view, Phys. Chem. Chem. Phys., 1999, **1**, 5633-5642.

[30] Madsen D., Thomsen C.L., Thögersen J., Keiding S.R., Temperature dependent relaxation and recombination dynamics of the hydrated electron, J. Chem. Phys., 2000, **113**, 1126-1134.

[31] Pépin C., Goulet T., Houde D., Jay-Gerin J.-P., Femtosecond kinetic measurements of excess electrons in methanol: substantiation for hybrid solvation mechanism, J. Phys. Chem., 1994, **98**, 7009-7013.

[32] Soroushian B., Lampre I., Bonin J., Pernot P., Pommeret S., Mostafavi M., Solvation dynamics of the electron produced by two-photon ionisation of liquid polyols. 1. Ethylene glycol, J. Phys. Chem. A, 2006, **110**, 1705-1717.

[33] Lian R., Crowell R.A., Shkrob I.A., Solvation and thermalization of electrons generated by above-the-gap (12.4 eV) two-photon ionization of liquid H_2O and D_2O, J. Phys. Chem. A, 2005, **109**, 1510-1520.

[34] Nordlund D., Ogasawara H., Bluhm H., Takahashi O., Odelius M., Nagasono M., Petterson L.G.M., Nilsson A., Probing the electron delocalization in liquid water and ice at attosecond time scale, Phys. Rev. Lett., 2007, **99**, 217406.

Chapter 4

Water radiolysis under extreme conditions. Application to the nuclear industry

Gérard BALDACCHINO and Bernard HICKEL

Introduction

Water radiolysis in the nuclear industry comes from the interaction of radiation (alpha, beta, gamma, recoil nuclei) with enough energy to interact with liquid water. As water is used as coolant in current Generation 2 and 3 nuclear reactors (PWR, BWR, EPR...), water decomposition is expected. This phenomenon was already recognized a century ago, and since then numerous chemists have succeeded in interpreting the physical and chemical mechanisms involved (Chapter 1) [1]. Nevertheless, water radiolysis under extreme conditions of temperature, pressure, pH and high Linear Energy Transfer (LET represents the loss of energy per unit length, LET = –dE/dx) is not fully understood due to its inherent rich and complex chemical reactivity.

In addition to understanding the interaction of radiation with water, the nuclear industry must obviously also take into account the excess production of molecular hydrogen and hydrogen peroxide, and control this excess in order to avoid explosive conditions and corrosion of the water circuitries. Due to the working conditions of the current reactors (T > 310 °C, P > 100 atm in Pressurized Water Reactor, PWR), it is mandatory to predict the evolution of the chemistry when submitted to high temperature and pressure. Nevertheless, a few experiments have shown that the linear Arrhenius law model is not applicable at temperatures above 250 °C. Hydrogen production overestimates have been necessary in order to render the safety control of these reactors as efficient as possible. Only in recent years have experiments been performed at high temperature, and fast (picosecond) processes after the ionization step have begun to be observed under almost supercritical conditions ($T > 374$ °C, $P > 21$ MPa). Temperature is one of the parameters suspected to affect the complex chemistry in water under radiation. Type and energy of the ionizing particles (which determine the LET), pressure and pH, are also known to affect the fast processes of recombination because they change the initial distribution and concentrations of the reactive species in the medium. Additionally, extreme conditions are also encountered in storage conditions of radioactive materials (high pH values, high pressure in porous materials...)(Chapter 8). These factors mean the fundamental aspects of water radiolysis represent a great and important challenge in the future of the nuclear energy industry.

The reactions of free radicals generated by the radiolysis of water (Chapter 1), such as hydrated electrons, hydroxyl radicals, superoxide radicals, hydrogen atoms, are now studied under extreme conditions using heavy ion beams of high LET generated by high energy cyclotrons and a high temperature up to and over the supercritical state ($T > 374$ °C, $P > 21$ MPa).

In future generations of nuclear reactors – especially supercritical water reactors (SCWR), 4th generation nuclear reactors and the ITER project (International Thermonuclear Experimental Reactor) – water should still be considered as a suitable coolant fluid, but it will be submitted to more extreme conditions of temperature and LET (high flux of neutrons). All contemporary studies show that it will be beyond reach to extrapolate the existing simulations to these new conditions without experimental determinations of essential parameters such as radiolytic yields and rate constants.

This chapter will focus on the influence of two essential parameters on the chemical mechanism of water under radiation: Linear Energy Transfer and temperature.

Influence of the type of radiation

In contrast to the situation that prevails in photochemistry, there is no compound transparent to high energy radiation. In water for example, the LET of high energy photons (such as gamma rays) is 0.2 keV *per* micrometer, while for visible light, it would be zero. In dilute aqueous solutions, the chemical changes observed on the solute are due indirectly to radicals generated from the solvent water, and not due to the direct radiation effect on the solute (Chapter 1).

The LET value is used to calculate the energy delivered to the material (the dose) but a similar value of LET can be attributed to different types of particles at different values of energy, with various resultant chemical effects. In order to identify the production yield of free radicals or molecules related to the type of ionizing particle (gamma rays, alpha rays or high energy heavy ions), other parameters should be used, such as the specific energy and the charge of the particle. The geometrical structure of the track takes greater and greater importance in the interpretation of the experimental and simulation results (**Fig. 1** gives an example of a track structure for an alpha ray of a few MeV). It has also been proposed that the initial energy density transferred to the matter might be an adequate parameter to predict the concentration level of molecular species (H_2 for example), as shown in recent publications based on pulse radiolysis experiments (**Fig. 2** is an example of the experimental set-up) [2].

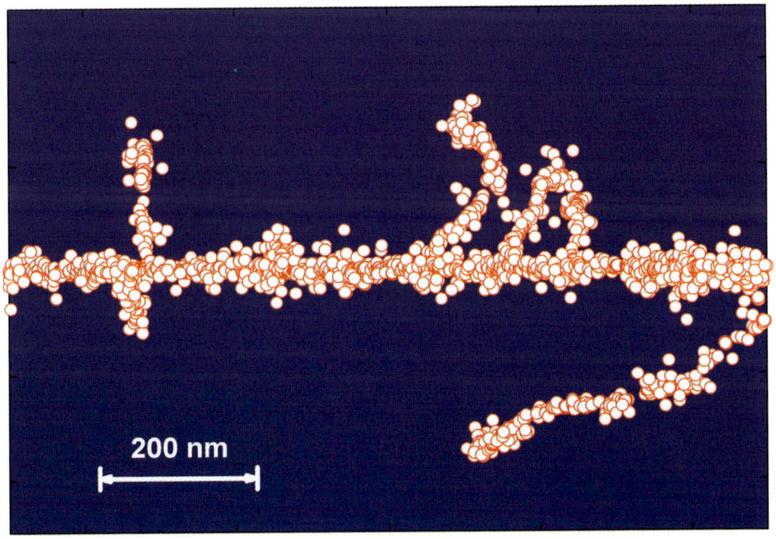

Figure 1 : *Simulated ionization track due to an alpha ray of a few MeV of energy coming from the left side in the «blue» water continuum. Each red circle is an ionization event. This local distribution, in the nanometer range, is the beginning of a complex chemistry. Ionizations occur mainly around the trajectory axis of this incident ion, and this area is named "core track". Some high energy electrons can be ejected and they can form their own track named "delta ray". When delta rays are sufficiently numerous (that depends on the incident ion energy and charge) a new area around the core can be named "penumbra". The penumbra has the characteristics structure of a "low LET area" because the ionizations are produced by high enrgy electrons.*

Decomposition and relaxation

When radiation interacts with matter, the energy involved is much higher than the bond energy of the molecules. Intuitively one would thus expect all chemical bonds to be

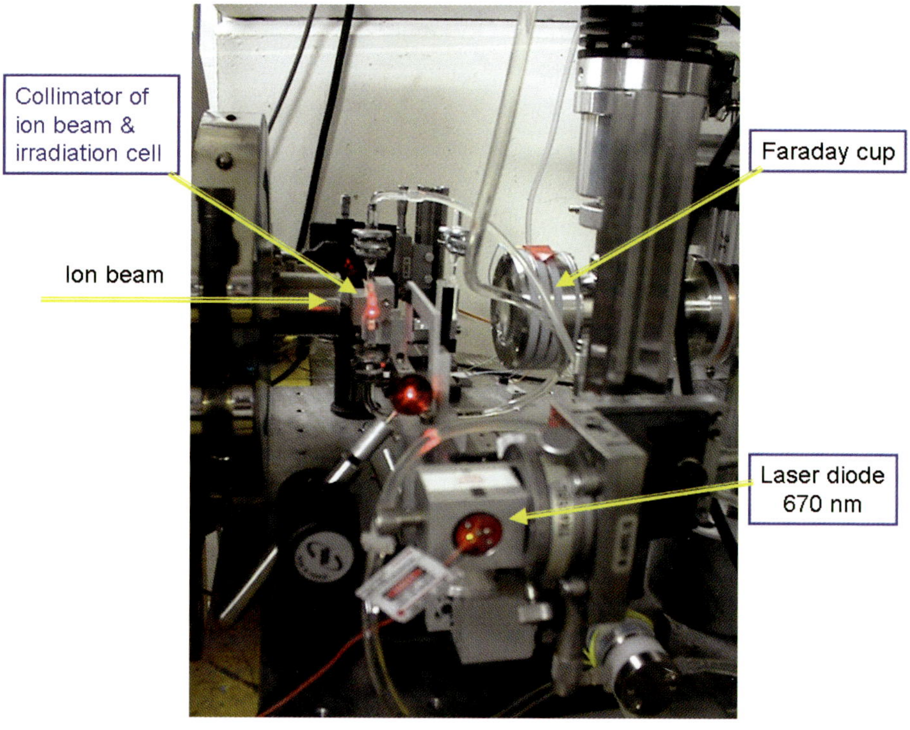

Figure 2 : *Picture of the experimental set-up using a red laser light to detect the hydrated electron in a flow cell by using the time-resolved absorption spectroscopy. High energy heavy ions were provided by the cyclotron GANIL, Caen, France. The safety protections against the laser light were taken off for the photography needing.*

immediately broken. This is however a common misconception since the duration of the interaction is less than 10^{-15} s and within this time span only ionizations and electronic excitations can occur. The bond scissions take place at a later stage during the relaxation of the system. All liquids are expected to be partly decomposed by radiation, and indeed they are, except some exotic species such as liquid sodium (where there is no bond to be broken after ionization) which is used to cool the breeder reactors. In the case of liquid hydrogen fluoride, the resultant compounds of decomposition (H_2 and F_2) probably react spontaneously to return to the state of HF.

Radiolysis of water: a puzzling case

The case of water decomposition under radiation is rather complex (Chapter 1). It has been known for a long time that, when carefully degassed and irradiated with low LET radiation such as X-rays or ^{60}Co gamma rays, pure water gives rise to almost no products, the amount of which remains at a very low steady state concentration. It seems that water decomposes at the very beginning of the irradiation and then becomes inert. But for high LET radiation such as alpha rays, water decomposition occurs with the formation of molecular hydrogen, hydrogen peroxide and molecular oxygen. When the water contains air or other solutes, water decomposition is always observed whatever the LET. In the presence of dissolved H_2, water decomposition is completely suppressed at low LET, but still occurs under alpha rays irradiation.

The explanation of this behavior, which has paramount importance to the nuclear industry, was given by Allen [3]. After World War II, Allen depicted a model of water decomposition under radiation that considers the production and consumption of H_2. The key role of $OH^•$, H_2 and O_2 involved in the chain as a carrier or a breaker is clear (see **Inset**). Within this chain reaction, the reaction between H_2 and O_2 (which is thermodynamically favorable) takes place in water at high temperature only in the presence of a catalyst such as copper or silver cations. In the radiolysis of water, the reaction can take place at room temperature in the presence of free radicals which form the molecular products H_2, H_2O_2 and O_2 at the first step inside the tracks or the spurs. Subsequently in the bulk of the solution, the free radicals which have escaped recombination in the tracks recombine as molecular products into water. The molecular products are formed in the nanosecond range and their recombination takes place in the millisecond range.

Inset : The series of reactions involved in the chain reaction depicted by Allen

First note that the global water radiolysis reaction can be written :

$$H_2O \xrightarrow{\text{radiation}} H^•, OH^•, e_{aq}^-, H_2, H_2O_2 + HO_2^• + H_3O^+ \quad (0)$$

We observe that O_2 is not the primary species in the water radiolysis; it is formed by secondary reactions. The key reaction which explains the unique role of H_2 is:

$$OH^• + H_2 \longrightarrow H_2O + H^• \quad (1)$$

This reaction converts an oxidizing radical $OH^•$ into a powerful reducing species, the $H^•$ atom. The first step is the reduction of O_2 in H_2O_2

$$H^• + O_2 \longrightarrow HO_2^• \quad (2)$$

$$HO_2^• + HO_2^• \longrightarrow H_2O_2 + O_2 \quad (3)$$

The second step is a chain reaction between $H^{\bullet}$ and H_2O_2

$$H^{\bullet} + H_2O_2 \longrightarrow OH^{\bullet} + H_2O \qquad (4)$$

followed by reaction 1 which gives back an $H^{\bullet}$ atom.

Whilst O_2 is still present, the chain reaction does not take place, because the rate constant of reaction 2, which is a chain breaker, is 200 times larger than the rate constant of reaction 4 which is a chain carrier.

There is another chain termination which is important to explain the paradoxical behavior of water radiolysis.

$$OH^{\bullet} + H_2O_2 \longrightarrow H^{\bullet} + HO_2^{\bullet} \qquad (5)$$

We can see that H_2O_2 is both a chain carrier (reaction 4) and a chain breaker (reaction 5). When its concentration increases too much it prevents the chain reaction by forming O_2 (reaction 5 followed by reaction 3).

The rate constants k_5 and k_1 have the same order of magnitude, so it is necessary to add a sufficient concentration of H_2 to insure that reaction 1 is always faster than reaction 5.

The net results of these reactions is

$$O_2 + H_2 + 2H^{\bullet} \longrightarrow 2H_2O \qquad (6)$$

Reaction 6 shows that we need at least two $H^{\bullet}$ radicals for one H_2 and O_2.

At low LET radiation like gamma and X-rays, this is easily achieved and the water behaves as if it was completely inert under irradiation. At high LET (alpha rays) the ratio of the molecular products over the free radicals increases and there are comparatively too few $H^{\bullet}$ to O_2 atoms, The recombination of H_2 and O_2 is thus incomplete, the chain reaction between H_2 and H_2O_2 cannot start and water decomposition is observed.

This set of reactions also clearly explains why water radiolysis takes place when some solutes are present. Most of the chemical species react with the $H^{\bullet}$ atom or the $OH^{\bullet}$ radical and prevent the chain reaction.

The special case of $HO_2^{\bullet}$

The particular case of superoxide radical ($HO_2^{\bullet}/O_2^{\bullet-}$) in water radiolysis comes from the property that its formation yield increases with LET, a behavior contrary to fast and natural recombination of other radicals in dense ionization tracks [4]. Logically, this increased radical recombination rate accelerates the production of molecular species (H_2 and H_2O_2) and decelerates the production of radicals. The superoxide species is probably formed in the track-core. However, it can easily escape the track to survive in the water bulk since its reactivity in water is low, mostly by a slow disproportionation (reaction 7),

$$HO_2^{\bullet} + O_2^{\bullet -} \xrightarrow{H_2O} H_2O_2 + O_2 + OH^- \qquad k = 9.7 \times 10^7 \text{ l mol}^{-1} \text{ s}^{-1} \qquad (7)$$

This reaction has several consequences in radiobiology concerning the sensitization of living cells submitted to heavy ion beams for radiotherapy (hadrontherapy). The efficiency of hadrontherapy with carbon ions mainly results from the property of local energy deposition in the Bragg peak region but can also be the consequence of molecular oxygen generation (reaction 7) in tumors which are known to be hypoxic cells. Some studies have tried to explain how the superoxide radical is formed in the tracks of heavy ions, and have proposed several possible mechanisms. A high local concentration of ionizations can favor reactions which can then interfere with geminate recombination. The production of $O^{\bullet}$ atoms has for a long time been suspected in the core track. Multiple ionizations of water molecules can be a source of $O^{\bullet}$ atoms because the huge energy deposited in the medium is considerably greater than the total energy needed to ionize the total number of water molecules along the ion track. This model has recently been exploited in Monte Carlo simulations [5] for which **Figure 3** presents a comparison between experimental and calculation results. There is a fair agreement between the few experimental results concerning $HO_2^{\bullet}/O_2^{\bullet -}$ and H_2O_2. Nevertheless, more exhaustive experiments are necessary to validate this model for values of LET higher than 1000 eV/nm.

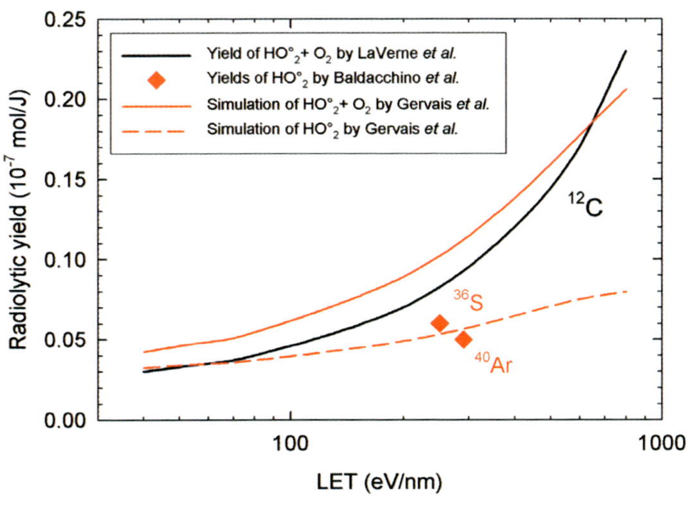

Figure 3: *Radiolytic yields of $HO_2^{\bullet}$ as a function of LET. Experimental data include the yield of O_2 due to a particular experimental method for determining the yield.* Carbon ions (black line) had various energies and then various LET values [4]. The red lines are from recent simulation results for $HO_2^{\bullet} + O_2$ [5] ; the red dash-line represents only $HO_2^{\bullet}$ for which two experimental data are available (red symbols obtained for Argon and Sulfur high energy ions, see [2] and references therein).

Application to nuclear reactors

In Pressurized Water Reactors (PWR) which are the type of reactors used by EDF for electricity production in France, it is important to avoid the formation of O_2 (and H_2O_2) arising from water radiolysis, since this leads to corrosion of the primary loop and increases the radioactivity of water. Minimization is achieved by adding H_2 and removing O_2. The other additives, boric acid and LiOH, used to absorb neutrons and regulate the water pH, do not react with the free radicals and cannot interfere with the chain reaction. The system works well owing to the intense flux of gamma radiation (low LET). The other types of radiation, fast neutrons and especially the alpha particles generated by the nuclear reaction between a thermal neutron and boron

$$^{10}B + n \rightarrow {}^{7}Li + {}^{4}He \qquad (^{4}He \text{ has an energy of 2.2 MeV}) \qquad (8)$$

have a much higher LET, and in the absence of gamma flux the water will be decomposed. This phenomenon is clearly visible in **Figure 4**. Water decomposition begins at a particular threshold of α/γ ratio. At a α/γ ratio of 8, a sharp change in the chemistry between alpha-ray radiolysis and gamma ray radiolysis is observed [6].

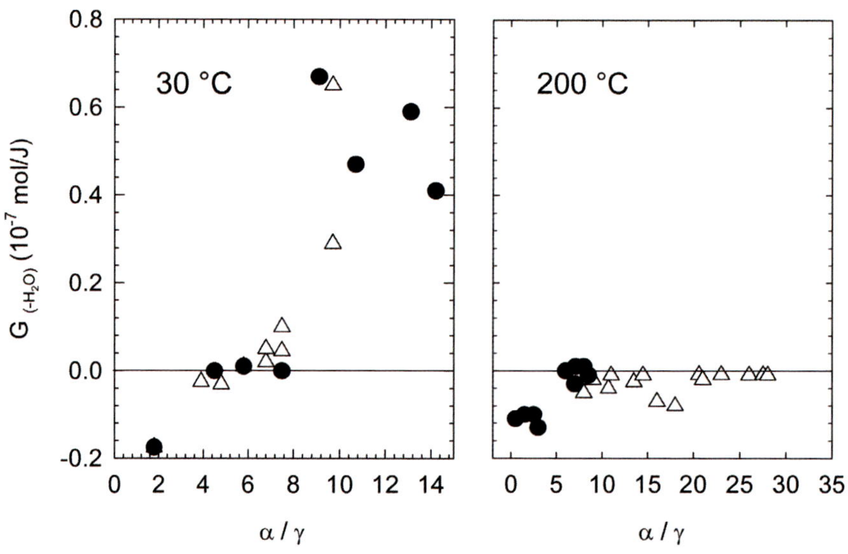

Figure 4 : *Water decomposition yield, G(-H_2O), as a function of the flux ratio alpha /γ irradiation (mixture of high and low LET particles). Increasing this ratio means increasing the LET value. The observed decomposition threshold is shifted to the right (higher LET) and disappears betwen 30 °C and 200 °C. The scale corresponds also to the concentration in ^{10}B. The difference between circle and triangle symbols is the initial enrichment in ^{10}B in the chemicals. Adding ^{10}B that absorbs neutrons under neutron/gamma irradiation produces alpha rays having high LET (200 eV/nm)* [6].

Water in the ITER reactor

A similar problem occurs for the future fusion reactor prototype, ITER, which is to be built in Cadarache, France (Project view in **Figure 5**). The first reactor wall will be cooled by water and again it is mandatory to avoid the formation of O_2 and H_2O_2. The nature of the radiation is different from that of a fission reactor and the radiolysis will be more severe, since 90% of the irradiation comes from high LET radiation (the 14 MeV fusion neutrons) and only 10% from low LET radiation (the gamma flux). We have performed experiments under a mixture of low and high LET radiation types to check whether the water radiolysis remains inhibited under the ITER conditions. By continuously varying the ratio of low to high LET radiation, we have observed a threshold at which water starts to abruptly decompose (**Fig. 4**). The position of the threshold depends on the temperature (**Fig. 4**, at 200 °C the decomposition disappears). Other conditions influence the water decomposition: LET, dose rate, temperature and the presence of impurities in water, such as copper cations which shift the threshold toward lower LET. In the ITER reactor, the irradiation conditions are fixed, and the only variable parameters are the concentrations of hydrogen and impurities in the cooling water.

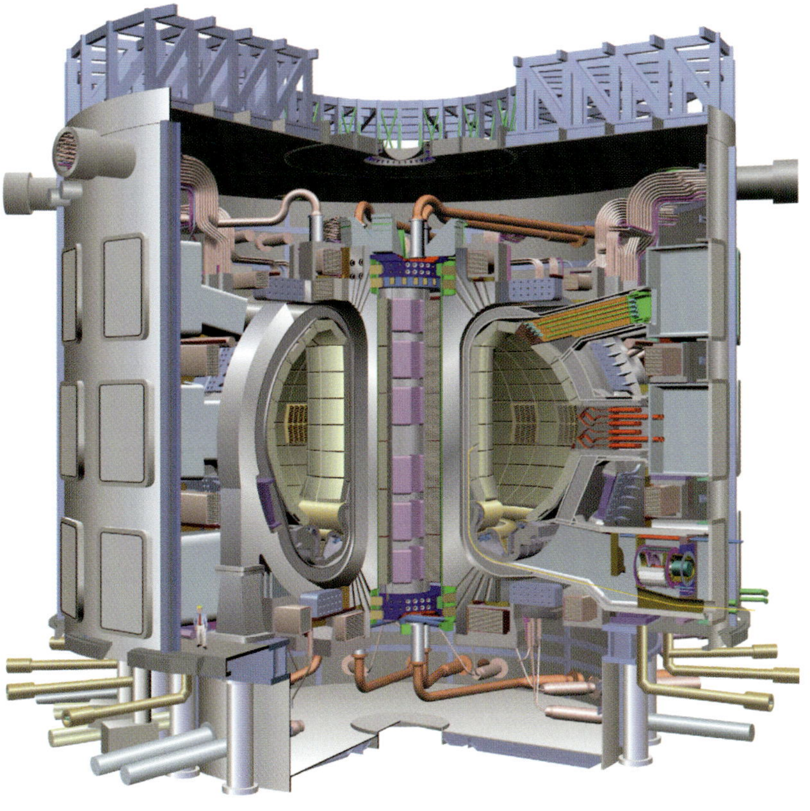

Figure 5 : ***View of the ITER project.*** *Water circuity will be submitted to a high flux of 14 MeV neutron and γ-rays. These radiations come from the fusion reaction in the plasma generated in the tokomak.*

We have shown that impurities such as copper cations have a great influence on the decomposition as it is presented in **Figure 6** by the production of hydrogen peroxide.

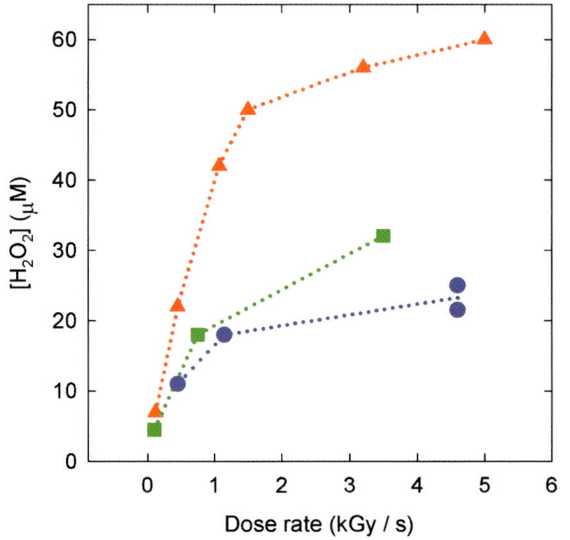

Figure 6 : *Influence of the dose rate and the copper cation concentration under irradiation with 1 GeV C^{6+} (LET = 26 eV/nm) at GANIL.* ●: *$[Cu^{2+}] = 0$* ; ■: *$[Cu^{2+}] = 1\ \mu mol\ l^{-1}$* ; ▲: *$[Cu^{2+}] = 6\ \mu mol\ l^{-1}$.*

Towards supercritical conditions

By increasing the temperature, the rate constants of the reactions occurring in the nanosecond/microsecond stage change according to the Arrhenius law up to 200-300 °C. The radiolytic yields of the radical species are similarly increasing with temperature. Recent publications tend to show that this increase is due to the acceleration of diffusion with temperature of the free radicals in the heterogeneous step in the pico-nanosecond range following ionization. **Figure 7** shows the main result of the single picosecond high temperature experiment yet performed. These experimental validations were obtained by pulse radiolysis (Chapter 1), using picosecond pulses of high energy electrons, delivered by the ELYSE picosecond electron accelerator in University of Paris-Sud, Orsay [7]. Below 300 °C the Arrhenius plot becomes non linear and the rate constants are no longer predictable. The challenge in 4th Generation nuclear plants of the SCWR project (Supercritical Water Reactor, see the scheme of the project in **Figure 8**) is to understand the chemical mechanisms in water irradiated under the extreme conditions of supercritical water (T > 374 °C, P > 21 MPa). Recent works presenting experimental results show a strong pressure dependence of the radiolytic yields [8]. We do not yet clearly understand whether this dependence is related to the change of the density of the medium or to the pressure activation of reactions.

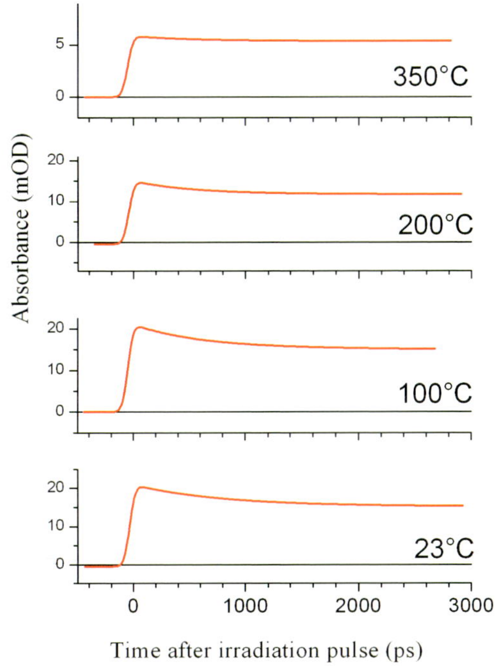

Figure 7 : ***Picosecond kinetics of hydrated electron recombination in the ionization spurs as a function of the temperature*** [7]. *Hydrated electron is produced in 15 ps pulse of 8 MeV electrons delivered by the electron accelerator ELYSE. The absorbance of 2 cm of pure water is analyzed by a laser at 790 nm. Due to the red-shift of the hydrated electron spectrum with increasing temperature, the absorbance maximum value at 790 nm decreases from 23 °C to 350 °C. In the time range of 3 ns, the recombination of hydrated electron appears less and less efficient with increasing temperature.*

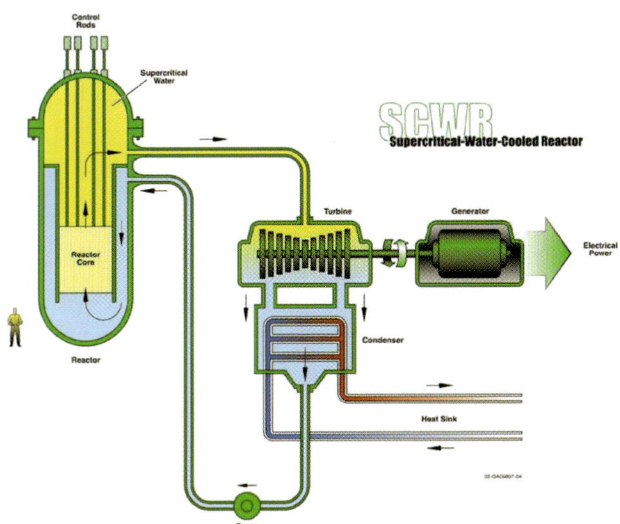

Figure 8 : ***Scheme of the SCWR project.*** *Supercritical water at about 510 °C and 25 MPa will be used as the coolant material.*

Challenge for the future

The challenge for the future is to explain radiation-matter interactions under a wide range of conditions of temperature, pressure, LET and pH. This will be approached by simulating and comparing interactions with experimental time-resolved results in the picosecond range. The realization of this challenge requires the development of increasingly sensitive detection methods and increasingly accurate simulations. It should be noted that, besides the nuclear industry which stands to benefit directly from the application of this research, the fields of radiobiology, medicine and astrobiology are also highly concerned by a better understanding of the mechanisms of this exotic chemistry.

References

[1] Mozumder A., Hatano Y. (eds), "Charged particle and photon interactions with matter. Chemical, physicochemical, and biological consequences with applications", Marcel Dekker, New York, 2004.

[2] Baldacchino G., Vigneron G., Renault J.-P., Pin S., Abedinzadeh Z., Deycard S., Balanzat E., Bouffard S., Gardès-Albert M., Hickel B., Mialocq J.-C., A nanosecond pulse radiolysis study of the hydrated electron with high energy ions with a narrow velocity distribution, Chem. Phys. Lett., 2004, **385**, 66-71.

[3] Allen A.O., "The radiation chemistry of water and aqueous solutions", Van Nostrand, New York, 1961.

[4] LaVerne J.A., Radiation chemical affects of heavy ions, *in* "Charged particle and photon interactions with matter. Chemical, physicochemical, and biological consequences with applications", Mozumder A., Hatano Y. (eds), Marcel Dekker, New York, 2004, 403-429.

[5] Gervais B., Beuve M., Olivera G.H., Galassi M.E., Rivarola R.D., Production of HO_2 and O_2^- by multiple ionization in water radiolysis by swift carbon ions, Chem. Phys. Lett., 2005, **410**, 330-334.

[6] Pastina B., Isabey J., Hickel B., Radiolyse de l'eau dans les réacteurs nucléaires à eau pressurisée. Effet de l'hydrogène, J. Chim. Phys., 1997, **94**, 226-229.

[7] Baldacchino G., De Waele V., Monard H., Sorgues S., Gobert F., Larbre J.P., Vigneron G., Marignier J.L., Pommeret S., Mostafavi M., Hydrated electron decay measurements with picosecond pulse radiolysis at elevated temperatures up to 350 °C, Chem. Phys. Lett., 2006, **424**, 77-81.

[8] Lin M., Katsumura Y., Muroya Y., He H., Wu G., Han Z., Miyazaki T., Kudo H., Pulse radiolysis study on the estimation of radiolytic yields of water decomposition products in high-temperature and supercritical water: use of methyl viologen as a scavenger, J. Phys. Chem. A, 2004, **108**, 8287-8295.

Part II

Radiation chemistry mechanisms and applications

Chapter 5

Molecular formation in the interstellar medium

Nigel J. MASON, Anita DAWES and Philip HOLTOM

Introduction

Ever since the development of consciousness human beings have been asking themselves that most basic question : where they came from ? The "answer" to this question has been a fundamental tenant of most religions, with each having a "genesis" story to explain how humanity came into being. With the emergence of science and our ability to explore our own solar system and look out into the universe (**Fig. 1**) the question of whether we are "alone" or if life has developed elsewhere has intrigued scientists and the public alike, with science fiction becoming one of the most popular genres of both literature and film. Hence, one of the greatest challenges of modern science is to determine how life began on Earth and whether the conditions for life to appear and develop are common across the Universe. Indeed, the discovery of life (even in its simplest form) on a planet other than Earth would be a defining moment in human history, while understanding the origins of life itself would be a crowning achievement of human scientific endeavour.

Today, for the first time, it is possible to develop a realistic, rigorous scientific programme to address these questions and begin to provide answers to these fundamental questions. Such a research programme has led to the development of a new scientific discipline, **astrobiology,** a necessarily interdisciplinary field of research bringing together chemists, physicists, biologists, as well as astronomers, geologists, engineers and even philosophers all with the common aim of both determining the origins of life and developing instrumentation to explore our own solar system and beyond for evidence of life.

Figure 1 : *The Heart and Soul Nebulas. Is the heart and soul of our Galaxy located in Cassiopeia? Possibly not, but that is where two bright emission nebulas nicknamed "Heart" and "Soul" can be found. The Heart Nebula, officially IC 1805 on the right, has a shape reminiscent of a heart . The nebulas are brightly coloured due to red light emitted from excited hydrogen gas. Deep in the nebulae forming from the dust and gas several clusters of young stars are visible. Such nebulae are also the chemical factories of many molecules including perhaps the precursors of life itself. [Credit : Richard Powell, Digitized Sky Survey, Palomar Observatory, STScI]*

The Origins of life

The fundamental laws of chemistry are believed to be universal, that is they are the same in any part of the universe. Thus in seeking to determine whether life has evolved elsewhere in the Universe we must seek to understand the chemical origins of life. The chemical ingredients of life are well known, the formation of self replicating molecules capable of expressing a genetic code (*e.g.* DNA) and molecules necessary for the containment of such molecules in a single entity, *e.g.* lipids forming the membrane of a cell. In addition, since life needs a medium in which to exchange chemicals, the presence of water is essential as a "universal solvent". The mechanisms by which such "prebiotic" molecules are formed is the topic of the rest of this brief review. Our understanding of how such prebiotic molecules were subsequently assembled on Earth to form the larger more complex cellular structures remains unknown with several theories having been developed but no experimental evidence to support any of these hypothesis has yet been provided. Indeed we remain a long way from being able to "create life in the lab".

In exploring the origins of life, we are faced with one of two possibilities: either the molecules of life were assembled on Earth or the building blocks of life were formed as part of the more general astrochemistry that gave birth to the solar system. The former leads to a more "Earth Centric" view of the universe suggesting that the Earth fulfilled some rather special criteria to allow molecules to assemble while the latter is more conducive to life being a universal phenomenon equally probable in any other solar system.

Knowing that organic molecules and microorganisms can be preserved as fossils, a record of the appearance of life on Earth and its early evolution, as well as the biosphere environment, should be preserved in rocks dating from the earliest epochs. The most salient problem, however, is to find such rocks. There are no rocks dating from the first 500 million years of Earth history because they have been destroyed by plate tectonics and other thermal processes. However there are examples of very well preserved rock formations *e.g.* 3.5 Ga-old in Australia and South Africa and the 3.8 Ga-old rocks of Isua, Greenland. These rocks, formed *just one billion years (1 Ga)* after the consolidation of the Earth, are believed to contain the fossilized remains of microorganisms, their colonies, biofilms and/or the geochemical signatures left by their metabolic activities.

Such fossil evidence itself provides scientists with a major challenge since it suggests that life was able to emerge very quickly from the earliest chemical conditions. Thus if the ingredients of life were formed on the Earth the chemical processes must have "simple" and capable of being sustained in what we believe to have been a very inhospitable regime, with no oxygen atmosphere and hence no ozone layer to protect the surface from biologically harmful UV irradiation, a surface prone to impact by meteorites and with active volcanism. Alternatively, the basic ingredients of life might have appeared formed as a result of the chemical processes that established our solar system and that are inherent in the formation of planetary systems.

In the 1950s in a pioneering experiment (**Fig. 2**) Stanley Miller and his student Harold Urey suggested that the molecules of life might have been formed in the early Earth's atmosphere. Placing a mixture of methane (CH_4), ammonia (NH_3), hydrogen (H_2), and water (H_2O) in a glass bulb – a mixture they thought was a good representation of the chemical constituents of the Earth's early atmosphere – they ran a continuous electric current through the system to simulate lightning storms, believed to be common on the early Earth. Analysis of the products of the discharged gas was performed using chromatography. After running the discharge for a week 10-15% of the carbon had been converted into new organic compounds including some 2% transformed into amino acids which in biology may be used to make proteins. This experiment demonstrated conclusively that molecular compounds, such as amino acids, which are essential to cellular life, could be made relatively easily by providing energy to a "chemical soup".

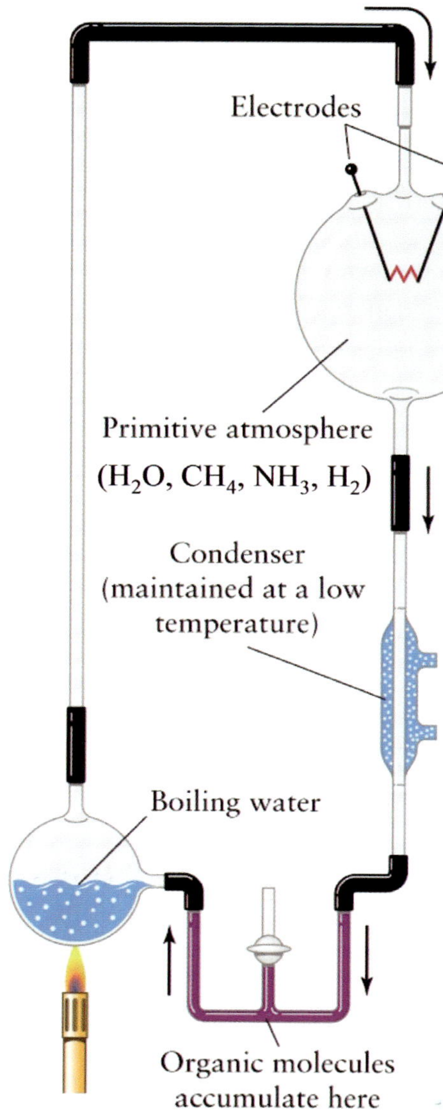

Figure 2 : *The Urey-Miller Experiment. The glass vessel to the lower left of the picture is used for the water reservoir and mimics oceans on the early Earth. An electrical discharge of a gas mixture mimicking the composition of the early Earth's atmosphere takes place in the large bulb at the top right. Molecules formed in the discharge may be condensed and are collected at the bottom of the system.*

However we now know that the Urey-Miller experiment is not in fact a good representation of the conditions of the early Earth since it is now believed that the early Earth's atmosphere was not reducing. Instead, scientists believe the atmosphere was full of oxidants, such as CO_2 and N_2. An oxidizing atmosphere is essentially neutral, and does not permit organic chemistry to occur. Furthermore while it is believed lightning storms were extremely common on the primitive Earth, they were not continuous as the Miller-Urey experiment portrayed. Thus it has been argued that, while amino acids and other organic compounds may have been formed, they would not have been formed in the amounts which this experiment produced. Hence, the time ranges necessary for the development of sufficient concentrations of such prebiotic molecules are too long to explain the emergence of life as witnessed by the fossil record – indeed some estimates suggest that, if the Urey-Miller hypothesis was to be adopted, then we would still be waiting for life to evolve on Earth.

So if these compounds were not created in a reducing atmosphere here on Earth as Miller and Urey suggested, then where did they actually come from? Perhaps they are formed in space itself as part of the process by which planets and stars are formed?

The Interstellar medium as a chemical factory

The interstellar medium (ISM) is the region of space between the stars (**Fig. 1**) that appears dark since gas and dust in these regions absorb the light from stars behind them. However, these regions still have very low densities (they constitute a vacuum far better than can be produced artificially on the surface of the Earth). From recent astronomical observations we now know that the interstellar medium is rich in molecules (**Table 1**) including vinegar (acetic acid), benzene and even simple sugars such as glycolaldehyde. Larger more complex organics, such as the simple amino acid glycine, may also be present but require the development of better instrumentation before they can be detected. The ISM is therefore a rich depository of prebiotic molecules and it is from the ISM that stars (and the planets) are formed.

Table 1· *Molecules detected in the Interstellar medium.*

2 atoms	SiN	HNC	C_2H_2·	HC_3N	**7 atoms**	$(CH_3)_2O$
H_2	SiO	HNO	CH_2D^+ ?	HC_2NC	C_6H	CH_3CH_2OH
AlF	SiS	MgCN	HCCN	HCOOH	CH_2CHCN	HC_7N
AlCl	CS	MgNC	$HCNH^+$	H_2CNH	CH_3C_2H	C_8H
C_2··	HF	N_2H^+	HNCO	H_2C_2O	HC_5N	**10 atoms**
CH	SH·	N_2O	HNCS	H_2NCN	$HCOCH_3$	CH_3C_5N (?)
CH^+	HD	NaCN	$HOCO^+$	HNC_3	NH_2CH_3	$(CH_3)_2CO$
CN	FeO ?	OCS	H_2CO	SiH_4·	c-C_2H_4O	$(CH_2OH)_2$?
CO	O_2 ?	SO_2	H_2CN	H_2COH^+	H_2CCHOH	H_2NCH_2-COOH Glycine ?
CO+	**3 atoms**	c-SiC_2	H_2CS	**6 atoms**	**8 atoms**	CH_3CH_2CHO
CP	C_3·	CO_2·	H_3O^+	C_5H	CH_3C_3N	**11 atoms**
CSi	C_2H	NH_2	NH_3	l-H_2C_4	$HCOOCH_3$	HC_9N
HCl	C_2O	H^{3+}.	c-SiC_3	C_2H_4·	CH_3COOH	**12 atoms**
KCl	C_2S	H_2D^+, HD^{2+}	CH_3·	CH_3CN	C_7H	C_6H_6· (?)
NH	CH_2	SiCN	**5 atoms**	CH_3NC	H_2C_6	**13 atoms**
NO	HCN	AlNC	C_5·	CH_3OH	CH_2OHCHO	$HC_{11}N$
NS	HCO	**4 atoms**	C_4H	CH_3SH	l-HC_6H· (?)	
NaCl	HCO^+	c-C_3H	C_4Si	HC_3NH^+	CH_2CHCHO (?)	
OH	HCS^+	l-C_3H	l-C_3H_2	HC_2CHO	**9 atoms**	
PN	HOC^+	C_3N	c-C_3H_2	NH_2CHO	CH_3C_4H	
SO	H_2O	C_3O	CH_2CN	C_5N	CH_3CH_2CN	
SO^+	H_2S	C_3S	CH_4·	l-HC_4H*.... (?)		

Underlined molecules are solid. AlNC : Aluminium Chloride. Those molecules marked with « ? » indicate molecules whose obervation is still to be confirmed.

The suggestion that the molecular building blocks of life could be formed in space is intriguing since such regions would seem to be rather unlikely places for the development of chemistry. The ISM is cold (temperatures of 10-30 K) and "empty" with pressures of less than 10^{-12} torr such that the probability for a collision between two compounds is low and, at such low temperatures, the "reaction rate" would be expected to be very low (hence in most industrial chemistry the reactants are heated to increase their reactivity). Nevertheless the detection of such molecules within the ISM makes it clear that these are chemically active zones. The solution to this apparent paradox is that the chemistry in the ISM is somewhat different from the conventional chemistry we observe on Earth, much of it being induced by radiation. The ISM contains several different sources of radiation, namely;

- **Light** (produced from stars) of which the ultraviolet is the most important for inducing chemistry .
- **Cosmic rays** – high energy ions about 89% of which are protons, 10% helium nuclei (alpha particles) and about 1% ions of the heavier elements. It is believed that most galactic cosmic rays derive their energy from supernova explosions, which occur approximately once every 50 years in our Galaxy.
- There are also sources of **gamma** and **X-rays**, the latter often being a signature of neutron stars.

Such radiation sources all have sufficient energy to break the chemical bonds of molecules in the ISM and thence produce both reactive radicals and ions capable of inducing further chemistry. In the gas phase much of the chemistry in the ISM is driven by ion-molecule reactions (**Fig. 3**). Such reactions are *barrierless* that is they require no energy to start the reaction rather once the reactants are brought together the reaction appears to occur spontaneously. Such barrierless reactions are also prevalent if one of the reacting species is a free radical (*e.g.* the hydroxyl radical $^{\cdot}OH$). Such reactions can therefore occur at low temperatures, indeed it has been noted that the reaction rate may actually increase at low temperatures.

Despite the success of ion-molecule and radical chemistry in explaining some of the molecular formation in the ISM, the wealth of molecules formed and the time scales upon which they develop can not be explained by gas phase alone. For example molecular hydrogen is the most abundant molecule in the ISM but there is no gas phase reaction scheme that can explain its formation. Furthermore, although complex hydrocarbons can be formed in such reactions, they may equally be destroyed by the radiation field. Indeed the more complex the molecular species is, often the more prone to destruction by ultraviolet light and cosmic rays it turns out to be.

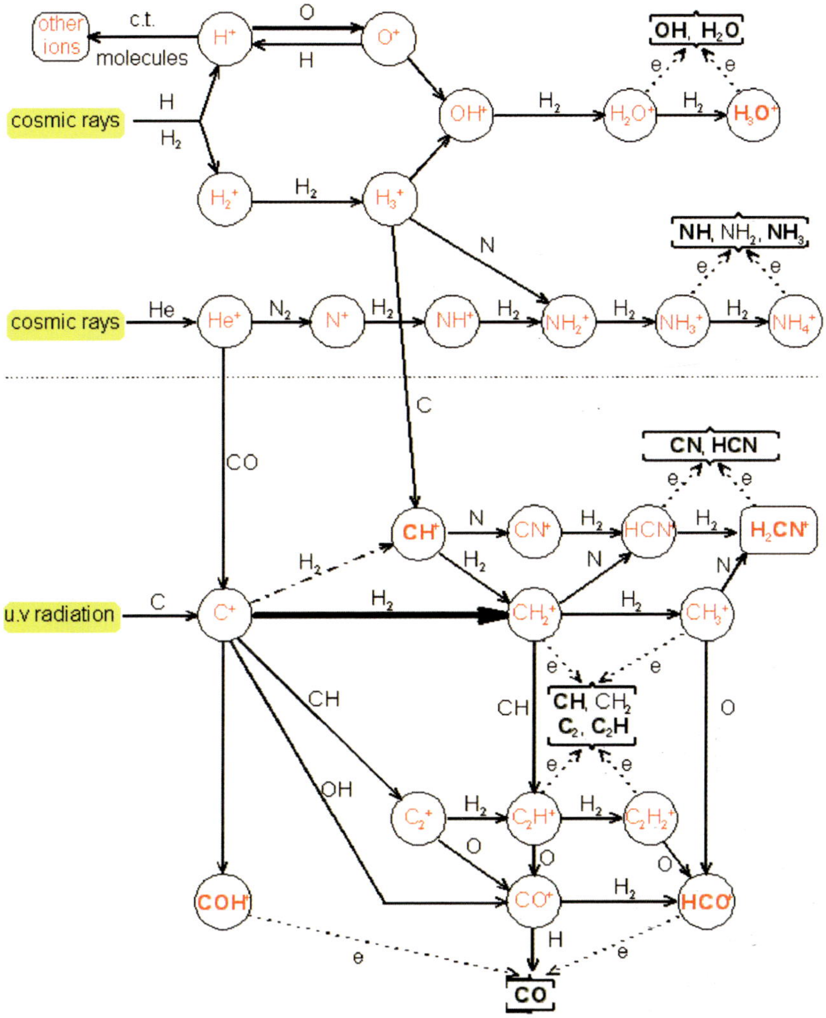

Figure 3 : *Schematic of ion molecule reactions induced by radiation in the interstellar medium. "e" indicates an electron-induced reaction.*

Thus, it is necessary to consider the role of the *dust* in the ISM. Dust comprises about 1% of the mass of the material in the ISM and can act as a surface upon which chemistry can occur (**Fig. 4**). Although we have yet to actually examine a piece of interstellar dust we believe it is similar to that found in the Solar System and recently collected by the Stardust mission. The dust is either carbonaceous or silicate in nature, comprising of small particles, typically sub-microns in size, probably with an irregular (fractal?) structure. Being so cold (around 10 K) the dust grains act as a depository for any gaseous molecules which "stick" to the surface. Hence $H^{\bullet}$ atoms may collide with the surface, and subsequent reaction between such $H^{\bullet}$

atoms is now believed to be the main mechanism by which molecular hydrogen is formed in the ISM, the product H_2 being desorbed back into the ISM as a result of subsequent heating of the grains (*e.g.* by grain-grain collisions). Other gaseous atomic and molecular species in the ISM may also collide with the dust slowly forming a complex multi component ice layer. It is this ice that provides the "high density" target within which chemistry may be induced by interstellar radiation allowing more complex molecules to be formed (**Fig. 4**).

Figure 4 : ***Molecular formation on interstellar dust grains.*** 1) *An ice layer is formed by accretion onto dust grains.* 2) *The icy layer is exposed to irradiation from the ISM.* 3) *Further ice is accreted on top of the irradiated layers.* 4) *Icy material and molecules are ejected from the ice into the gas phase through heating events such as shocks and grain–grain collisions.*

Molecular formation in astrochemical ices

Experimental studies of the formation of molecules within astrochemical ices are now being conducted in several international laboratories. A typical apparatus is shown in **Figure 5**. An ultrahigh vacuum chamber is required (typically less than 10^{-9} torr) to mimic the ultralow pressures found in space within which is mounted a cryostat, cooled by liquid helium to temperatures typical of the ISM (10 K). Gases are then introduced into the vacuum chamber and deposited on a sample mounted off the cryostat. Irradiation of the ice may be provided by a variety of sources: ultraviolet light may be produced by ultraviolet lamps or by synchrotron radiation (such as that to be provided by the French *Soleil synchrotron*; synchrotrons are also a bright source of X-rays (*e.g.* the ESRF at Grenoble) while cosmic rays may be produced by ion accelerators (*e.g.* Van de Graaff accelerators). After irradiation the products of the subsequent chemistry may be detected using a variety of techniques including: (i) infrared spectroscopy, which can detect molecules in the ice itself and (ii) mass spectrometry heating the ice after irradiation until the products are desorbed.

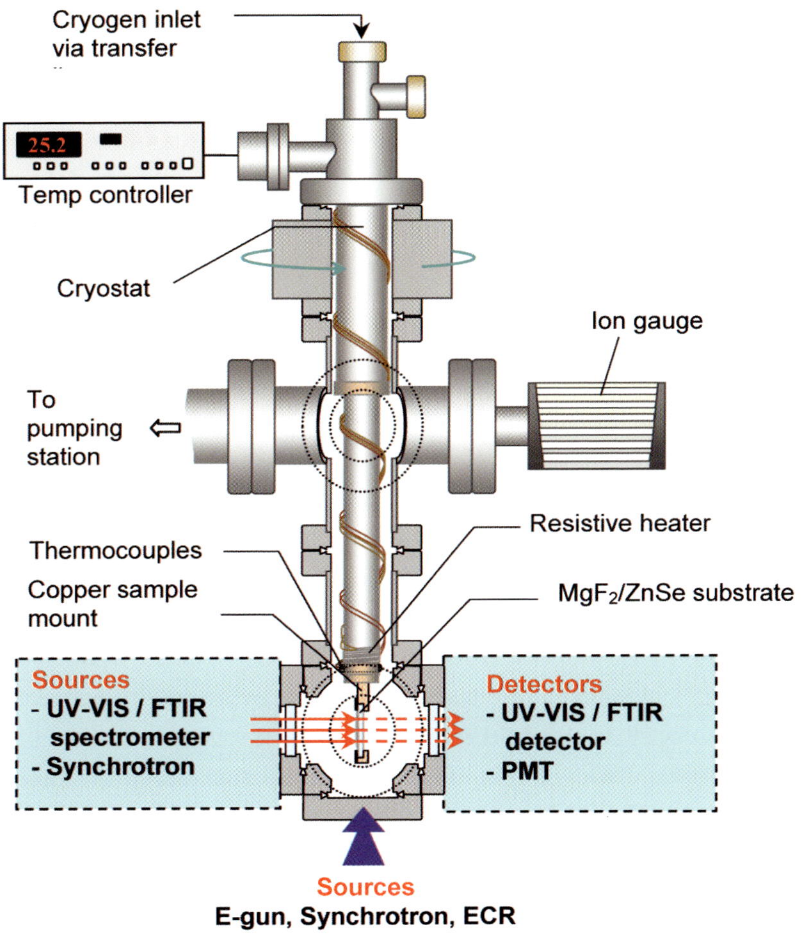

Figure 5: *A typical experimental apparatus used to explore astrochemistry, comprising an ultrahigh vacuum chamber to mimic low pressures in space, a cryostat to prepare surfaces at very low temperatures, an irradiation source to provide energy to induce chemistry and a detection system (e.g. Fourier transform infrared spectroscopy or FTIR) to determine what molecules are formed in the astrochemical processes.*

Recent research has shown that the structure of the ice (morphology) depends not only on the deposition temperature but also on the deposition time with slow deposition forming crystalline structures at temperatures where during fast deposition amorphous structures are produced. Ice morphology plays a vital role in the chemistry that can occur after irradiation of the ice mixture. A densely packed ice may slow migration of radicals through the ice whilst an ice full of pores and voids may allow faster diffusion and provide a larger surface area upon which reactions can occur. At present we do not know the morphology of the ice on an interstellar dust grain and this highlights the difficulty in undertaking experiments that aim to mimic conditions in the ISM. However, it should be remembered that in space deposition rates are very slow – a few molecules per year or less – far slower than any rates we can achieve in the laboratory. Similarly, in all the laboratory experiments undertaken to date, the bulk ice surfaces used are much larger than those found in space being $> 1\ cm^2$, whilst the typical area on a dust grain will only be a few μm^2 such that thermal effects and the role of surface compared to bulk chemistry may be quite different. Nevertheless using such experimental systems made it possible to demonstrate that in the ISM irradiated ices may induce sufficient chemistry to form complex molecules that can form the "building blocks of life".

In 2002 two different research groups, one at the NASA AMES laboratory in the USA (M P. Berstein *et al.*), the other at Leiden University in the Netherlands (G.M. Muñoz-Caro *et al.*), studied the irradiation of a mixture of ices using UV light. The NASA Ames group used a 20:2:1:1 ice mixture of H_2O: CH_3OH:NH_3:HCN, the Leiden group a 2:1:1:1:1 ice mixture of H_2O:CH_3OH:NH_3:CO:CO_2. In both cases, after several hours of irradiation at temperatures of 10 K, amino acids, 6 of which are protein constituents, glycine, alanine and serine were formed. More recently experiments using electrons to irradiate simpler binary mixtures of CO_2 and CH_3CN (methylamine) or CO_2/NH_3 have also shown that glycine is readily formed in simple astrochemical ices (P.D. Holtom *et al.*). These experiments demonstrate that the formation of amino acids in the cold regions of space is not only possible but may occur with a high probability. Thus the ISM may form a large depository of molecules that may subsequently be assembled to form the building blocks of life such as the DNA bases.

Meteorites and panspermia

Demonstrating that it is possible to form molecules in the ISM that may subsequently assemble to form the building blocks of life on Earth (or any other planet) is not in itself evidence that the origins of life lie in the ISM. It is also necessary to know how such molecules might be transported to the planet. Our knowledge of how stars and planets form relies on physical/chemical models but we believe we know the basic mechanisms. Within the ISM some disturbance (*e.g.* a shock wave from a nearby exploding star) triggers the collapse of the dust cloud and its surrounding gases to form a nebula within which dust coalesces and eventually a star is born. Close to this new Sun solar heat may vaporise ices and prevent lightweight elements, like hydrogen and helium, from condensing but further from the Sun it may still be cold enough for ices to remain intact in the form of comets or meteorites. Hence comets and meteorites may be a repository of molecules formed in the ISM capable of then impacting planets and depositing their "prebiotic" material on an early Earth. Evidence for this is found in the chemical composition of meteorites found on Earth. On September 28, 1969, a meteorite fell over Murchison, in the state of Victoria, Australia (**Fig. 6**). While only 100 kilograms were recovered, analysis of the meteorite has shown that it is rich with amino acids. Indeed over 90 amino acids have been identified by researchers to date. Nineteen of these amino acids are used in biological processes on Earth. The Murchison meteorite (and others) has therefore demonstrated that the Earth may have acquired some of its prebiotic material by "planetary infall", a hypothesis that is often described as "panspermia".

Figure 6 : *The Murchison meteorite which landed at Murchison, Australia on September 28, 1969. Over 100 kilograms of this meteorite have been collected. Classified as a carbonaceous chondrite, type II (CM2), this meteorite is likely to have come from a comet. More than 92 different amino acids have been identified within the Murchison meteorite, nineteen of these are also found on Earth such that many believe that such meteorites provided the 'seeds' of life on the early Earth.*

Conclusion

How life evolved on Earth and whether life can (and has) evolved elsewhere in the universe remains one of the greatest unanswered questions of modern science. Our understanding of how irradiation in the interstellar medium can induce chemistry in the most inhospitable regions of the universe and form complex molecules is an essential first step into our understanding where the molecules of life were (are) formed. A combination of experiment, modelling and astronomical observations provides us with much information on the chemical processes in the ISM, and suggests that life's molecular ingredients may be formed throughout the universe. What remains unknown is how these prebiotic molecules (amino acids, hydrocarbons, sugars) were assembled to form proteins, the nucleotide bases – guanine adenine, cytosine, thymine and uracil – and thereafter the self replicating molecules RNA and DNA. This represents today's great challenge for those interdisciplinary scientists who are developing the new research field of astrobiology.

Acknowledgements

The authors would like to take this opportunity to thank several research agencies for their own work in astrochemistry: in the UK the CCLRC, EPSRC, NERC and PPARC Research councils, The Royal Society and The British Council as well as the Open University. Such research is necessarily international and we therefore whish to thank the EU Framework programme and the European Science Foundation for their support in conducting research and for integrating the European Community in common research projects.

References

Gilmour I., An Introduction to Astrobiology, Cambridge university Press, 2004.

Shaw A., Astrochemistry: From Astronomy to Astrobiology, John Wiley and Sons Ltd., 2006.

Chela-Flores J., The New Science of Astrobiology : From Genesis of the Living Cell to Evolution of Intelligent Behaviour in the universe, Kluwer Academic Publishers, 2004.

Bernstein M.P., Dworkin J.P., Sandford S.A., Cooper G.W., Allamandola L.J., Racemic amino acids from the ultraviolet photolysis of interstellar ice analogues, Nature, 2002, **416**, 401-403.

Holtom P.D., Bennett C.J., Osamura Y., Mason N.J., Kaiser R.I., A combined experimental and theoretical study on the formation of the "aminoacid" glycine (NH_2CH_2COOH) and its isomer ($CH_3NHCOOH$) in extraterrestrial ices, Astrophysical J., 2005, **626**, 940-952.

Muñoz Caro G. M., Meierhenrich U. J., Schutte W. A., Barbier B., Arcones Segovia A., Rosenbauer H., Thiemann W. H.-P., Brack A., Greenberg J.M., Amino acids from ultraviolet irradiation of interstellar ice analogues, Nature, 2002, **416**, 403-406.

Chapter 6

Water remediation by the electron beam treatment

Salvatore S. EMMI and Erzsébet TAKÁCS

Introduction

Water shortage and security are becoming a major national and regional priority because of water-intensive lifestyles, rapid industrialization, urbanization, and agricultural intensification worldwide. Therefore the treatment of municipal and industrial wastewater as well as of drinking water is a prime task for environmental engineering. Let's consider, for example, the problems encountered with halogenated organic compounds. The ubiquitous nature of halogenated organics, from trichloroethylene to the wood preservative pentachlorophenol, has spawned considerable discussion about the impact of such halogenated organics in the environment. Chlorophenols are a serious health concern in drinking water treatment, as they can be present in the source water or be formed as by-products of chlorine-based disinfection methods. An ample literature reports on effective degradation of chlorophenols and other toxic pollutants such as dioxins, polychlorinated biphenyls (PCBs), and benzofuran derivatives [1-3]. Another class of compounds, emerged as a focus of attention because of their impact on the environment, is that of dyes and pigments, which are the single largest group of industrial chemicals. This chapter briefly surveys the electron beam (EB) Advanced Oxidation Process, reporting some recent results on the decompositions of a dye (Apollofix-Red) and of a pesticide (carbofuran). Two successful applications of the technology on the industrial scale are also presented.

Advanced Oxidation Processes (AOPs)

A large class of Advanced Oxidation Processes (AOPs), such as UV-peroxide, ozonation, Fenton and Fenton-like processes, photocatalysis, electrokinetics and sonolysis have been effectively employed in order to decompose hazardous compounds in water [3-5]. All these methods are based on the generation and use of hydroxyl radicals as the primary oxidant for the degradation of organic pollutants (**Fig. 1**). **Table 1** reports the reduction potentials of popular radical species relevant to wastewater remediation. From that it can be seen that $^{\cdot}OH$ has the highest oxidative power. Being also a very fast reactant, $^{\cdot}OH$ can attack and oxidize most of the hazardous compounds dissolved in water. Electron beams in the MeV range, and to a lesser extent γ-rays, are excellent tools to produce $^{\cdot}OH$ radicals by operating the lysis of water induced by radiation (radiolysis).

Table 1. *Reduction potentials of radical species relevant to wastewater remediation* ($Ox + e^- \rightarrow Red$).*

Species (alphabetic order)	Couple	pH	Std Red Pot (1)
Chlorine (Cl_2)	$Cl_2/Cl_2^{\cdot-}$		0.42 .. 0.60
Chlorine dioxide ($ClO_2^{\cdot}$)	$ClO_2^{\cdot}, H^+/HClO_2$		~ 1.30
Chlorine radical anion ($Cl_2^{\cdot-}$)	$Cl_2^{\cdot-}/2Cl^-$		2.09(2)
Hydrogen peroxide (H_2O_2)	$H_2O_2, H^+ / H_2O, {}^{\cdot}OH$	7	0.46 .. 0.87
Hydroxyl radical (1) ($^{\cdot}OH$)	$^{\cdot}OH / OH^-$		1.9
Hydroxyl radical (2) ($^{\cdot}OH$)	$^{\cdot}OH, H^+/ H_2O$	acid	2.7
Hypochlorous acid (1) (HClO)	$HClO, H^+ / H_2O, Cl^{\cdot}$		- 0.46
Hypochlorous acid (2) (HClO)	$HClO/ Cl^-, {}^{\cdot}OH$		- 0.04
Oxide radical ion ($O^{\cdot-}$)	$O^{\cdot-}, H_2O/ 2 OH^-$		1.78
Ozone (O_3)	$O_3 / O_3^{\cdot-}$	11-12	1.01
Perhydroxyl radical (1) ($HO_2^{\cdot}$)	$HO_2^{\cdot}, H^+ / H_2O_2$	0	~ 1.50
Perhydroxyl radical (2) ($HO_2^{\cdot}$)	$HO_2^{\cdot} / HO_2^-$		0.79
Superoxide radical ($O_2^{\cdot-}$)	$O_2^{\cdot-}, H^+/ HO_2^-$		1.00

A more positive reduction potential for the couple Ox/Red, means that Red is a more powerful oxidant. When protons are involved, the reduction potential of the half-cell varies with pH.* **1) *Wardman P., J. Phys. Chem., Ref. Data, 1989,* **18**,*1637-1755.* **2**) *Atkins P.W., Physical Chemistry, Fifth Edition, Oxford University Press (1994).*

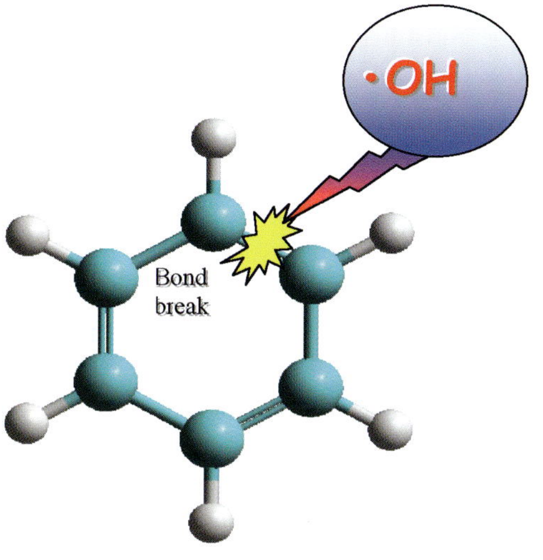

Figure 1 : *Hydroxyl radicals decompose pollutants. Typical ˙OH attack to an aromatic ring.*

Water radiolysis by electron beam

A beam of electrons, travelling through water, forms a swarm of highly reactive radicals: hydrated electron (e_{aq}^-), hydroxyl radical (˙OH) and hydrogen atom (˙H) [6] (see G.V. Buxton, Chapter 1). To perform an AOP, e_{aq}^- and H atoms should be neutralized to poorly reacting species or, better still, transformed into a further amount of ˙OH radicals. This is achieved by adding one or more scavengers to the system. **Figure 2** outlines a variety of systems which characterize different EB operating systems. Among these, a straightforward oxidative degradation process is represented by the saturation of water with N_2O above pH 3: N_2O captures the strong reducing e_{aq}^- and converts it stoichiometrically into ˙OH. As a result of this method, practically only ˙OH remains, except for a negligible quantity of H˙. The plainness of this system is also suitable for mechanistic studies with pulse radiolysis, *i.e.* the chemical kinetics application of EB (see next page).

The rapidity with which the EB treatment initiates the decomposition of pollutants may be promptly checked by reasonably assuming that *i*) ˙OH reacts with rate constants in the range 10^6-10^9 l mol^{-1} s^{-1}, and *ii*) pseudo-first order conditions are maintained throughout the process *(i.e.* ˙OH reacts quantitatively with the pollutant). Therefore a pulse of electrons depositing 10 Gy of energy in N_2O saturated water triggers the decomposition of ~5 x 10^{-6} mol/l of pollutant. In an ordinary accelerator facility, sweeping a flow of water with pulses of 1 microsecond duration at a repetition rate of 100 pulses *per* second, 0.5 millimol/l of compound per second are immediately modified. Certainly, the calculation of the production speed of the oxidizing species allows only a rough assessment of the efficiency of an AOP.

In fact, the whole rehabilitation is affected by a wide range of factors, among which a primary role is played by the variety and nature of pollutants, and the required quality of water. As a matter of fact, industrial plants demonstrate the ability of EB to treat and recycle millions of litres of wastewater per day (see later), a goal well above the reach of the other technologies.

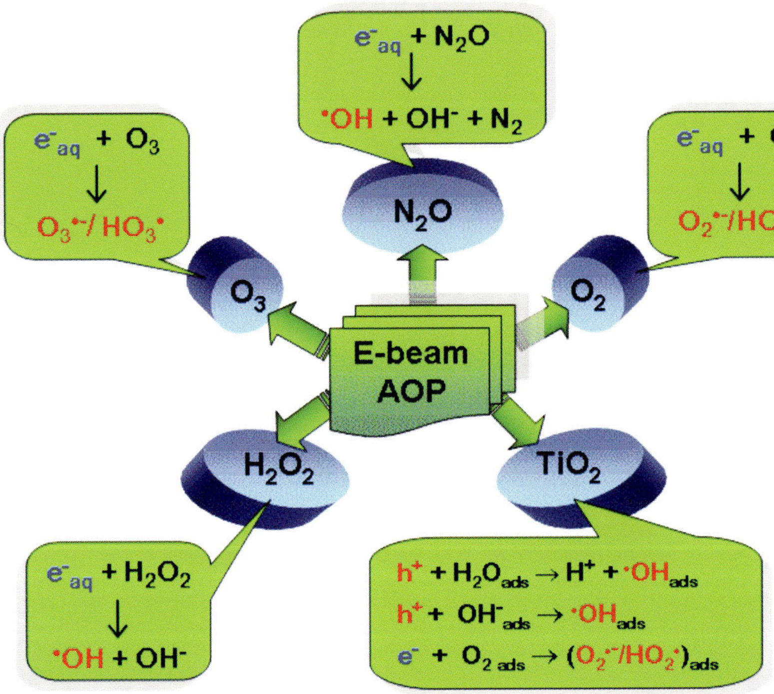

Figure 2 : *Homogeneous and heterogeneous aqueous systems used to produce ˙OH and other oxidizing radicals in typical E-beam AOP experiments. Any system can be employed separately or in association with the others. Oxidation chain carriers are red tagged.*

To exemplify the destruction of a pollutant, the decomposition of a hydrocarbon is illustrated in **Figure 3**: the right side represents the radical production systems, while on the left the most important reactions following ˙OH attack are given. In **Figure 3**, ˙OH operates its oxidizing action by abstracting a hydrogen atom from the hydrocarbon chain. Besides water, an alkyl radical is formed from the pollutant: this radical starts a series of reactions, such as peroxidation, fragmentation, further H abstraction, disproportionation, molecular growth etc. ˙OH may also initiate an oxidative pathway by accepting an electron (electron transfer), or by adding to a π electron system (bond breaking). The latter is the most common pathway in the presence of an aromatic ring, and it will be described later by means of two representative pollutants.

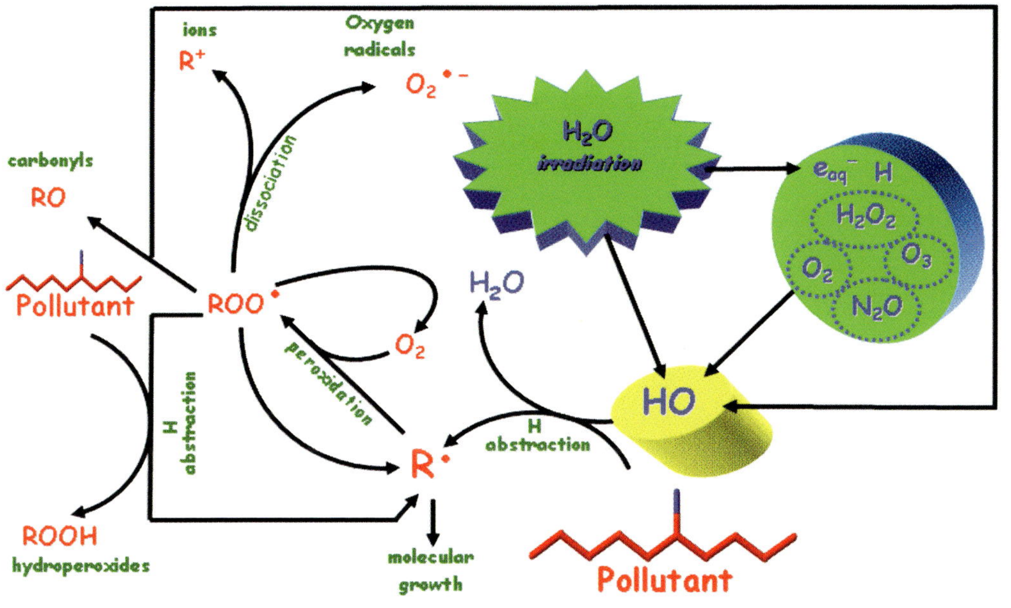

Figure 3 : *Oxidative degradation of a hydrocarbon by E-beam AOP. Irradiated water produces $^{\bullet}OH$ both directly and by the mediation of dissolved scavengers (right side). $^{\bullet}OH$ oxidizes the hydrocarbon by H abstraction. The resulting alkyl radical starts a recursive chain degradation giving rise to new carbon- and oxygen-centered radicals, carbonyls, carbocations, hydroperoxides, polymers, etc. (left side).*

Pulse radiolysis

The simultaneous presence of pollutants of different nature in wastewater gives rise to a very complex kinetic degradation scheme. The kinetic parameters of each reaction constitute the basic data needed to design an efficient degradation process. A suitable tool to make such kind of investigation is pulse radiolysis (Chapter 2) coupled with a kinetic spectrometry detection system [7]. Aiming to achieve meaningful results, reactions with water radicals are studied separately as outlined below (Chapter 1):

- In N_2 or Ar saturated solutions, between pH 3 and 11 the reactions take place with $^{\bullet}OH$, $H^{\bullet}$ and e_{aq}^{-}.

- Below pH 2 in N_2 or Ar saturated solutions, nearly equal amounts of the $^{\bullet}OH$ radicals and $H^{\bullet}$ atoms are present. This is because aqueous electrons are converted to $H^{\bullet}$ by reacting with the hydroxonium ions:

$$e_{aq}^{-} + H_3O^{+} \rightarrow H^{\bullet} + H_2O \qquad (1)$$

– $H^{\bullet}$ atoms reactions with solutes can be carried out alone below pH 2 in N_2 or Ar saturated solutions containing 0.2 – 1 mol l^{-1} tert-butanol.

In this system, tert-butanol scavenges $^{\bullet}OH$ and forms poorly reactive radicals:

$$^{\bullet}OH + (CH_3)_3COH \rightarrow {}^{\bullet}CH_2(CH_3)_2COH + H_2O \quad (2)$$

(The reaction between $H^{\bullet}$ and tert-butanol is slow).

– Reductions by e_{aq}^{-} are performed in N_2 or Ar saturated solutions above pH 3 and in the presence of 0.2 – 1 mol l^{-1} tert-butanol (there is a small contribution from the $H^{\bullet}$ atom reactions). Radicals formed from tert-butanol are also present.

– In air or oxygen saturated solutions, the reactive species are the $^{\bullet}OH$ radical, and the couple $O_2^{\bullet-}/HO_2^{\bullet}$ (superoxide radical anion/perhydroxyl radical). The couple is originated by scavenging e_{aq}^{-} and $H^{\bullet}$ with O_2:

$$e_{aq}^{-} + O_2 \rightarrow O_2^{\bullet-} \quad (3)$$
$$H^{\bullet} + O_2 \rightarrow HO_2^{\bullet} \quad (4)$$

and is in acid/base equilibrium with a pK_a value of 4.8

$$O_2^{\bullet-} + H_3O^{+} \rightarrow HO_2^{\bullet} + H_2O \quad (5)$$

– Solutions saturated with O_2 and containing tert-butanol are used for studying the reactions of the $O_2^{\bullet-}/HO_2^{\bullet}$ acid-base radical pair.

The degradation of pesticides: carbofuran

Carbofuran ($C_{12}H_{15}NO_3$), is a carbamic insecticide and nematocide vastly employed in North America and in some European countries to protect maize, rice, alfalfa, onion, garlic, potatoes, etc. However, it is toxic by contact and ingestion, and lethal over the level of 11 mg/kg (LD_{50}, rats). The Maximum Contamination Level (MCL) for drinkable water, as established by EPA (USA), corresponds to 0.18 μmol/l.

Several AOPs have been proposed to operate carbofuran decomposition, and many of them have demonstrated their effectiveness. The choice of the more appropriate system will consider parameters like the rapidity of action and economical factors.

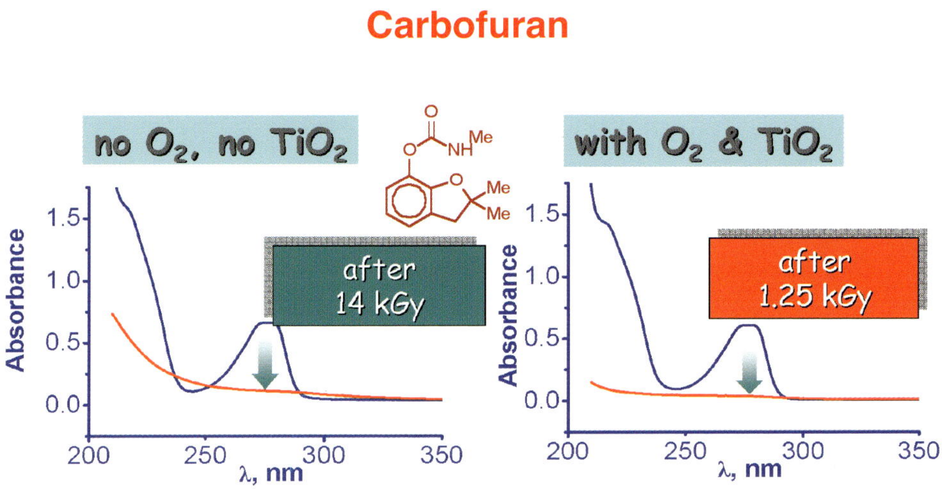

Figure 4 : *Gamma-Radiolytic degradation of a pesticide carbofuran at 200 μmol l^{-1} and the synergistic effect of a metal oxide (TiO_2).* Left*) : carbofuran in deaerated atmosphere and in the absence of catalyst.* Right*) : carbofuran in oxygenated atmosphere and in the presence of TiO_2. In the latter system the dose necessary to decompose 200 μmol l^{-1} carbofuran decreased by an order of magnitude.*

The example below refers to a recent study by pulse and gamma radiolysis [8]. A pH ≈ 6 has been generally used. Hydroxylation mainly concerns the aromatic ring of carbofuran (**Fig. 4**); in fact, the first intermediate shows the absorption of a cyclohexadienylic type radical in the region 280-330 nm. The hydroxylation attack to carbofuran

$$^{\cdot}OH + \text{carbofuran} \rightarrow {}^{\cdot}CF\text{-}OH \quad (6)$$

proceeds with a rate constant $k_{OH+CF} = 6.6 \times 10^9$ l mol^{-1}s^{-1}. In the presence of oxygen, the carbofuran cyclohexadienylic radical adds oxygen (peroxidation) with a rate constant $k_{perox} \approx 10^7$ l mol^{-1}s^{-1}. As measured after gamma radiolysis, oxygen participation makes the global decomposition three times more efficient than in anaerobic solutions; the efficiency is further enhanced by adding a 1% amount of TiO_2. Typically, under these conditions a sample containing carbofuran 200 μ mol l^{-1} is completely decomposed by 1.25 kGy of γ-radiation dose, *i.e.* an order of magnitude more efficiently than without O_2 and TiO_2 (**Fig. 4**).

TiO_2 catalysis

TiO_2 catalysis proceeds through the production of $^{\cdot}OH$ radicals at the solid-liquid interface. The mechanism of action lies on the separation of charge upon absorbing a radiation of energy higher than 3.2 eV, *i.e.* below 380 nm (UV photons, E-beam and γ-rays): holes (h^+) are left in the valence band (VB) and electrons (e^-) are promoted to the conduction band (CB).

It appears that TiO_2 particles in water are widely hydroxylated. Therefore holes escaping annihilation may migrate to the surface and oxidize adsorbed water molecules and hydroxyl ions. Oxygen adsorbed at the surface captures electrons preventing their recombination with h^+, and therefore favouring the yield of $^{\bullet}OH$ radical (**Fig. 5**).

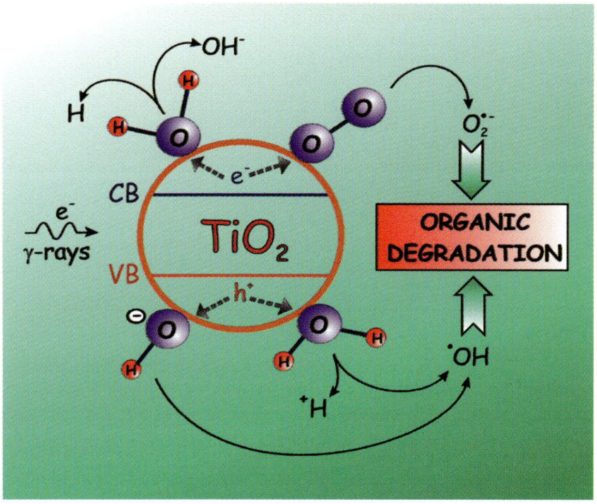

Figure 5 : *Formation of oxidizing radicals at the solid-liquid interface of TiO_2 and water. Radiation promotes a separation of charge in the catalyst. H_2O, OH^-, and O_2, adsorbed at the surface are converted to radicals by electrons and holes which migrate to the interface. Oxidative degradation is greatly enhanced (see also* ***Fig. 4****).*

Decoloration of dyes : Apollofix-Red

In dilute dye solutions all the intermediates of water radiolysis, *i.e.* hydrated electron (e_{aq}^-), hydroxyl radical ($^{\bullet}OH$) and hydrogen atom ($^{\bullet}H$) induce the decomposition of the solute.

The degradation of the dye can be easily followed by taking decoloration curves. This is usually done by preparing appropriate solutions, irradiating them by increasing the dose stepwise and by taking the UV-VIS absorption spectra after each irradiation [9]. The decoloration is due to the destruction of the color centers. The color fading with dose is illustrated in **Figure 6**.

The upper part shows the spectra of aqueous N_2O saturated solutions of a reactive dye (Apollofix-Red), before and after EB irradiation with different doses (0-3 kGy), while the lower part shows the photographs of samples treated in the same way. The chromophore center of this dye is an azo group connecting a benzene ring with a naphthalene part. Decolored products are formed when the extensive conjugation is destroyed, for instance when the conjugation through the –C–N=N–C– bridge is lost. End-product analysis by High Performance Liquid Chromatograph (HPLC, diode array detection was used for this measurement) also showed a sharp decrease in the concentration of the starting compound (**Fig. 7**).

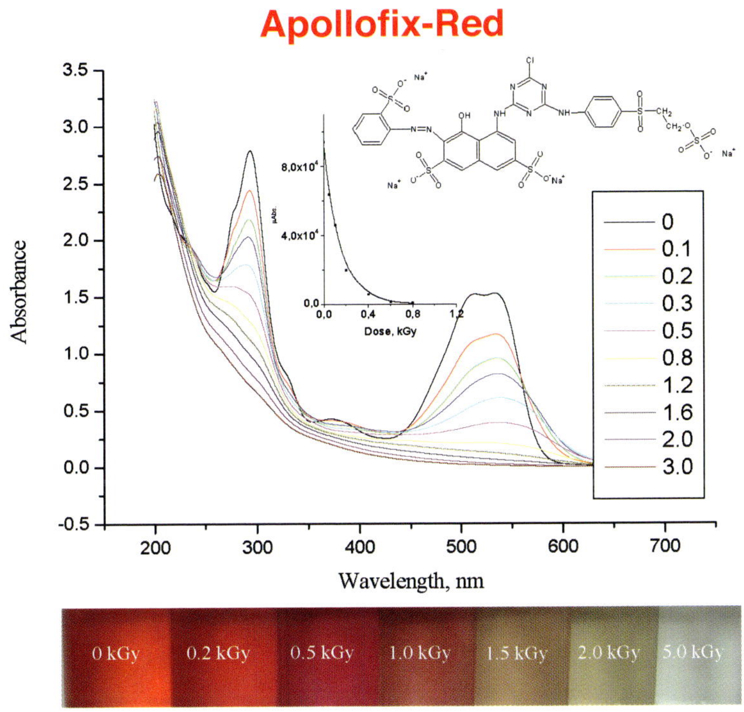

Figure 6 : *Dye decoloration. Aqueous N_2O saturated solutions of a dye (Apollofix-Red 0.25 µmol l^{-1}) are E-beam irradiated with increasing doses: the reaction with* $^{\bullet}OH$ *destroys the color centers. This is shown both by a progressive weakening of the absorbance of samples (spectra) with dose (0-3 kGy) and also by a progressive increasing of their transparence (photographs below the spectra). The* ***inset*** *shows the decoloration-dose curve at 530 nm.*

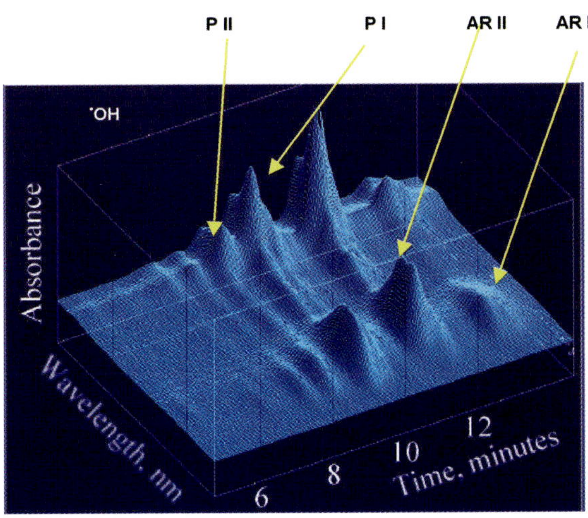

Figure 7 : *Three-dimensional HPLC chromatogram showing the separation of degradation products in N_2O saturated dye (Apollofix-Red AR-28 at 0.25 mmol l^{-1}) solution irradiated with 0.6 kGy dose. Destruction of AR-28 molecules and formation of degradation products in* $^{\bullet}OH$ *radical reactions as measured by HPLC with diode array detection. AR I and AR II are used for AR-28 and for its hydrolysed form, respectively, PI and PII are products.*

The mechanism is not as straightforward as it may appear: in the first step, in fact, after ˙OH radical addition to any aromatic ring, a new colored product is formed, *i.e.* the cyclohexadienylic type radical may reform the conjugated system in disproportionation reactions [7,10]. The process is exemplified with a substituted benzene (**Scheme 1**). Using high doses, and especially in the presence of O_2 or O_3, aldehydes and organic acids are formed from the ˙OH-dye adducts and finally decomposed to CO_2 and H_2O (see upper part of **Scheme 1**).

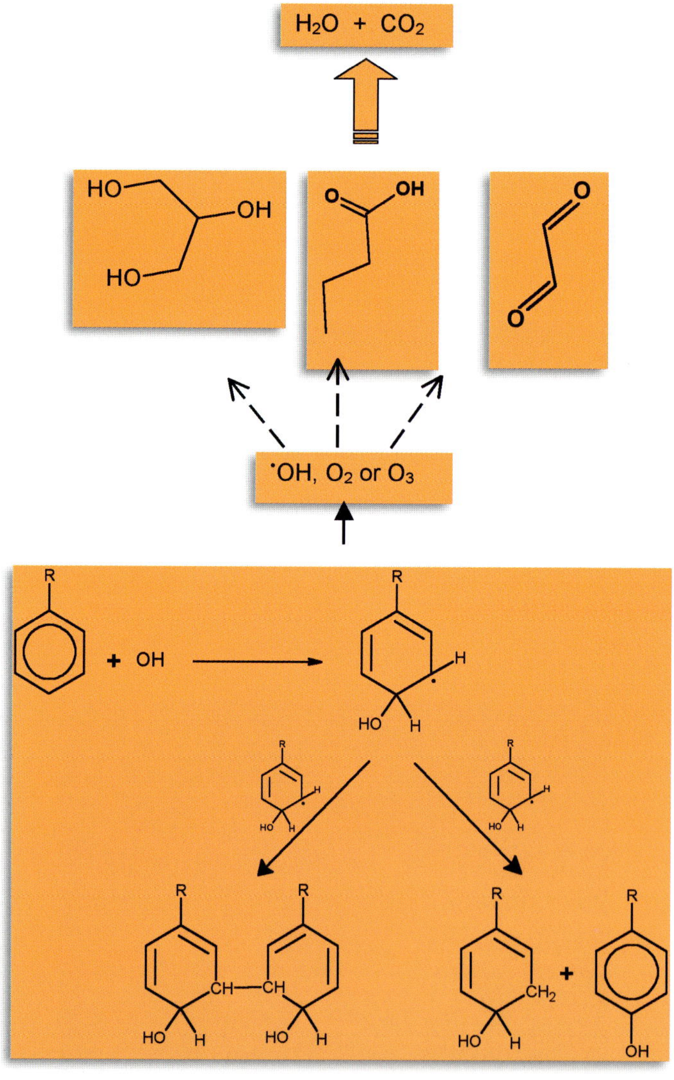

Scheme 1 : *Decomposition outline of the aromatic part of Apollofix-Red dye, modeled with a substituted benzene (lower part) and achieving its complete mineralization (upper part).* A disproportionation reaction between two cyclohexadienylic radicals regenerates the aromatic ring structure in a phenol derivative. The R-group can be an alkyl group or a halogen atom.

Electron accelerators and operative conditions

At present radiation technology is a popular method for sterilization, polymerization, polymers and semiconductors modification, but not for environmental remediation. Actually, while various new features of irradiated materials were worthy of industrial investments, the improvement given to the quality of water and air was not economically rewarding. In the past indeed, the requirement of high initial investments constituted a restraint to the E-beam implementation in the environmental field. Nowadays, however, accelerators are becoming cheaper while freshwater is more and more precious, such that radiation rehabilitation and recycling of wastewater can be profitably considered by both public authorities and investors.

To see that, let us focus on a couple of factors that mainly determine the economy of the EB remediation process: *i.e.* a **high power** (by which high volumes of water can be rehabilitated daily), and the **efficiency** (which establishes the transfer rate of the accelerator nominal power into useful energy). To this respect, it is instructive to learn from **Table 2** that DC and UHF machines (Chapter 2) best match with these requirements. In the DC type machines, electrons are accelerated by a direct-current field, while in the UHF type, acceleration occurs across an electromagnetic field oscillating at few hundreds MHz. These accelerators achieve high powers coupled with moderately-high efficiencies, and therefore represent the best choice for wastewater treatment. At present, the linear accelerators, which are based on a microwave field, show efficiencies and power below the others, such that they are not suitable for environmental purposes.

Table 2. *Features of accelerators for environmental processing.*

Type	Accelerating field	Efficiency range (%)	Power (kW) (up to)	Energy range (MeV)
DC	electrostatic	60-80	400	0.1-5
UHF	100-200 MHz	25-50	700	0.3-10
Linear	1.3-3 GHz	10-50	150	2-10

Considering some practical aspects, it has to be remarked that the EB is best used as a pre-treatment step to integrate a conventional bio-, and physico-chemical plant. Under such conditions a dose of 2 kGy allows the destruction of most organic pollutants with unmatched success even in case of bio-refractory compounds. Simultaneously, disinfection of wastewater is achieved. When only disinfection is required, the dose may be reduced to 0.2 kGy, resulting in more than 95% inactivation of the total coliform bacilli without any previous chlorination. An appropriate machine for these processes should produce 1-2 MeV electrons, while the output power can be within the 40 to 400 kW range, depending on the flow rate required. Studies showed that uniform dose distribution is achieved by transversally pumping a thin layer of effluent under the beam area, as illustrated in **Figure 8**.

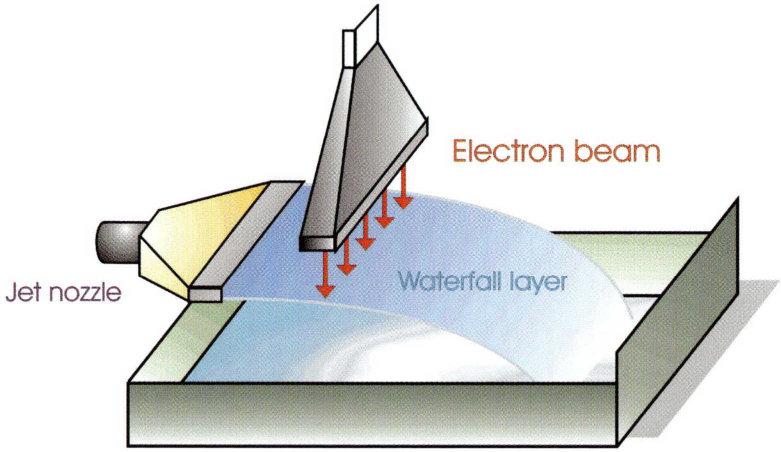

Figure 8: *Optimal flow system irradiator geometry. Penetration of 2 MeV electrons in water is 1 cm.*

Industrial and municipal treatments

Data on the reactivity of various classes of toxic substances and of dyes with $^{\bullet}OH$ and other radicals were made available by means of pulse radiolysis (see above). Oxidative degradation in complex systems could then be simulated with numerical methods and tested on pilot plant scale. Finally, some fully operating industrial plants demonstrated the viability of EB processing both on the environmental and economical aspect.

Among these plants, the wastewater rehabilitation plant of the Voronezh synthetic rubber plant (Russia) is worth to mention [2]. Here, two electron accelerators (0.7-1 MeV energy and 50 kW power), can treat up to 2,000 m^3 of wastewater per day with the specific goal to transform the non-biodegradable emulsifier Nekal into a biodegradable form. A further decomposition is achieved with biological methods.

Another industrial plant was put into operation in December 2005 for the EB treatment of dyeing wastewater (**Fig. 9**) in Korea at Daegu Dyeing Industrial Complex (DDIC). DDIC includes about hundred factories, the majority of which has equipment for dip dyeing, printing, and yarn dyeing [11, 12]. The production requires high consumption of water and emits large amount of highly colored industrial wastewater. The whole facility treats up to

80,000 m^3 wastewater /day. From this volume, 10,000 m^3/day are pre-treated presently by EB, using a 1 MeV - 400 kW - accelerator combined with bio-treatment facilities.

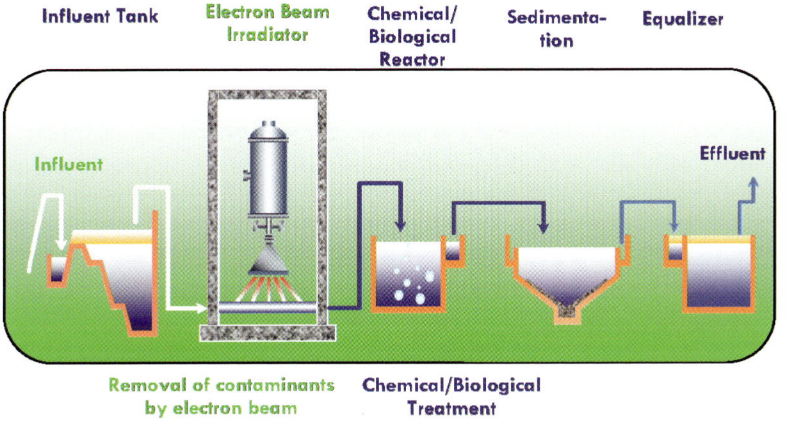

Figure 9: *Combined wastewater treatment at DDIC* *(from Bumsoo Han originals, EB-TECH Co., Ltd., Daegu, Korea, on permission of the author).*

An instructive case study, illustrating different operational conditions in the Mediterranean area, was carried out by Lopes [13]. The author made a predictive modelling on the implementation of EB to the urban wastewaters of the city of Palermo (Sicily). At present the city of Palermo is served by two depurator plants. One of them, called "Corsairs water", is under development to accomplish the treatment of 300,000 m^3/day. Goal of Lopes's investigation aimed at assessing the disinfection line of the depurator by EB. The process was then modelled on a daily treatment of 33,000 m^3 with 10 MeV electrons and a dose of 0.2 kGy, produced by one or two 190 kW accelerators of the UHF type, *i.e.* the Rhodotron TT300 (IBA).

Cost evaluation

Cost evaluations for EB treatment may vary, mainly depending on accelerator voltage and power, on the cost of electricity, and on the reuse assigned to restored water (*i.e.*, on the degree of pollutant decomposition or disinfection). Among other factors, the cost for generating a unity of power is of primary importance: 1 kW is cheaper for high power accelerators, *i.e.* it may cost 20 k$ for a 40 kW accelerator, but only 5 k$ for a 400 kW machine. Nonetheless, a realistic and updated evaluation is available from the experience at the Daegu plant above [12] and from the mentioned study at the University of Palermo [13]. Let us prospect two typical cases at Daegu, both using a 1 MeV-400 kW-accelerator (cost 2 – 2.5 million US$), running 365 days/year. Case A: 100,000 m^3/day of wastewater from a municipal plant are rehabilitated for irrigation and industrial use with a dose of 0.2 kGy (disinfection only). Case B: 10,000 m^3/day of textile wastewater are rehabilitated for self-recycling with a dose of 1-2 kGy (dyes decoloration and disinfection). In Case A, the capital investment per m^3 of effluent would be around 0.11-0.12 US$, while the operating cost would be *ca.* 0.03 US$. In case B, costs increase 7-8 times, being respectively 0.82 US$/$m^3$ and 0.24 US$/$m^3$ (**Tab. 2**). These financial quotations include investments for the industrial plant (accelerator, shielded room, water reactor, design, and others), interest, depreciation, and electricity (80% efficiency at a cost ≈ 0.05 US$/kWh) and do agree with the actual costs after 12-15 months of operation. Costs for land, research and development, and for authority approval are not included. At the same site, EB treatment combined with bio-plant leads to ≈ 20% savings in operating costs, compared to bio-plant alone, reduces the residential time, and moreover yields water of better quality.

In the Palermo case [13], one accelerator (cost ≈ 4 million €) working almost at full capacity (180 kW), or two accelerators at half-power, are conceived. A comprehensive economical model reached conclusions matching with the Daegu budget plan: precisely, in Palermo, each m^3 could cost 0.17 € employing one accelerator, or 0.28 € with two accelerators (**Tab. 3**). The latter system, although more expensive, is still acceptable as it could minimize risk of service interruption for possible failure. However, a pilot plant with only one accelerator should be a wise starting system, as has been done in Daegu.

Table 3. *Operational features and cost of EB treatment at DDIC (Daegu, Korea) and Corsairs Water (Palermo, Sicily).*

Plant	Accelerator					Aim	Status	Dose, (kGy)	Water treated (m^3/day)	Total cost (/m^3)
	Model	Type	Energy, (MeV)	N. of Acc	Operating power (kW)					
DDIC	ELV-12	DC	1	1	400	disinfection	tested in pilot plant	0.2	100 ,000	0.15 US$
			1	1	400	industrial recycling	operative	2.0	10,000	1.06 US$
Corsairs Water	Rhodotron TT300	UHF	10	1	180	disinfection	study	0.2	33,000	0.17 €
			10	2	90 each	disinfection	study	0.2	33,000	0.28 €

Concluding remarks

In the preceding sections, the reader got acquainted with a few basic concepts on the remediation of water pollution by injecting energy in water and its latest developments. The Electron beam treatment is a physico-chemical method which pursues the rehabilitation of polluted waters by the oxidative action of the ˙OH radical, and we saw that the methods for ˙OH generation distinguishe various Advanced Oxidation Processes.

The EB treatment is founded on the excitation and dissociation of water by high energy electrons. Energetic electrons are produced by devices which operate similarly to a TV set and, in the same way, are switchable on and off at the touch of a button. Exactly like in the TV operation, no radiation is stored in the environment. Wastewater can be rapidly and safety rehabilitated to be reused, for example, in industrial cooling and washing cycles, fire fighting, street washing, green park and horticultural irrigation. EB is particularly suitable for treating medium-large volumes of effluents from industries, hospitals, municipal plants, and animal-breeding. It does not give rise to concern for the environment, as it minimizes or rules out the addition and stock of chemicals. Furthermore, it works at room temperature and atmospheric pressure. Electrons penetrate in the depth of water even in case of turbidity (not possible by UV methods), and generate very high concentration of ˙OH in a fraction of a microsecond.

The versatility of EB is further experienced by the ease with which radical kinetics is governed; in fact, the concentration of ˙OH can be modulated on that of pollutants by easily tuning the beam current.

Conventional rehabilitation treatments are inadequate when wastewaters include bio-resistant chemicals, *e.g.* the carcinogenic polychlorobiphenyls (PCB's) and aromatic compounds. In this case a pre-treatment with EB converts bio-refractory compounds to bio-degradable ones and increases their solubility in water, finally reducing cost and residential time. Therefore, the design of a cost effective rehabilitation plant conceives the usage of EB as a method for improving the biodegradability of hazardous compounds, rather than for achieving their complete mineralization (*i.e.* production of CO_2 and water). Complete mineralization is, in fact, a rather costly objective and may not always prove to be necessary for remediation. On the other end, conversion of a pollutant into a useful non-toxic product (*e.g.*, fuel or polymer) may be alternatively a desirable goal. However, if complete mineralization is the ultimate purpose of wastewater treatment, radio-catalysis may be associated with other radiolysis methods, granting a synergistic effect to the process. An environmental friendly metal oxide, such as TiO_2 for instance, may be used in this case. Besides the benign chemical action, the EB kills viruses and bacteria and should be regarded as the method of choice for water security, *i.e.* a rapid rehabilitation of bio-contaminated waters, in the eventuality of environmental accidents, acts of terrorism or vandalism.

Recent industrial achievements and projects showed the EB method to be both economically rewarding and effective in protecting the environment, and as such was granted credit by the IAEA [14]. Finally the Electron Beam technology should be regarded both as a trustworthy and an innovative process to fulfil a sustainable chemistry.

References

[1] Getoff N., The Role of Peroxyl Radicals and Related Species in the Radiation-Induced Degradation of Water Pollutants, *in* "Environmental Applications of Ionizing Radiation", Cooper W.J., Curry R.D., O'Shea K.E. (eds), Wiley, New York, 1998, 231-246.

[2] Pikaev A.K., Electron Beam Purification of Water and Wastewater, in "Environmental Application of Ionizing Radiation", Cooper W.J., Curry R.D., O'Shea K.E., Wiley (eds), New York, 1998, 495-506.

[3] Pera-Titus M., Garcia-Molina V., Banos M.A., Gimenez J., Esplugas S., Degradation of chlorophenols by means of advanced oxidation processes: a general review, Appl. Catal. B. Environment, 2004, **47**, 219-256.

[4] Legrini O., Oliveros E., Braun A.M., Photochemical process for water treatment., Chem. Rev., 1993, **93**, 671-698.

[5] Kamat P.V., Meisel D., Nanoparticles in advanced oxidation processes. Current Opinion in Colloid and Interface, Science, 2002, **7**, 282-287.

[6] Buxton G.V., Greenstock C.L., Helman W.P., Ross A.B., Critical review of rate constants for reactions of hydrated electrons, hydrogen atoms and hydroxyl radicals ($^{\bullet}OH/^{\bullet}O^{-}$) in aqueous solution, J. Phys. Chem. Ref. Data, 1988, **17**, 513-886. Updated version : www.rcdc.nd.edu.

[7] Wojnárovits L., Takács E., Emmi S.S., Reactivity differences of hydroxyl radicals and hydrated electrons in destructing azo-dyes, Radiat. Phys. Chem., 2005, **74**, 239-246.

[8] Emmi S.S., De Paoli G., Takács E., Pálfi T., Electron beam remediation of wastewater. The hydroxylation of carbofuran, 24th Miller Conference on Radiation Chemistry, Le Londe les Maures, France, 10-15 September 2005, **64**.

[9] Solpan D., Güven O., Takács E., Wojnárovits L., Dajka K., High-energy irradiation treatment of aqueous solutions of azo dyes: steady-state gamma radiolysis experiments. Radiat. Phys. Chem., 2003, **67**, 531-534.

[10] Wojnárovits L., Takács E., Irradiation treatment of azo dye containing wastewater : An overview, Rad. Phys. Chem., 2008, **77**, 225-244.

[11] Han B., Ko J., Kim J., Kim Y., Chung W., Makarov I.E., Ponomarev A.V., Pikaev A.K., Combined electron-beam and biological treatment of dyeing complex wastewater. Pilot plant experiments, Radiat. Phys. Chem., 2002, **64**, 53-59.

[12] Han B., Kim J.K., Kim Y.R, Salimov R.A., Kuknasov N.K., Nemytov P.I., High power accelerators for environmental application, IAEA-TECDOC-1473, IAEA, Vienna , 2005, 119-124, and personal communication, April 2007.

[13] Lopes V., Nuclear Technologies Case Studies in the Civil and Industrial field, Laurea magistralis Thesis, 2005-2006, University of Palermo, 106 pp.

[14] IAEA, Radiation processing of wastewater, in Radiation processing: environmental applications, IAEA, Vienna, 2007, 25-44, and references quoted therein.

Chapter 7

Metal clusters and nanomaterials

Jacqueline BELLONI and Hynd REMITA

Introduction

In the past, tiny particles of matter were widely used as pigments in paints, inks, cosmetics and stained glass. In the 19th century, the properties of colloids attracted increasing interest and the specific colours of metal colloids were explained by Mie (1908). Fogging of photographic plates, that is reduction of silver ions into atoms and finally into tiny metal clusters in the emulsion of silver halides, enabled W.C. Roentgen to discover X-rays in 1895 and H. Becquerel uranium rays in 1896. This effect is still used in some dosimeters to detect ionizing radiation and more generally in radiography. Due to the ionization of liquids and the production of solvated electrons, e_s^-, and of $H^•$ atoms (Chapter 1), metal ions in solution are likewise reduced by these strong reducing agents into lower valency ions, and then into atoms. Indeed, Marie Curie used to compare the radiolytic processes to an "electrolysis without electrodes", since the primary pair electron-solvent cation is formed in the bulk solution instead of at the surfaces of separated electrodes. After coalescence, the atoms give rise to clusters containing an increasing number of atoms.

However, such atoms and the smallest clusters (or oligomers), observed after the radiation-induced generation, appeared to be produced to a lesser extent than expected and were easily corroded, even if the corresponding bulk metal presented a noble character [1]. In other words, small clusters of, for example, gold, silver, copper, platinum, etc., exhibited a new reactivity under conditions where the metals have none. We solved this apparent violation of thermodynamics in 1973 by assigning to both the atoms and the smallest clusters a reduction potential much more negative than for the bulk metal and depending on the number of atoms they contain [1]. This peculiar behaviour was assumed to be due to their extremely divided state and to discrete energy levels, different from the band structure of large metal crystals. This new concept was confirmed in 1977 by A. Henglein's calculations of the reduction potential of the silver atom in aqueous solution, found to be 2.6 V more negative than the bulk metal [2], and by measuring oxidation rate constants of the atom and the dimer with several electron acceptors. Since then, due to their deep penetration and the homogeneous production of strong reducing species, ionizing radiation has been used as a particularly powerful tool, in the steady and the pulse regimes, to observe at room temperature the early and transient steps of metal atom formation, nucleation and growth into clusters, in the depth of a solution or of the pores of a support and to determine their nuclearity-dependent properties [3]. Based on these results, which were obtained under various conditions, the nucleation and growth mechanisms involving competition between several processes could be established. Their understanding is a guide to control the final size, shape and structure of mono- or multi-metallic clusters. The mechanisms are also inspiring the improvement of some processes for various applications where they play an important role.

Metal ion reduction and atom coalescence

The atoms are produced in solution by radiation-induced reduction of the metal ion precursors without any other added electron donor. The species arising from the radiolysis of water, solvated electrons e_{aq}^-, and $H^•$ atoms, are indeed the strongest reducing agents of this wet process. They easily reduce at room temperature all metal ions, possibly complexed by a ligand, down to the zero-valent state (**Fig. 1**). In contrast, sibling radicals which are also formed in radiolysis, such as $^•OH$ radicals in water, are able to oxidize the ions or the atoms into a higher oxidation state. To prevent this reaction, a scavenger of $^•OH$ radicals, such as a secondary alcohol or formate anion is added to the solution. The secondary radicals formed, $H_3C\text{-}^•COH\text{-}CH_3$ or $CO_2^{•-}$ radicals, respectively, are also strong reducing species (Chapter 1). To prevent oxidation by di-oxygen or corrosion by protons of the easily oxidisable atoms, the solutions are de-aerated and, for non noble metals [4], made slightly basic.

Such reduction reactions have been observed directly by pulse radiolysis for several metal ions. Most of the reduction steps have been observed and their rate constants determined. **Figure 1** presents the example of Ag^+ reduction observed by pulse radiolysis coupled with time-resolved spectrophotometry. The evolution of the optical absorption spectrum in the successive fast steps is recorded just before and after the short electron pulse delivering the irradiation dose, as in a movie filming the fast cascade of reactions initiated by the pulse. Provided the transient species have specific absorption bands (or colours), they can be identified and the detailed mechanism of their reactions and the corresponding rate constants is deduced from the variation of their absorbance which is related to their concentration. The broad and intense band in the near infra-red at the end of the pulse in **Figure 1** is due to hydrated electrons. They react readily here with Ag^+ giving rise to Ag atoms (λ_{max} = 360 nm) at the same time as the electron absorbance decays [2]. The binding energy between two metal atoms is stronger than the atom-solvent or atom-ligand bond energy. Therefore, the atoms dimerise on encounter. The bonding between atoms or clusters with unreduced ions is also strong and these association processes are fast, the first one being that of atoms with excess ions leading to Ag_2^+ (λ_{max} = 310 nm) (**Fig. 1**). The atom absorbance vanishes while that of Ag_2^+ increases. Then, Ag_2^+ dimerise into Ag_4^{2+} (λ_{max} = 270 nm) which progressively coalesce by a cascade of processes into growing oligomers and clusters. At nuclearity larger than n = 13, the spectrum of yellow clusters (λ_{max} = 380 nm), due to a plasmon resonance of electrons confined in the particule, is fully developed [2] (**Fig. 1**).

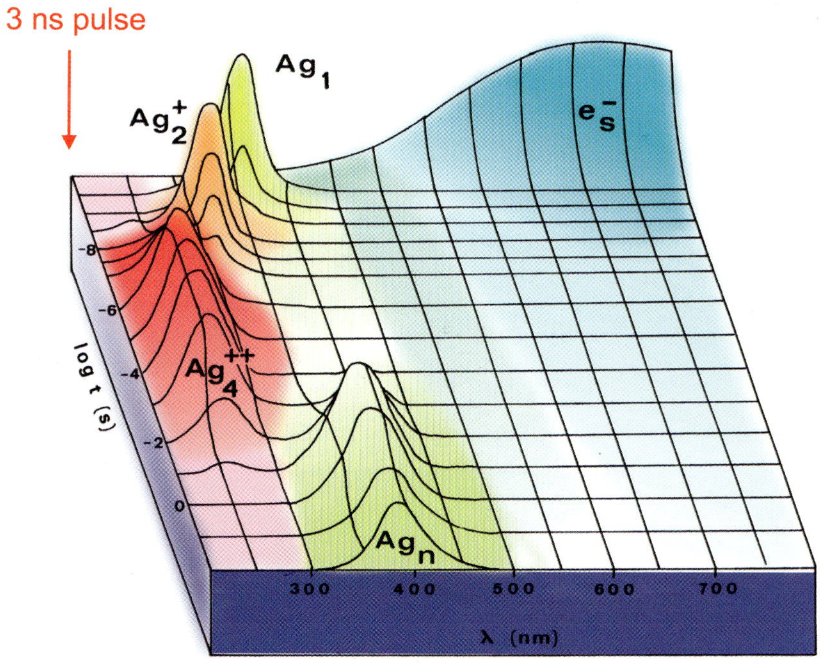

Figure 1 : *Evolution with time (logarithmic scale) after a 3 ns electron pulse of the optical spectra of transients formed in an aqueous solution of silver ions as observed by pulse radiolysis.*

Because the radiation-induced reducing agents, solvated electrons and radicals are generated randomly in the bulk of the sample and in the vicinity of metal ions, the atoms and clusters are formed with a rather homogeneous distribution throughout the solution with a remarkable homodispersity. Note that in the pulse regime (at high dose rate), all reducing species are produced and scavenged within a short time, followed by the steps of coalescence of atoms separately created (**Fig. 2a**) [6]. In contrast, in the continuous irradiation regime (at low dose rate), the association of M^+ ions with atoms, as in the charged dimers M_2^+ and clusters M_{n+1}^+, and the subsequent coalescence processes are faster than the production rate of the reducing radicals (**Fig. 2b**). Therefore, the reduction of M^+ ions occurs mostly *in situ* on clusters M_{n+1}^+ already formed. As a consequence, the new atoms formed are not isolated but contribute to the growth of a smaller number of preformed nuclei. In fact, the final nuclearity *n* and the cluster radius, which is proportional to $n^{1/3}$, are found to be systematically larger by steady than by pulse irradiation. When a chemical agent generally chosen as an electron donor D is added, the reduction potential ($E°D^+/D$) is not negative enough to reduce directly isolated metal ions to atoms (see below). Thus, it essentially reduces ions adsorbed on the nuclei generated by radiolysis and acting as seeds, a development process which results in still larger clusters (**Fig. 2c**) [7].

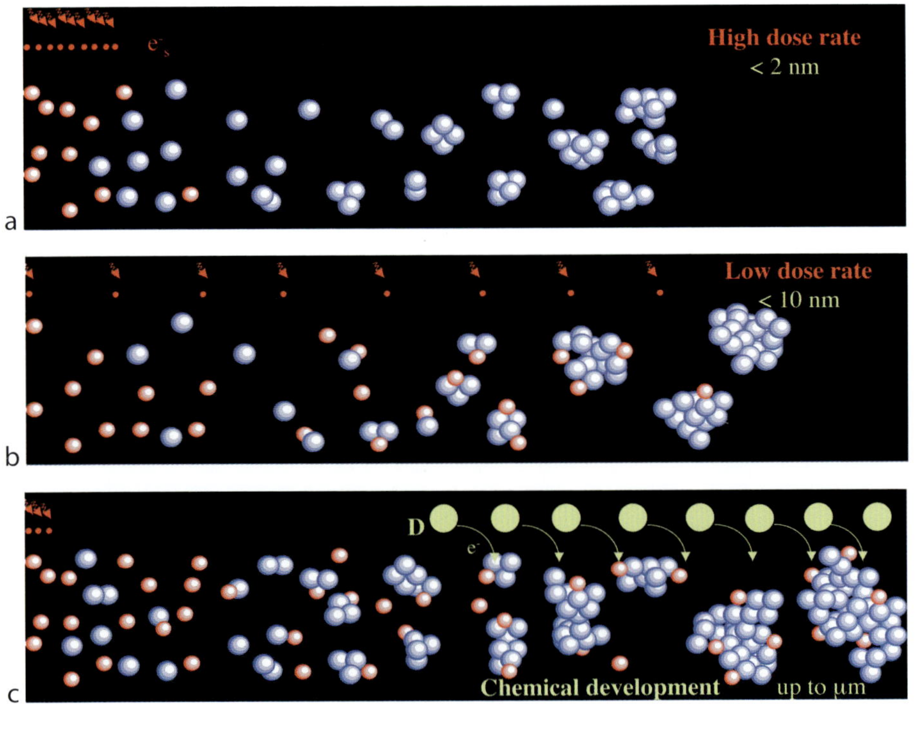

Figure 2. *Nucleation and growth of clusters generated by radiolytic radicals at high (a) and low (b) dose rates, without or with an electron donor D (c). The stabilizing effect of the added polymer prevents exclusively coalescence beyond a certain limit of nuclearity, but does not prevent successive ion and electron transfers (from the radicals at low dose rate and from the donor D), which lets the cluster develop up to much larger sizes than at high dose rate.*

Cluster stabilization

The radiolytic method of cluster synthesis, which is a bottom-up approach, can be used in various environments (**Fig. 3**). The control of the final size depends on the limitation applied to coalescence beyond a certain nuclearity, and to reduction *in situ* on nuclei (limitation favoured at high dose rate without chemical donor as shown in **Figure 2a**). For free clusters such as nanocolloids in solution, the coalescence may be limited by a polymeric molecule acting as a cluster stabilizer. Functional groups with high affinity for the metal, such as carboxyl groups, ensure the anchoring of the molecule at the cluster surface, while the polymeric chain protects the cluster from coalescing with the next one through electrostatic repulsion or steric hindrance and thus inhibits at an early stage further coalescence. Note that the polymer protects the cluster from growth by coalescence but not from development by metal ion addition and electron transfer *in situ* from radiolytic radicals or chemical donors (**Figs. 2b** and **2c**). The polymer should not chemically reduce the ions fixed on the clusters so as to prevent their growth by the development process (**Fig. 2c**). Poly(vinyl alcohol) (PVA) for example generally fulfils this criterion (**Fig. 3a**), in contrast to stabilizers also often used as the electron donor in some chemical reductions.

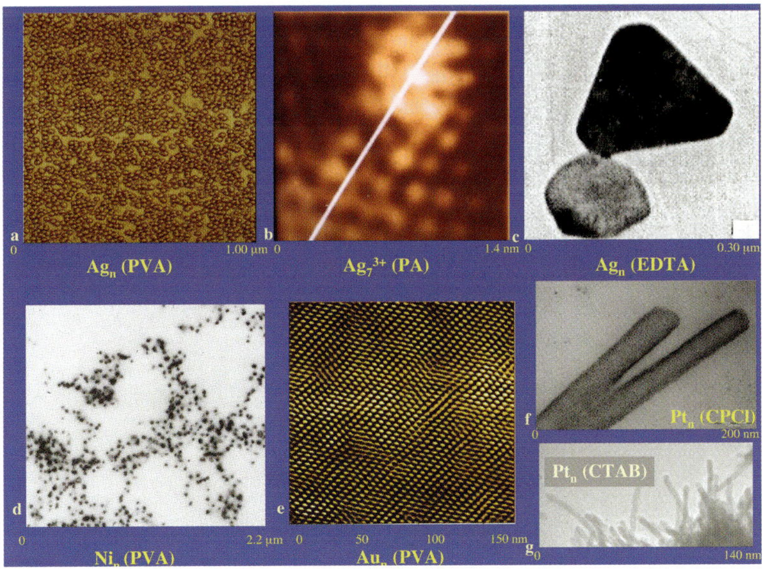

Figure 3: ***Radiation-induced metal clusters.*** (a) *Silver nanoclusters stabilized by PVA (10 nm).* (b) *STM imaging of a single cluster of the blue sol of silver oligomers* Ag_7^{3+} *formed by* γ*-irradiation (n = 4).* (c) *Clusters of* Ag_n *partially reduced by irradiation and then chemically developed by EDTA. (100 nm large and 15 nm thick).* (d) *TEM bright-field image of* Ni_n*, PVA clusters (5 nm).* (e) *Two-dimensional self-assembled array of gold clusters (PVA) on mica with remarkable homodisperse size (5 nm).* (f) *Monocrystalline Pt nanotubes with CPCl (10 nm diameter and a few 100 nm long).* (g) *Pt nanorods with CTAB (3-4 nm thick and 20-40 nm long).*

The final size of the clusters depends on the metal; it decreases as the initial polymer/ion concentration ratio increases and with PVA lies in the nanometre range (1-10 nm). The radiolytic method has been used for the synthesis of a great number of noble and non noble metal nanocolloids in various solvents, water, alcohols, liquid ammonia, etc. Their intense colours, due to a Mie resonance in light absorption by the electron pool of the particle, depend on the surrounding medium, on the cluster size and shape, and on the metal (**Fig. 4**): in water, yellow for Ag_n spheres of a few nm ($\lambda_{max} = 380$ nm), purple for Au_n ($\lambda_{max} = 520$ nm), pink for Cu_n ($\lambda_{max} = 570$ nm), and brown for the others (λ_{max} in the UV). For such small sizes, no light scattering is observed and the solutions are very clear. Cobalt and nickel clusters are ferromagnetic [4] (**Fig. 3d**). As a consequence of the initial homogeneous distribution of atoms, the size distribution of the clusters is remarkably narrow, as shown in **Figures 3a, 3d,** and **3e**. Due to their high homodispersity, they can even arrange in a two-dimensional self-assembled array when the droplet of the nanometre colloidal sol is dried on mica surface (**Fig. 3e**). Phase imaging in tapping mode AFM (atomic force microscopy) permits discriminating between silver or gold clusters and the amorphous polymer stabilizing the nanoparticles [7]. In the absence of an added radical scavenger, the polymer PVA can be cross-linked under irradiation during the simultaneous reduction of metal ions (silver, palladium, nickel, etc.). Finally, after drying, the clusters formed are trapped in a thin polymeric film, which presents the specific optical absorption band of the metal nanoparticles. In the case of ferromagnetic nickel nanoparticles, the film can be moved by a magnetic field [6].

Figure 4 *: Nanoclusters synthesized by irradiation of water solutions and stabilized by PVA. From left to right: copper, silver, and gold clusters. The colors of ultradivided metals are known from long. In the Middle Ages, they were produced in fused glasses under reducing conditions to produce stained glasses.*

Another polymer, sodium polyacrylate (PA), displays the property to stop at low dose the coalescence at only a few atoms, as shown by pulse radiolysis (**Fig. 3b**) [8]. Such oligomers of silver Ag_7^{3+} or platinum Pt_{5-7} were observed by STM (Scanning Tunneling Microscopy) (**Fig. 3b**) and by their optical absorption spectrum (λ_{max} = 800 nm for blue silver and 540 nm for pink nickel oligomers) assigned to a strong metal-acrylate ligandation. The silver and platinum oligomers stabilized by PA are long-lived even in air, but nickel-PA oligomers are spontaneously oxidized by water within one day [3].

Some ligands, such as CN^- for gold or silver clusters, are also able to stabilize, without added polymer, small size particles, probably because they are strongly linked to the cluster, thus protecting them from coalescence by electrostatic repulsion. However, the reduction potential of these clusters is still more negative than without CN^- and they are extremely sensitive to oxygen [6]. Carbonyl clusters, such as Chini clusters $[Pt_3(CO)_6]_m^{2-}$ with m = 3-10, can be easily obtained by radiation-induced reduction of metal ions under CO atmosphere [3]. The synthesis is selective and m is controlled by adjusting the dose (high doses yielding low m values). The colours (or optical absorption spectra) and redox reactivity also change with the nuclearity.

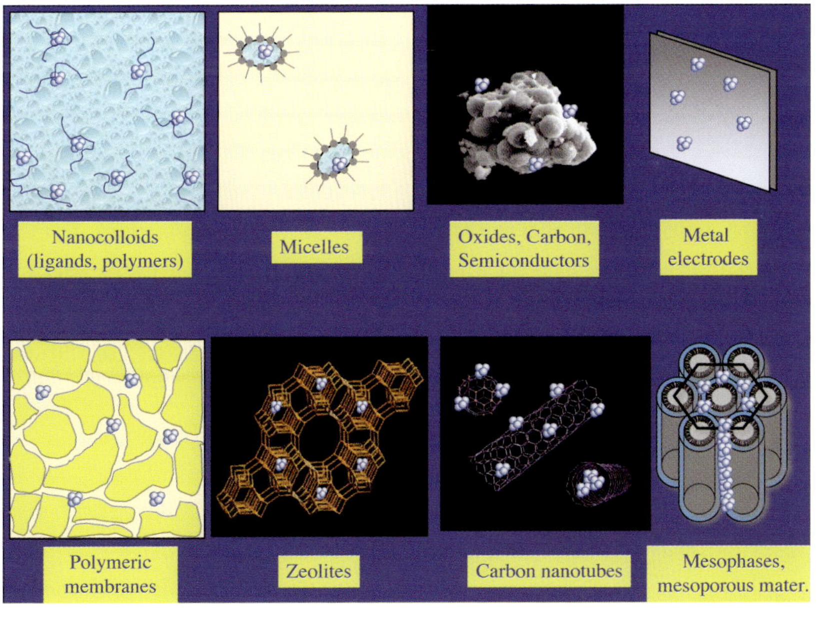

Figure 5 : *Clusters generated by irradiation in various hard (oxides, semiconductors, metals, zeolites, electrodes, carbon nanotubes) or soft matrices (liquids, micelles, polymeric membranes, mesophases).*

Clusters supported on solids or confined in various structures, hard or soft matrices (**Fig. 5**), are produced by letting the ion solution diffuse into the matrix pores or by mixing the ion and surfactant solutions [3, 6]. The support is often charged superficially with a layer of counter ions of opposite charge in the liquid phase. Metal ions of the same charge can thus exchange with these counter ions. Due to the penetrating properties of ionising radiation, reducing radicals are produced in the vicinity of these ions and yield the metal atoms in tight interaction with the support. At room temperature, the coalescence is restricted to the region, surface or intra-pore, where the atoms are fixed, and the sintering is limited by the strong interaction with the host material.

Another challenge in cluster synthesis is the control of the shape of the nanoparticles which governs their properties. One-dimensional metal nanostructures such as nanowires and nanorods are very attractive because of their unique electrical, optical, magnetic and mechanical properties, and because of their potential applications in nanodevices. Surfactant molecules self-assemble under certain conditions in water into a large variety of morphologies, including micelles and liquid crystalline phases such as lamellar or hexagonal mesophases. Such mesophase structures provide a class of useful and versatile templates for generating 1D or 2D nanostructures in relatively large quantities. A flexible approach has been developed for the radiolytic synthesis of nanostructured materials of various shape and size such as metal nanofibres or lamellae (and also polymer nanowires), using mesophases with hexagonal, cubic or lamellar symmetry as soft templates [9] (**Fig. 5**). Metal nanorods of a few nm diameter and a few μm length, and lamellae of a few nm thick and tens of nm^2 in area have been obtained.

In the Pt-doped hexagonal mesophase formed from CPCl (cetyl pyridinium chloride), platinum ions are adsorbed at the surface of the surfactant cylinders. They are reduced radiolytically into a metal layer as a nanotube of around 10 nm diameter and a few hundred nm long (**Fig. 3f**). Extraction of all these nanostructures is achieved by dissolution of the soft template using alcohol. This possible easy extraction constitutes a marked advantage over the synthesis in hard templates, such as mesoporous silica or carbon nanotubes, the dissolution of which is more hazardous for the metal nanostructures.

Because of a preferential binding to a single face of clusters, some surfactants are known (CTAB, cetylmethyl ammonium bromide, for example) to drive the growth of metal nanocrystals unidirectionally. Single monocrystalline Pt nanorods, monodisperse in diameter (3-4 nm) and 20-40 nm long, were recently obtained by orientated coalescence of spherical seeds of 3-4 nm radiolytically produced in the presence of CTAB surfactant [10] (**Fig. 3g**).

Note that solvated electrons can also be produced by photodetachment of electrons from certain anions or photoionization of molecules using UV excitation. Then the mechanism of metal ion reduction is expected to be quite similar to the radiolytic processes of **Figures 2a, 2b** and **2c**. However, the matrices used as cluster hosts are generally not transparent to light, and the reduction is often restricted to the surface.

Nuclearity-dependent reduction potential of clusters

The most important changes in cluster reactivity occur, as for other properties, at low nuclearity. However, due to the spontaneous coalescence, the small oligomers are generally short-lived and observable only by pulse radiolysis and time-resolved detection in the course of their coalescence as in **Figure 1**. Likewise, the determination of their reactions is only possible by kinetics methods, studying the influence of an added reactant of known reduction potential on their usual coalescence processes. Depending on the reactant, metal clusters M_{n+1}^+ may behave as electron donors or electron acceptors. In the first case they are oxidized into the ions (some examples, such as spontaneous oxidation by protons or oxygen, are given above).

When clusters are electron acceptors, the electron coming from a donor may reduce the metal ion fixed on the cluster and the cluster nuclearity is incremented by one unit (**Fig. 2c**). The donor D is produced by the same pulse as the atoms M^0 and is observed by a change in the absorbance at a specific wavelength. The remarkable feature of the kinetics signal at the donor wavelength is that the absorbance remains constant after the pulse for a rather long induction time [7]. Nevertheless, during this time, millions of encounters between the donor and isolated Ag^+, Ag_2^+ or Ag_4^+ have occurred without reaction. Actually, the electron transfer requires that $E^0(M_n^+/M_n)$, which increases with n during the spontaneous coalescence, becomes higher than the reference potential $E°(D^+/D)$. The reduction potential of the reference donor thus creates a threshold for n, that is a critical nuclearity n_c. For $n < n_c$, it is observed indeed that the coalescence occurs as in the absence of D, and that isolated ions and charged clusters can not be reduced except by radiolytic species. However, when $n \geq n_c$, D starts to decay at the end of the induction time due to the transfer of an electron to M_n^+. Simultaneously, the clusters grow autocatalytically by successive reduction of adsorbed ions to atoms and addition of a supplementary ion. Once formed, a supercritical cluster behaves indeed as a growth nucleus. The sequence of alternate reactions of ion fixation and reduction makes the cluster reduction potential more and more favourable for the transfer, so that an autocatalytic growth is observed up to the total consumption of D or M^+. Actually, n_c is deduced from the time-dependent donor concentration variation and from a numerical simulation taking into account the competitive processes of coalescence and electron transfer [7, 11]. By changing the reference potential in a series of reduction monitors (such as reduced forms of sulfonato-propyl viologen, methyl viologen or naphthazarin), the dependence of the cluster potential on the nuclearity has been obtained, for example for silver and copper clusters. The potential $E°(Ag_n^+/Ag_n)_{aq}$ in water increases with n as shown in **Figure 6** [11]. A value of the reduction potential is reached asymptotically at the nuclearity around $n = 500$ (radius $\approx$ 1.25 nm). More generally, the reduction potential of any metal atom $E°(M^+/M^0)$ is expected to be quite negative, and the oligomer potential to be lower than that of the bulk metal with a similar variation as in **Figure 6**.

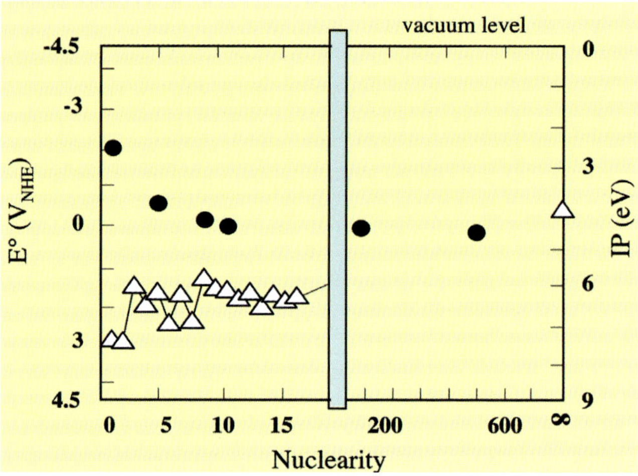

Figure 6 : *Size-dependence of the reduction potential E° of silver clusters in water (●) and of the ionization potential IP of silver clusters in the gas phase (Δ). The reduction potentials refer to the normal hydrogen electrode one which is at 4.5 eV above vacuum [11].*

The same phenomenon of a required critical nuclearity should exist as well in chemical reduction by donors having a reduction potential much less negative than radiolytic species. It is indeed well known that the reduction of a silver salt occurs essentially on the walls of a vessel where it forms a silver mirror. In contrast to free ions which are not reducible, the ions adsorbed on the walls present a potential higher than $E°(D^+/D)$.

The potential $E°(Ag_n^+/Ag_n)_{aq}$ represents the energy difference required to ionize the cluster in water compared to the ionization energy of the normal hydrogen electrode, which is 4.5 eV relative to vacuum. Thus, a comparison can also be made between the nuclearity effects on $E°(Ag_n^+/Ag_n)_{aq}$ and on the ionization potentials IP (Ag_n) of bare silver clusters in the gas phase [12] (**Fig. 6**). It is obvious from this figure that the variations of E° and IP do exhibit opposite trends *vs. n* for the solution and the gas phase, respectively. The origin lies indeed in the solvation free energy of the cation Ag_n^+ which in water assists the ionization of Ag_n. The difference between the ionization potentials of bare and solvated clusters decreases with increasing *n*, and corresponds fairly well to the solvation free energy of the cation Ag_n^+ deduced from the Born model.

An important consequence of the nuclearity-dependence of the reduction potential concerns the cluster nucleation/growth process itself. As shown in **Figure 2c**, the potential of most chemical donors D is not negative enough to start a homogeneous reduction of isolated ions in the bulk with the potential E° (Ag^+ / Ag_1) = – 1.8 V_{NHE} [2] (**Fig. 6**). They contribute instead by reducing *in situ* ions adsorbed on supercritical preformed particles (or on walls) (E° (Ag_n^+ / Ag_n) > E° (D^+ / D)) to develop larger ones. In the example of **Figure 3c**, the ligand ethylenediamino tetracetate (EDTA) protects the clusters from coalescence during the first step of irradiation. They are spherical with a size of 10-15 nm. However, if a large part of the ions are not radiolytically reduced but adsorbed on the former clusters, they are slowly reduced after days by EDTA in a post-effect as in **Figure 2c**, whereas solutions of silver ions complexed by EDTA are stable. The growth is orientated and favours the 111 surface, giving pellets of 100-150 nm [6]. They are bluish-grey and a second band centred at 1100 nm appears in their optical absorption spectrum. These pellets are particularly efficient for surface enhanced Raman scattering (SERS) detection of single molecules adsorbed on Ag. Similarly, gold clusters formed by partial radiolytic reduction of a solution of gold ions $AuCl_4^-$ are very slowly developed by reduction of adsorbed excess ions by PVA to form homodisperse cubic crystallites which range in size from 10 to 500 nm, depending on the respective parts of radiolytic and chemical reduction [6].

Bi- and multi-metallic clusters

Clusters containing two or more metals are of high interest for various applications, particularly catalysis, and their physical and chemical properties can be adjusted according to the composition [3, 6]. In alloyed clusters, the different metals are all present among the surface atoms of the cluster and can accelerate the successive steps of the catalyzed reaction.

Radiolytic reducing agents are strong enough to reduce rapidly the ions of any metal. However, as shown by pulse radiolysis of mixed solutions, further electron transfer from less noble atoms to the ions of the more noble metal may lead to a privileged reduction of the latter, then to the reduction of the former in a shell coating the initial noble metal core, resulting in a core-shell structure (**Fig. 7 top**). The optical absorption band due to the surface plasmon is in these cases progressively shifted in mixed solutions irradiated by γ-rays (a few kGy h^{-1}) from that of the noble metal core to that of the coating less noble metal. But, at high dose rate (delivered for example by pulses of an electron beam with a few kGy s^{-1}), the reduction of all ions can be achieved rapidly (**Fig. 7 bottom**) before any inter-metal electron transfer can occur. In that case, the coalescence occurs between the various atoms irrespective of their potential but according to their abundance, and an alloyed structure of the cluster is found having the same composition as the initial ion solution [3, 4]. This structure is checked by the evolution of the optical absorption spectrum, by X-ray photon spectroscopy (XPS) and X-ray analysis at increasing dose, and by electron scattering patterns.

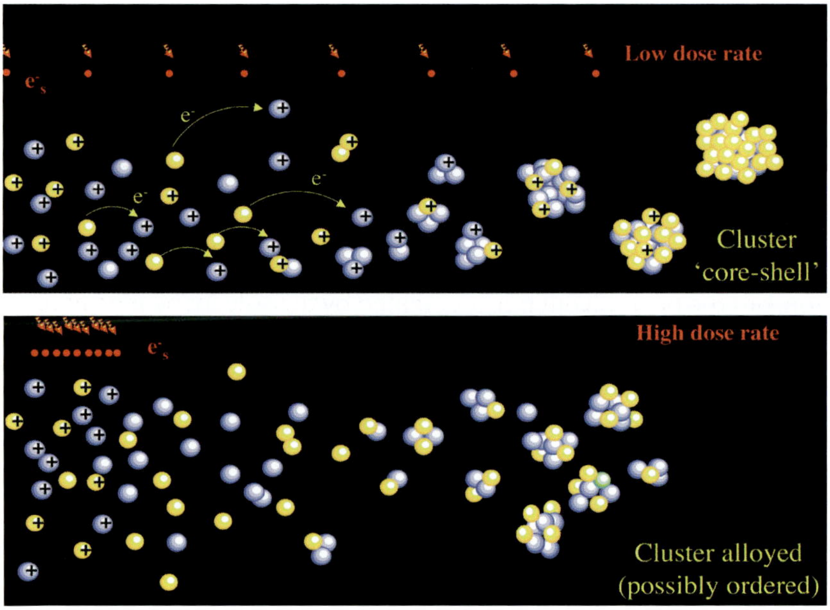

Figure 7 : *Scheme for the influence of the dose rate on the competition between the inter-metal electron transfer and the coalescence processes during the radiolytic reduction of mixed metal ion solutions.* Top) *Low dose rates favour core-shell segregation of the metals in the cluster due to electron transfer from less noble metal atoms (in yellow) to noble metal ions (in blue). The noble metal atoms thus constitute the core and are coated by the other metal.* Bottom) *High dose rates favour alloying of the metals by sudden reduction of all the ions, followed by the coalescence of atoms according to their abundance. The cluster size is also smaller (**Fig. 2**).*

Depending on the rate of inter-metal electron transfer competing with the dose rate-dependent coalescence [6], alloying is spontaneous, even at low reduction rate, as for Cu_3Pd, CuPd, NiPt, CuAu, AgPd and AgPt [3, 4], whereas it is only obtained at high dose rate, as for AuAg, AuPt and AuPd [3].

Alloyed multi-metallic clusters have been synthesized in the same way [3, 6]. However, characterization of the alloyed structure is still more difficult than for bimetallic clusters. The alloyed character is inferred rather from their catalytic properties, for example, which are enhanced when they are produced at high compared to low dose rate.

Optical limitation

At low light fluence, clusters behave like other molecules and their optical absorbance is independent of the exciting light fluence because they relax rapidly (**Fig. 8**) [13]. However, at very high light fluence ($\geq$ 0.9 J cm^{-2}) delivered by a short laser picosecond pulse (30 ps), the absorbance of gold clusters increases suddenly, particularly for large clusters (15 nm mean radius), indicating a non linear optical behaviour. The metal cluster nanosol behaves as an ultra- fast optical limiter or shutter, which can be used as a shielding against powerful lasers. From the analysis of the relaxation kinetics, which depends on the energy confinement (fluence, pulse duration, cluster size), two kinds of mechanisms are found: in the first one, faster than 1 ns, the gold cluster cumulates the energy absorbed, vaporizes and expands, so that scattering of light causes a sudden decrease of the direct light transmission [13] **(Fig. 8)**. Then, in the second mechanism, after energy transfer to the solvent, the cluster relaxes by condensation, but the new solvent bubble created by the solvent heating around the cluster now expands, and also behaves as a secondary scattering centre up to the solvent bubble cooling and the energy dissipation to the bulk. The cluster size relaxation is not strictly reversible, and the cluster seems to undergo fragmentation during the 20 first pulses. But the results are then reproducible, and an optical limitation with half efficiency is reached. For medium fluences, the first mechanism is negligible.

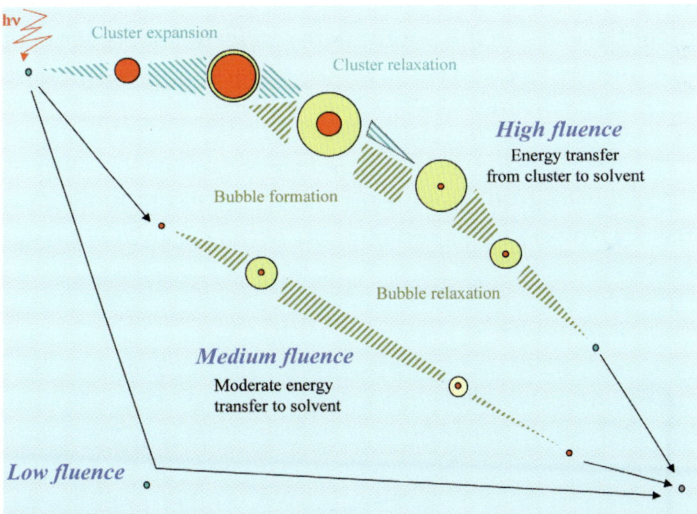

Figure 8 : *Non-linear optical effects. Scheme for the formation of two types of scattering centres at increasing fluences of a laser pulse used to excite gold clusters.* High fluence : *vaporized and expanded metal clusters limit light transmission immediately after the pulse and are replaced, after energy transfer to the solvent, by solvent bubbles acting as secondary scattering centres.* Medium fluence : *only solvent bubbles are formed.* Low fluence : *light transmission depends linearly on the intensity.*

Catalytic properties

One of the important applications of metal clusters is catalysis. The clusters are fixed on a support and their role is to accelerate the reaction between reactants. A catalytic process generally involves repetitive multistep reactions where reactants are transformed into products, and where the catalyst, in spite of its transient interactions with the reactants and products, is steadily regenerated. The catalysts should offer a large specific area in order to accelerate the access of reactants to the active sites. Ultra-divided metal clusters are thus particularly efficient in a number of reactions. However, as in the autocatalyzed growth described above (**Fig. 2c**), the catalyzed reaction is controlled not only by kinetics (large area), but also by thermodynamics (size-dependent reduction potential) [14] (**Fig. 9**). These aspects are often overlooked, although they concern each elementary step of the mechanism. Actually, when an overall electron transfer reaction between a donor D and an acceptor A is catalyzed by metal clusters **(Fig. 9**), they have the capability to play the role of an intermediate electron relay due to their reduction properties: in a repetitive cycle, alternately M_n acts as a donor faced to A, and M_n^+ as an acceptor faced to D [14]. Very small amounts of M_n^+/M_n are thus able to transform large concentrations of D/A into D^+/A^-. This mechanism will reach the optimized efficiency provided $E°(D^+/D) < E°(M_n^+/M_n) < E°(A/A^-)$ (**Fig. 9**). This thermodynamic condition also implies that the size of the cluster is small enough to correspond to a potential value in the appropriate range, between two thresholds imposed by D and A. Otherwise, for a given system D/A, too a large size cluster, (with $E°(A/A^-) < E°(M_n^+/M_n)$), *a fortiori* a bulk metal, will be unable to transfer the electrons to A, and too a small one, (with $E°(M_n^+/M_n) < E°(D^+/D)$), will be irreversibly corroded by transferring electrons to A without receiving electrons from D. Alloying between two or more metals permits one to adjust the potential according to the requirements of the successive catalyzed steps.

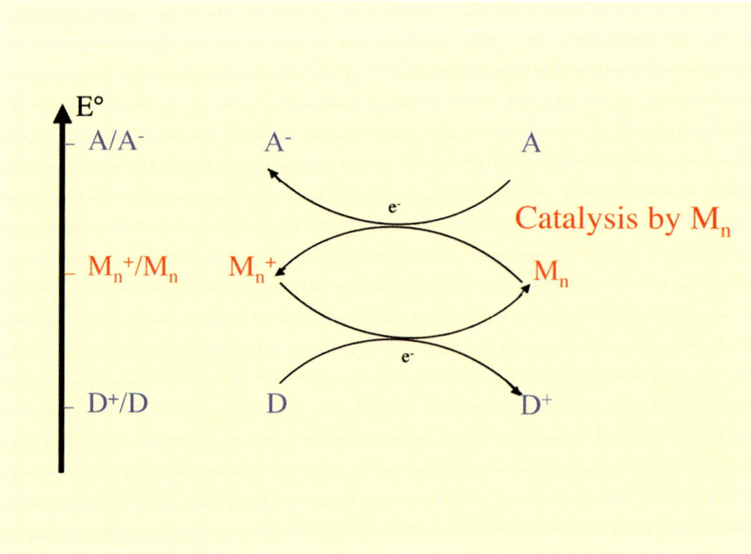

Figure 9 : *Scheme of the mechanism of a catalytic electron transfer involving metal clusters as a relay. The thermodynamic conditions to be fulfilled are that the cluster reduction potential is higher than the donor D and lower than the acceptor A potential. This implies an optimized range for the cluster size value.*

The mechanism of some catalyzed reactions of electron transfer has been studied by pulse radiolysis, for instance at the surface of colloidal TiO_2 particles [14]. Due to their very small and homodisperse size and to the strong interaction with the support, radiation-induced metal clusters are remarkably efficient in various reactions. For example, the radiolytic method has been used to graft metal nanoaggregates onto anodes or cathodes involved in the electrochemical chlorine-soda process. In both cases, significant overpotentials are usually measured on unmodified electrodes. Radiation-induced bimetallic nanoparticles (such as Pt-Ru and Ni-Ru), once grafted onto bulk metal electrodes (Ti or Ni), display a remarkable electrocatalytic efficiency. As a synergistic effect, a drastic decrease of the overpotential is observed when Pt and Ru are alloyed in atomic ratio 2:1 [3]. Similarly, the electrochemical behaviour of transparent SnO_2 counter electrodes grafted with Pt, Ru or Rh clusters approaches that of bulk Pt electrodes. Platinum $Pt_x(CO)_y$ and trimetallic Pt-Ru-Sn alloyed clusters, strongly anchored to the surface of carbon fibres and powders with high loadings up to 60 Pt wt.%, are very active as electrocatalysts for methanol oxidation in fuel cells [3] while Pt-Co clusters are very efficient for oxygen reduction.

Radiation-generated platinum and platinum-tin supported on α-Al_2O_3 or CeO_2 appeared as particularly active catalysts in the oxidation of vehicle exhaust gases (deNOx reactions). The temperature-dependence of the methane conversion in the vapo-cracking reaction ($CH_4 + H_2O \longrightarrow CO + 3\ H_2$) has been studied over radiolytic catalysts Ni/α-Al_2O_3. The catalytic tests indicate that the reaction starts at a low temperature with a very high selectivity. When complexed $Ni(NH_3)_6^{2+}$ ions adsorbed on CeO_2 are reduced radiolytically at room temperature, highly dispersed Ni^0 and intermetallic phases NiCe and Ni_2Ce are detected. The catalytic activity in the benzene hydrogenation reaction is remarkably high and the total conversion into cyclohexane is achieved in a markedly low temperature range [15].

Silver Photography

Silver photography is based on the formation, during the exposure, of the latent image made of silver clusters distributed at the surface of each silver halide crystal embedded in the gelatine and containing from 0 to 10 atoms.

Although the environment of the clusters is different in solutions and at the surface of silver halide crystals, we proposed to extend the same growth mechanism, which was demonstrated by pulse radiolysis for Ag_n^+ clusters free in solution, to a theoretical explanation of the development process in photography [7] (**Fig. 10**). The development occurs in both cases at the interface between an aqueous solution of an electron donor and a silver cluster acting as an autocatalytic growing site alternately accepting electrons and silver ions.

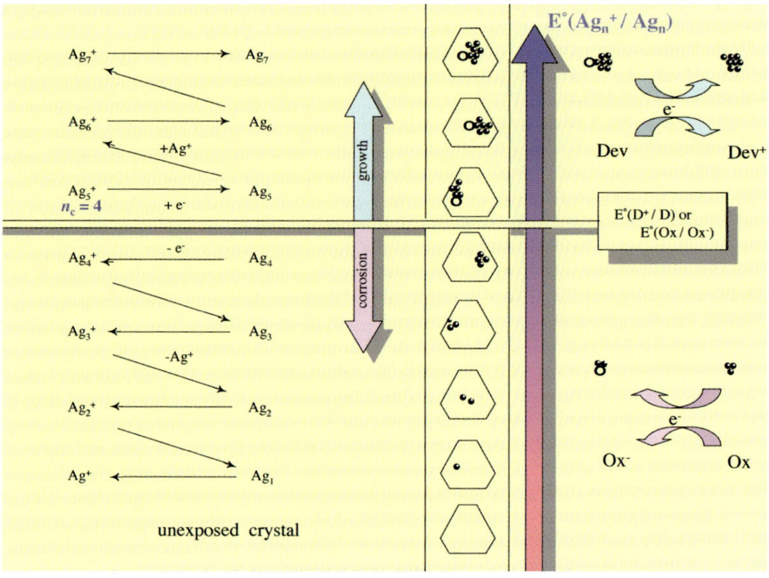

Figure 10 : *Photographic development mechanism. The reduction potential, E°(Ag_n^+/Ag_n), of the latent image clusters, when in contact with a solution, increases with the number of atoms n. Therefore a nuclearity threshold for development is created by the redox potential of the developer E°(D^+/D). Above the critical nuclearity n_c, the potential E°(Ag_n^+/Ag_n) is higher than E°(D^+/D), and alternate electron transfer toward Ag_n^+ and Ag^+ adsorption on Ag_n allows the cluster to grow autocatalytically. On the contrary, when E°(Ag_n^+ / Ag_n) is lower than E°(D^+/D), corrosion of subcritical clusters takes place by oxidizing molecules, such as D^+ or Ox [7].*

In fact, the various aspects of the catalytic growth revealed by the kinetic studies of solutions mimic the characteristics of the development known experimentally to photographers, namely the existence of a critical nuclearity as a lower limit for the development, and the dependence of its value on the reduction potential of the developer [7]. According to the mechanism established in solution, the discrimination induced by the developer is explained as the consequence of a quantum-size effect on the reduction potential of the silver nuclei which, at the aqueous interface, does increase with n: the critical nuclearity n_C, as a lower limit, is determined by the threshold imposed by the first one-electron reduction potential of the developer (n_C = 3-5 for usual developers) **(Fig. 10)**. Sensitization by ions of a noble metal such as gold generates alloyed clusters having a more positive potential. Only supercritical oligomers with a potential higher than that of the developer act as growth nuclei. The model is now widely accepted.

Another important application to photography of our understanding of the cluster formation mechanism in radiation chemistry is the enhancement of the sensitivity of photographic emulsions **(Fig. 11)** [16]. The primary effect of the photon absorption by the silver halide crystals of the emulsion during the exposure is indeed to produce one

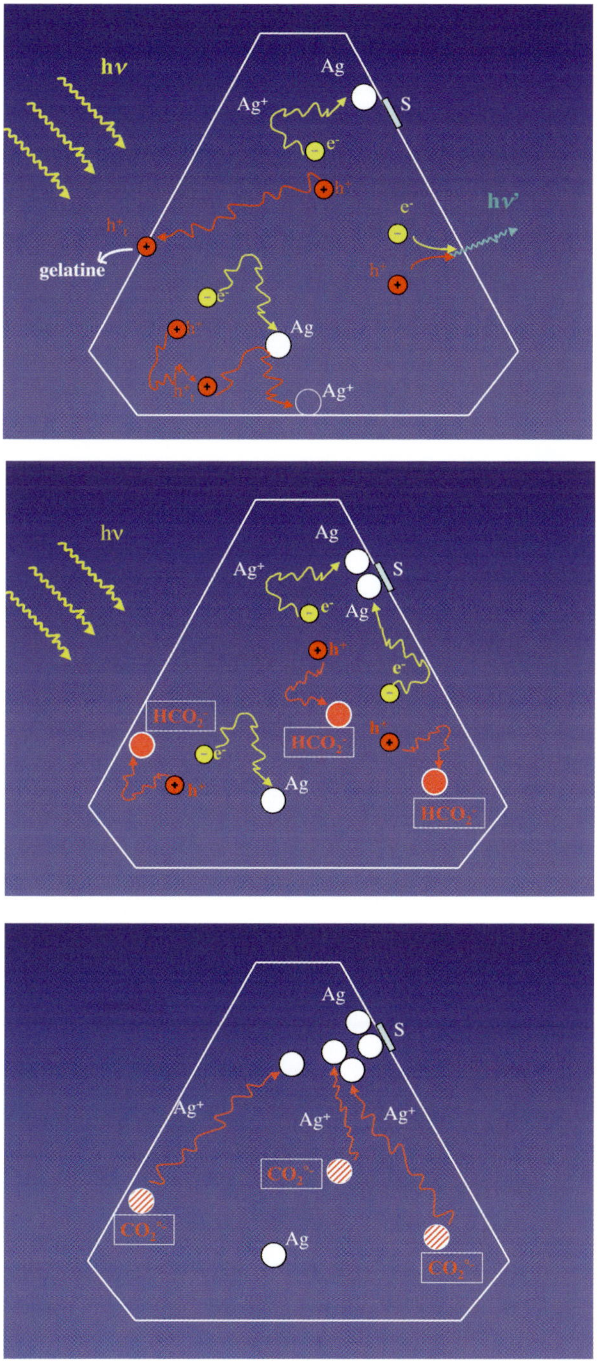

Figure 11 : ***Photographic latent image formation in undoped and formate-doped and gold-sulfide (S) sensitized AgBr crystals.*** Top: *undoped crystal with electron-hole recombination.* Centre: *formate-doped crystal and hole scavenging step by formate (HCO_2^-).* Bottom: *formate-doped crystal and delayed reduction step of additional silver ions by carboxyl radicals $CO_2^{\bullet-}$* [16].

electron-hole pair (e^--h^+) *per* photon absorbed. However, a fraction only of the holes (20%) diffuses to the crystal surface and is eliminated by gelatine. Indeed, the effective sensitivity decreases markedly due to the very fast recombination of part of the initial pairs with production of fluorescence before electrons reduce silver cations, and to the oxidation by holes of the newly formed atoms (**Fig. 11 top**). This analogy with $^{\bullet}OH$ oxidation processes occurring in irradiated solutions prompted us to use the same scavenging method to inhibit the fast electron-hole pair recombination and the silver atom oxidation by the holes. When the silver halide crystal has been doped during the precipitation with formate ions (a low amount, $HCOO^-/Ag^+ = 10^{-6}$, is sufficient), the holes produced simultaneously with electrons during the exposure are indeed immediately scavenged by the formate ions HCO_2^- (**Fig. 11 centre**) and produce carboxyl radicals $CO_2^{\bullet-}$ (**Fig. 11 bottom**). The fluorescence arising from the recombination is suppressed and the quantum yield is enhanced from 0.20 for the undoped reference up to 1 atom *per* photon for the doped emulsion after immediate developing. This means that the loss of electrons by recombination is totally suppressed. The yield is still doubled to an effective quantum yield of 2 atoms *per* photon when the development is delayed, because a supplementary reduction of Ag^+ by the strong reductant carboxyl radicals $CO_2^{\bullet-}$ occurs within 15 minutes (**Fig. 11 bottom**). This remarkable overall enhancement by ten-fold is applicable to all processes based on silver imaging, black and white and colour photography, radiography or holography [16].

Conclusion

Radiation chemistry methods have been shown to be of high potentiality to generate small- size monodisperse metal clusters, and to study the dynamics of nucleation and growth of clusters, mono- and multi-metallic, from the monomers to the stable nanoparticle. Their structure, core-shell or alloy, is governed by the dose rate used. Pulse radiolysis provides the means to observe directly by time-resolved techniques their reactivity, especially to determine during the growth their nuclearity-dependent properties, such as the optical absorption spectrum and the reduction potential. These are of crucial importance for the understanding of the mechanism of the cluster growth itself, in the radiation-induced as well as in the chemical or photochemical reduction processes. This knowledge is exploited to guide the synthesis of the clusters and the control of their size and shape. These specific properties have important applications in several fields such as electronics, optical limitation, catalysis, and photographic latent image formation and development.

References

[1] Delcourt M.-O., Belloni J., Capture de précurseurs de l'hydrazine par les ions Cu^{+} au cours de la radiolyse de l'ammoniac liquide, Radiochem. Radioanal. Lett., 1973, **13**, 329-338.

[2] Henglein A., The reactivity of silver atoms in aqueous solutions (a γ-radiolysis study), Ber. Bunsenges. Phys. Chem., 1977, **81**, 556-561. Tausch-Treml R., Henglein A., Lilie J., Reactivity of silver atoms in aqueous solution. II. A pulse radiolysis study, Ber. Bunsenges. Phys. Chem., 1978, **82**, 1335-1340.

[3] Belloni J., Mostafavi M., Remita H., Marignier J.-L., Delcourt M.-O., Radiation-induced synthesis of mono- and multi-metallic clusters and nanocolloids, New J. Chem., 1998, **23**, 1239-1258.

[4] Marignier J.-L., Belloni J., Delcourt M.-O., Chevalier J.-P., New microaggregates of non noble metals and alloys prepared by radiation-induced reduction, Nature, 1985, **317**, 344-345. Belloni J., Marignier J.-L., Delcourt M.-O., Minana M., Non noble metal catalytic microaggregates, a method for their preparation and their application in the catalysis of the photoreduction of water., US Pat. 1986, **4**,629,709. Belloni J., Marignier J.-L., Delcourt M.-O., Minana M., New mono or multimetal microaggregates, a method for their preparation and their application in the catalysis of the photoreduction of water, US Patent 1987, 4745094.

[5] Janata E., Henglein A., Ershov B.G., First Clusters of Ag^{+} Ion Reduction in Aqueous Solution, J. Phys. Chem., 1994, **98**, 10888-10890.

[6] Belloni J., Mostafavi M., Metal and Semiconductor Clusters, in: Studies in Physical and Theoretical Chemistry 87. Radiation Chemistry : Present status and future trends, Jonah C.D., Rao M. (eds), Elsevier, 2000, 411-452.

[7] Mostafavi M., Marignier J.-L., Amblard J., Belloni J., Nucleation dynamics of silver aggregates. Simulation of the photographic development process, Radiat. Phys. Chem., 1989, **34**, 605-617. Belloni-Cofler J., Amblard J., Marignier J.-L., Mostafavi M., La photographie révélée, La Recherche, 1990, **212**, 48-57. Belloni-Cofler J., Amblard J., Marignier J.-L., Mostafavi M., The principles of photographic development, Endeavour, 1991, **59**, 2-9. Belloni-Cofler J., Marignier J.-L., Amblard J., Mostafavi M., Guyon S., Roussi G., The Magic of the Image, 26 min film, CNRS, 1989.

[8] Mostafavi M., Keghouche N., Delcourt M.-O., Belloni J., Ultraslow aggregation process for silver clusters of a few atoms in solution, Chem. Phys. Letters, 1990, **167**, 193-197. Remita S., Orts J.M., Feliu J.M., Mostafavi M., Delcourt M.-O., Chem. Phys. Letters, 1994, **218**, 115-121.

[9] Surendran G., Tokumoto M., Pena dos Santos E., Remita H., Ramos L., Kooyman P.J., Santili C.V., Bourgaux C., Dieudonné Ph., Prouzet E., Nanomaterial Synthesis in Highly Swollen Hexagonal Phase Templates, Chem. Mat., 2005, **17**, 1505-1514.

[10] Krishnaswamy R., Remita H., Impéror-Clerc M., Even C., Davidson P., Pansu B., Synthesis of single-crystalline Platinum nanorods within a "soft" crystalline surfactant-Pt(II) complex, Chem. Phys. Chem. 2006, **7**, 1510-1513.

[11] Khatouri J., Mostafavi M., Amblard J., Belloni J., Ionization potential of clusters in liquids., Z. Phys. D., 1993, **26**, 82-86.

[12] Jackschath C., Rabin I., Schulze W., Electron impact ionization potentials of gold and silver clusters Me[n] n ≤ 22, Z. Phys. D, 1992, **22**, 517-520; G. Alameddin, J. Hunter, D. Cameron, Kappes M.M., Electronic and geometric structure in silver clusters, Chem. Phys. Letters, 1992, **192**, 122-125.

[13] François L., Mostafavi M., Belloni J., Delaire J., Optical limitation of gold clusters. Mechanism and Efficiency, Phys. Chem. Chem. Phys., 2001, **3**, 4965-4971.

[14] Khatouri J., Mostafavi M., Belloni J., Kinetics of electronic transfer in solution catalyzed by metal clusters, in Photochemistry and Radiation Chemistry : Complementary methods for the study of electron transfer. Wishart J., Nocera D. (Eds), Adv. Chem. Series, 1998, **54**, 293-314. Belloni J., Nucleation, growth and properties of nanoclusters studied by radiation chemistry. Application to catalysis, Catal. Today, 2006, **113**, 141-156.

[15] Keghouche N., Chettibi S., Latrèche F., Bettahar M. M., Belloni J., Marignier J. L., Radiation-induced synthesis of α-Al_2O_3 supported nickel clusters : Characterization and catalytic properties, Radiat. Phys. Chem., 2005, **74**, 185-200. Chettibi S., Keghouche N., Wojcieszak R., Boudjennad E.H., Belloni J., Bettahar M.M., Catalytic properties of CeO_2- supported nickel clusters synthesized by radiolysis, Catal. Today , 2006, **113**, 157-165. Redjala T., Remita H., Apotolescu G., Mostafavi M., Thomazeau C., Uzio D., Bimetallic Au-Pd and Ag-Pd clusters synthesized by gamma or electron beam radiolysis and study of the reactivity/structure relationships in the selective hydrogenation of buta-1,3-diene, Gas and Oil : Science and Technology, 2006, **61**, 789-797.

[16] Belloni J., Treguer M., Remita H., De Keyzer R., Enhanced yield of photoinduced electrons in doped silver halide crystals, Nature, 1999, **402**, 865-867. De Keyzer R., Treguer M., Belloni J., Remita H., Photosensitive silver halide element with increased photosensitivity, US patent 6,436,625, Aug. 20, 2002.

Chapter 8

Water radiolysis in cement-based materials

Pascal BOUNIOL

Introduction

Cement-based materials, that are widely used in nuclear industry, concern two main fields of applications:

1/ matrices of coating or coverage of radioactive wastes of low and medium level activities within primary or composite waste-forms (**Fig. 1**);

a b

Figure 1 : *Examples of cement-based matrices used in the conditioning of radioactive wastes :* a) *concrete container accommodating 4 primary waste forms ;* b) *mortar for embedding metallic wastes.*

2/ concretes for structural purposes or for radiation shielding within installations, which are able to be exposed to strong irradiation (silos and casemates for reactor vessel or burned fuels storage and high activity wastes).

The relatively moderate cost, great availability and flexibility in their use, the wide available composition and the high physico-chemical inertia (*e.g.* buffer medium, resistance to fire and irradiation) cause an inevitable interest in the use of cement-based materials. On the other hand, cement-based materials such as grouts, mortars and concretes constitute complex mediums of which some ingredients can prove weak with age, when the nature and the combination of diverse stresses such as drying, leaching, chemical attack, mechanical effort, irradiation, etc. are considered. Typically in this case, the free water is the major cause of the encountered problems, when the preservation of the concrete properties and integrity is the main concern (concept of durability). In the presence of internal- or external-source of irradiation, the action of ionizing radiation results, in particular, in a decomposition of water and mainly a production of gaseous dihydrogen. Besides the ageing mechanism, which is mainly related to the heating and drying (the direct radiation effects are indeed negligible), the radiolysis phenomena raise a real safety problem regarding the risk of internal pressurization or gas explosion in ionizing environment. In order to adopt the best countermeasures in industrial situation (*e.g.* ventilation of the buildings to provide a maximum of 3 to 4% H_2 in the air) or to try to inhibit the radiolysis while intervening on the process through formulation of material, addition of specific products, it is always advisable first to evaluate the level of risk with an adequate accuracy. The aim is, beyond the rough estimation of the source term H_2 alone, to describe correctly the radiation chemistry within the interstitial liquid in the material, as well as the satellite radiolysis phenomenologies by taking account of the greatest number of elementary mechanisms. A description as complete as possible of the radiolysis phenomena in a macroscopic and heterogeneous system (*e.g.* mineralogy and porosity resulting from the hydration of cements, transport of gases in porous medium, transport of the radiation and energy deposition, heterogeneous chemistry gas/solution/solid phases, etc.) is necessary for the development of an operational model of behaviour. Though not developed in this study, a high level of coupling exists between these various phenomena and radiation chemistry. We will mainly focus on this last point in the following developments.

The cement based matrix: an original medium

Independent of the material implemented, the binder phase ensuring cohesion of the aggregates and mechanical rigidity comprises:

- anhydrous cement (mixture of mainly calcium silicates and aluminates);

- hydration products of these minerals, essentially amorphous silicates and crystallized $Ca(OH)_2$ (**Fig. 2**):

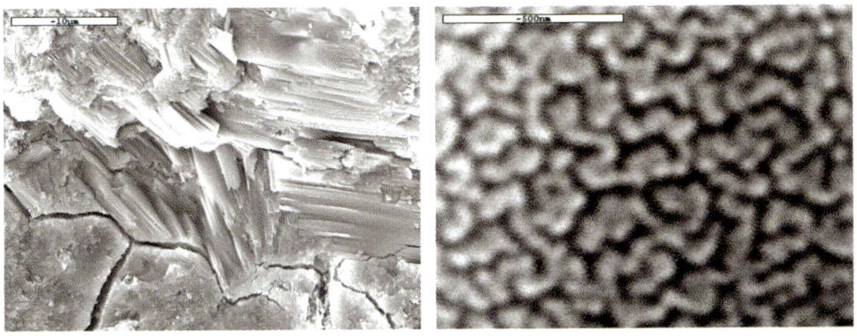

Figure 2 : *Hardened paste of hydrated Portland cement observed with MEB* : a) *tabular crystals of* $Ca(OH)_2$ *(portlandite) on fissured calcium silicate hydrate substratum* ; b) *zoom on the intrinsic porosity of calcium silicate hydrate.*

- a multimode porosity, including fine pores of just a few nanometers as well as coarser capillary sections up to hundreds of nanometers, occupied by residual liquid and air (**Fig. 3**):

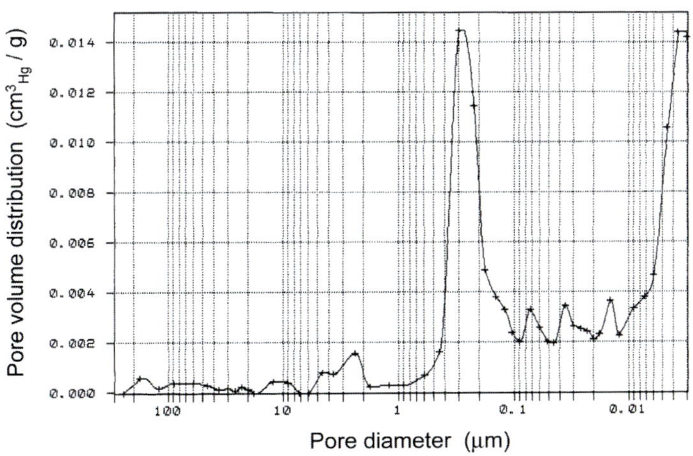

Figure 3 : *Distribution of pore sizes of an hydrated Portland cement paste 4 months old (initial mass ratio water/cement of 0.4) obtained by mercury intrusion. The pore family related to the calcium silicate hydrate (some nm) and that of capillary porosity (a few hundreds of nm) can be easily distinguished. The finest porosity is completely saturated with interstitial liquid while the capillary porosity is partially.*

All these make cement matrix a composite, multiphasic and very heterogeneous material, that is difficult to be studied. In addition, cement matrix has also an evolutive character due to the progress of hydration, modifying the microstructure in its early ages (from a few months to several years). From a physical point of view, the amorphous character of the main hydrated phase containing numerous defects allows the material to adapt to various conditions of temperature and irradiation. In addition, the presence of a porosity accessible to gases ensures an effective transport of radiolytic dihydrogen, essentially by gas diffusion. This means that, when H_2 is evacuated, it is replaced by air in which the O_2 fraction interferes with the radiolytic processes. From a chemical point of view, the presence of alkaline elements in cements (K and Na) makes the pore liquid in contact with the solid phases to be far from being pure water, since it is an aqueous solution of pH higher than 13 with an ionic strength close to 0.2 mol/l. The important proportion of crystallized $Ca(OH)_2$ (portlandite) with an already high balanced pH (12.45 at 25 °C) provides this medium with a reinforced buffer power, with respect to which the chemical variations induced by radiolysis remain overall very limited.

In the case of a concrete structure, the presence of steel reinforcements can further complicate the system by causing a local heterogeneity since they constitute in terms of dose rate (driving of the radiolysis) or by the influence of iron ionic species on radiolysis at the interface with the cement paste. Although still not very well-studied, the resulting radiolysis-corrosion coupling (as a function of temperature) is identified as the only aspect which could possibly affect the concrete durability under radiation [1].

Effect of cement chemistry on radiolysis

Even if radiolysis does not, or only very weakly, influence the chemistry of the cement medium, the reverse assertion is not true. The radiolysis of the interstitial liquid only relates to the solvent (H_2O) which gives approximately 18 species, well-known for most of them (Chapter 1). Due to the very high pH, only the basic forms will play a role. Thus, the equilibria listed below implying the main radiolytic species are displaced towards the right :

$$HO_2^{\bullet} + OH^- \leftrightarrow O_2^{\bullet -} + H_2O \qquad pK_a = 4.80 \qquad (1)$$
$$H^{\bullet} + OH^- \leftrightarrow e_{aq}^- \qquad pK_a = 9.60 \qquad (2)$$
$$H_2O_2 + OH^- \leftrightarrow HO_2^- + H_2O \qquad pK_a = 11.68 \qquad (3)$$
$$^{\bullet}OH + OH^- \leftrightarrow O^{\bullet -} + H_2O \qquad pK_a = 11.90 \qquad (4)$$

Before leading to the formation of these various species in the solution at the homogeneous stage, the high concentration of OH^- ions interferes with the primary production processes

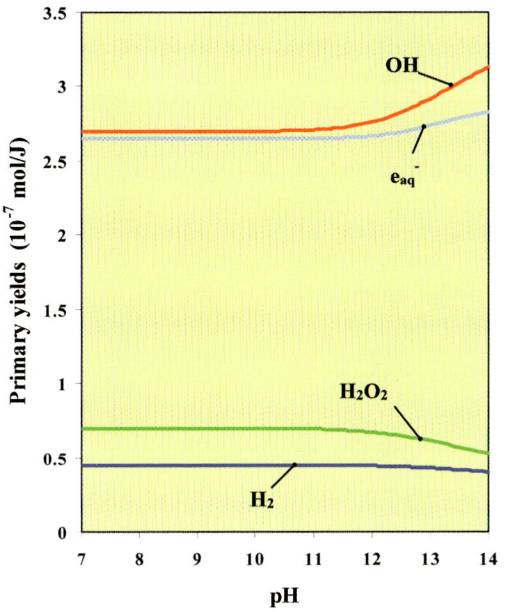

Figure 4 : *Evolution of primary yields for the main species of the water radiolysis (gamma radiation) according to the pH at 25 °C.* The pH of the cement based matrices, beyond 13, is in the most basic zone.

within the radiolytic spurs due to a higher encounter probability. **Figure 4** shows thus that between pH 12 and 14, the primary yields of $^{\cdot}OH$ and e_{aq}^- radicals increase significantly, whereas the primary yields of H_2 and H_2O_2 decrease more moderately. At the end of approximately 10^{-6} s, about sixty secondary reactions between species resulting from the primary stage initiate the already known mechanisms involved in the radiolysis of water (Chapter 1) with a specific orientation, due to the very high basicity of the medium :

– Respective attacks of H_2 by the oxidizing radical $O^{\cdot -}$, and of HO_2^- by e_{aq}^- are faster in comparison with the homologous reactions constituting the chain-reaction highlighted by A.O. Allen [2] at neutral pH. This results in a persistence of that chain in alkaline medium under an alternative form, which is theoretically more effective to recycle H_2 and inhibit the water decomposition:

initiation :

$$O^{\cdot -} + H_2 \rightarrow e_{aq}^- \qquad k = 1.2 \times 10^8 \, l \, mol^{-1} \, s^{-1} \qquad (5)$$

propagation :

$$e_{aq}^- + HO_2^- \rightarrow O^{\cdot -} + OH^- + H_2O \qquad k = 3.5 \times 10^9 \, l \, mol^{-1} \, s^{-1} \qquad (6)$$

– The peroxide and superoxide species, which naturally tend to accumulate, are prone to very slow reactions of disproportionation (reactions (7), (8)) ($< 1\ \mathrm{l\,mol^{-1}\,s^{-1}}$). These reactions are also the cause of the secondary dioxygen formation (always less compared to H_2), that occurs at long time. Among these reactions, specially the three following can be listed, the last one (mixed type) being a basic equivalent of the Haber-Weiss reaction :

$$H_2O_2 + HO_2^- \rightarrow O_2 + OH^- + H_2O \quad (7)$$
$$O_2^{\bullet-} + O_2^{\bullet-} \rightarrow O_2 + O_2^{2-} \quad (8)$$
$$HO_2^- + O_2^{\bullet-} \rightarrow O_2 + OH^- + O^{\bullet-} \quad (9)$$

Like in neutral medium, the presence of dioxygen partially interferes with the chain-reaction. That is to say that in a closed system, safe from air, a regulation of the radiolysis leading to a steady state is always possible by a mass action mechanism relating to the whole reaction system. This also means that, in the presence of a strong reductant inducing a ratio of initial concentrations such as $[H_2] / [O_2] > 2$, the chain-reaction is strongly activated, causing O_2 to disappear and resulting in a low and stable concentration for H_2. This mechanism is similar to that implemented in the pressurized water nuclear reactors where a partial pressure of H_2 is applied in order to obtain a slack pressure and an anoxic medium (Chapter 4). Under these conditions, at any time the amount of H_2 produced is equivalent to the amount of H_2 consumed.

With a very high pH, the second peculiarity of the cement medium lies in the presence of important amounts of calcium in the form of portlandite $Ca(OH)_2$. Paradoxically, the content of calcium in solution appears rather moderate, approximately $3 \times 10^{-3}\ \mathrm{mol\,l^{-1}}$ at 25 °C. However, this might be of a secondary importance compared to the fact that portlandite provides a permanently available and quasi-inexhaustible source of calcium. This aspect can be considered as interesting since a new solid compound, the calcium peroxide octahydrate, can be formed due to the increase in the peroxide concentration during the radiolysis [3]. With a very low solubility ($K_s = (Ca^{2+})(O_2^{2-})(H_2O)^8 = 2.5 \times 10^{-11}$), $CaO_2.8H_2O$ precipitates indeed in contact with the portlandite as soon as the solubility product K_s is reached (heterogeneous precipitation not requiring a supersaturation), and grows gradually with the consumption of the portlandite. When the two solid phases coexist, the equilibrium relations with the pore solution already buffered in term of pH result in obtaining a mineralogical super-buffer where portlandite and $CaO_2.8H_2O$ respectively control the concentrations of calcium and peroxide in the solution. The concentration of the whole of the species, and especially of O_2 and H_2, becomes constant by propagation to all other equilibria, which results in a second mode of radiolysis regulation within the interstitial liquid. Precipitation is however conditioned by a strong production of radiolytic peroxide (*i.e.* by a high dose rate (**Fig. 5**)).

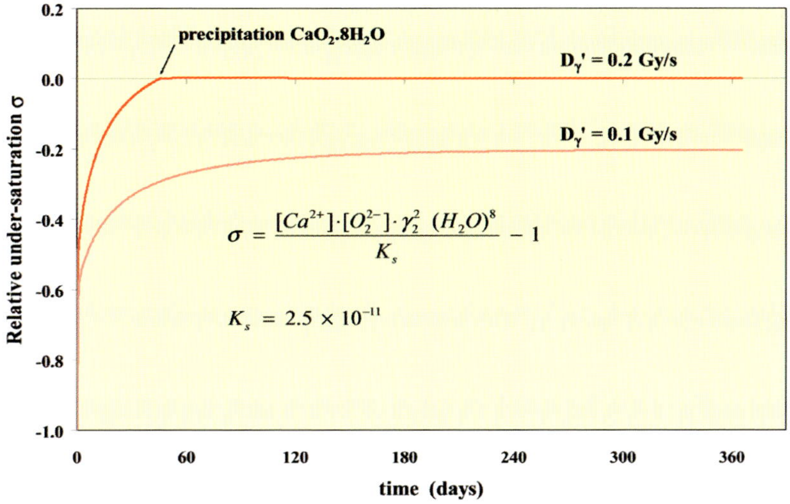

Figure 5 : *Evolution of the relative under-saturation index σ according to the duration of gamma irradiation of a Portland cement paste for two dose rates (simulation). The solubility product K_s is reached for the highest dose rate that causes an earlier regulation of the radiolysis in the pore solution.*

The effect of the cement material porosity on radiolysis

Another set of relations of the heterogeneous type is also established between the interstitial liquid and the gaseous phase present in the material porosity. At the end of the hydration process of a Portland cement paste having an initial mass ratio of water/cement < 0.42 (index generally associated to a good quality), the gaseous phase occupies approximately 25% of the total porosity in a closed system. Taking into account the low solubility of H_2 and O_2 initially formed in aqueous phase by radiolysis, the latter passes then very easily into the gaseous phase of the porous volume, thus having significant effects on the global behaviour of the system :

- in term of quantity of matter, H_2 and O_2 happen to be mainly in gaseous phase: (about 57 and 93 molecules remain in solution, respectively, for 1000 in the gas phase at 25 °C) ;

- in the gas phase, H_2 is not attacked by the oxidizing radicals while in the solution the relative enrichment in O_2 favours the radiolytic decomposition of water and subsequent formation of hydrogen ; therefore, H_2 accumulates in the gas phase ;

- the accumulation of H_2 gas strongly attenuates the efficiency of the regulation mechanisms of the solution radiolysis previously described ; reaching a global equilibrium remains still possible but after a longer duration, since the heterogeneous equilibrium for gas/solution (Henry's law) moreover introduces an additional "resistance".

The saturation degree of the pores obviously constitutes a control parameter of the radiolysis. When increased, it leads more directly to equilibria in solution, and becomes more efficient in the regulation of the phenomenon. In contrast, the regulation mechanisms based on a high H_2 concentration in solution are not very active and that causes production and accumulation of gaseous dihydrogen in the system. The opening of porosity to the atmosphere (open system) results in an alternative for this case: the expansion volume of radiolytic gases becomes infinite and then, the decomposition of water occurs indefinitely. In this last case, the fast evacuation of H_2 by gas diffusion is accompanied with the penetration of O_2 in exchange. This would also interferes with the regulating chain-reaction of the "Allen" type (reactions (5), (6)). In the macroscopic waste form level, the operational consequences that arise due to previously mentioned issues, lead to conflicting situations in gamma radiolysis: in a closed system the total pressure increases up to a moderate plateau value, while in an open system the total pressure remains lower but ensures a constant H_2 release.

From a more fundamental point of view, the porous medium also raises very interesting questions on the effect of water confinement within the pores in nanoscale. Thus, whatever the topological characteristics the network may be, it is legitimate to think that the tight mixture of liquid and solid phases with a considerably developed interface impacts not only the modes of energy deposition due to the irradiation, but also the probabilities of meeting of the various species in solution. This last aspect is being studied through the modification of the primary yields of the water radiolysis in porous media model (glasses with monodisperse porosity). The first results clearly confirm the effect of the confinement with a very significant increase in the primary production of the molecular species H_2 and H_2O_2 [4]. One of the hypotheses about this behaviour proposes that, if the characteristic pore size is lower than the mean free diffusion path of the primary radicals, the latter are forced to recombine, giving more molecular products. This effect starts to be less significant when the pore size increases and then disappears when it becomes much higher (a few tens to a few hundreds of nanometers) than the size of the radiolysis spurs. In theory, the interference of porosity with the chemical reactivity should also impact all the reactions in solution, consecutive to the primary stage, with a more or less important effect on the kinetic constants, depending on the size and the charge of the species. This field of interest should be developed in the future. For the moment, it does not warrant significant corrections to the current model describing the radiolysis in cement based-materials in which primary yields and kinetic constants still result from determinations in solution, in the absence of porous environment. In any case, it will be supposed to bring an answer to certain questions of operational relevance: do primary yields still keep a meaning in porous environments ? Would the super-buffer character of the cement medium and the presence of a gas porosity be able to attenuate the range of the elementary mechanisms highlighted in porous environment ?

Global phenomenological approach and simulation

The combination of the various phenomena involved in radiolysis allows us to propose a model based on the previously evoked knowledge: primary yields, secondary reactions, initial concentrations, dose rate, solid-solution and solution-gas exchanges [5]. A more accurate description of the system moreover results in studying it, according to determining operational parameters such as the pH, the saturation degree of the pores, the radiation source decay and the temperature. When the relations existing between the various phenomenological compartments are examined (**Fig. 6**), it is seen that in terms of internal pressure or quantity of released H_2, the global result of radiolysis in a cement matrix cannot be simplified to a simple addition of elementary phenomena.

In a system characterized by a strong heterogeneity on various scales and by a large variety of mechanisms, the global behaviour proceeds, on the contrary, from the complexity and the numerous couplings. The most obvious demonstration for this results in the production of responses, at best non linear, indeed with thresholds. **Figure 6** shows thus that, starting from various initial parameters (intensity and type of the irradiation, composition of the interstitial liquid), the evolution possibilities of the radiolysis are multiple, with more or less broad loops according to operating conditions. The two solid-liquid and liquid-gas heterogeneous equilibria, the impact of which is very strong because of quasi-direct feedbacks on the medium, can be recognized in particular on the upper part of the diagram (related to the radiolytic chemistry). With an operating mode in an open system, the dimension of transport is superimposed with a feedback at the same time less direct and more complex on the reactional medium (smaller H_2 solubilization, contamination by entering O_2). In all cases, it is interesting to emphasize that the equilibrium pressure in the porosity plays the central role in these feedbacks, more than the solution concentration. This leads to consider that the cement matrix could exhibit a global behaviour more or less similar, independently of the nature of the mechanisms in solution.

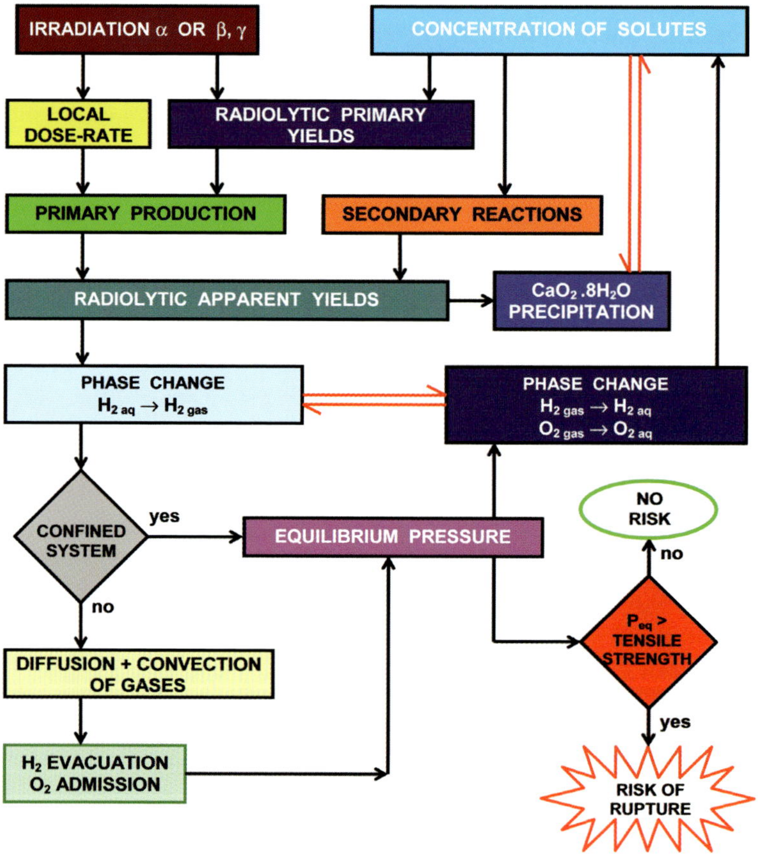

Figure 6 : *Physico-chemical parameters and their couplings involved in the radiolysis of cement-based materials at the level of a waste form. Heterogeneous equilibria (solid solution and solution-gas) are located by a double red arrow.*

An example of simulation of the radiolysis in a closed and partially saturated cement medium illustrates perfectly the non-linearity of the behaviour for various dose rates (**Fig. 7**). Hence, this kind of simulation shows that :

- partial pressure of H_2 in the porous network is not proportional to dose rate D', except at the beginning of irradiation ;

- an equilibrium can be reached by the action of different mechanisms (by precipitation of $CaO_2.8H_2O$ for D' = 0.2 and 0.4 Gy/s ; by mass action for D' = 0.1 Gy/s) ;

- a close equilibrium plateau can be obtained under different conditions of irradiation and mechanisms (note the important delay of setting to equilibrium in the case of the mass action).

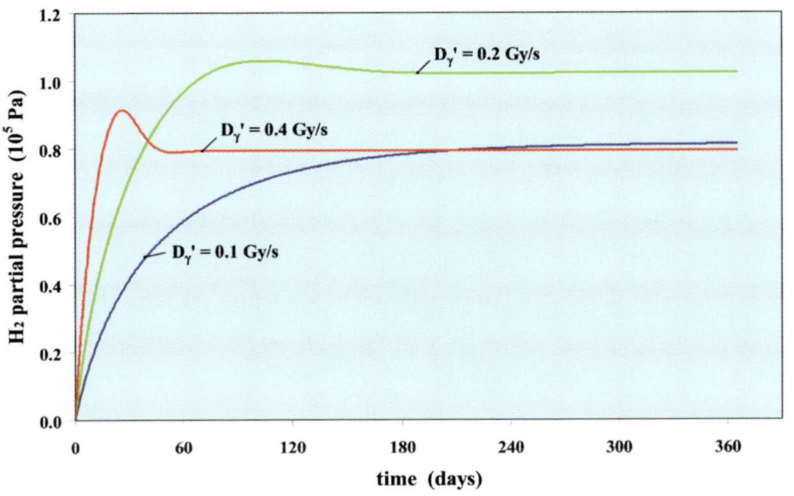

Figure 7 : *Compared evolutions of the H_2 partial pressure in a Portland cement paste with various gamma constant dose rates (simulation for a degree of pore liquid saturation equal to 0.75). Dose rate of 0.1 Gy/s is not sufficient to cause the precipitation of $CaO_2.8H_2O$ and an earlier regulation of the radiolysis in the medium.*

Although simulations happen to be the most economic and fastest way to carry out the evaluation of the H_2 source term in various cement-based materials, it is advisable to perform periodical experiments, preferably following an important conceptual modification, in order to check certain predictions and also to fix certain parameters of the model. These experiments generally are related with the test-bars of cement material irradiated in closed vessel or the material poured within air-tight mini-containers (**Fig. 8**).

Figure 8: *Radiolysis experiments on cement pastes leading to know the total pressure and the composition of gases according to the duration of gamma irradiation :* a*) sealed glass phial around a test-bar 4 × 4 × 16 cm ;* b*) mini-containers out of stainless of 500 cm^3 (allowing a quasi-integral filling of cement paste) with outlets for $P(H_2)$ measurement in line.*

The follow-up of the time dependence of the total pressure and the H_2 partial pressure, supplemented by a complete analysis of the gas space by chromatography, allows comparison between experiment and simulation reasonably under well-defined operating conditions. One should keep in mind that a badly characterized porosity or an inaccurate dose rate constitute the most common causes of divergence, with the strongest impact.

The currently selected model is based on strong assumptions of homogenization (*e.g.* dose rate, average concentrations and gradients) and can be described with an 1-D geometry. This model enables us to calculate the total and partial pressures to a satisfying order of magnitude as well as a minority presence of O_2 even safe from the air, and also a similar evolution of the long-term behaviour (1 year and more).

Besides the fundamental studies which will determine if the cement matrix can be still regarded as an overall homogeneous medium in spite of the incidence of the porous environment on the primary phenomena and the reactions between species, the prospects for the model improvement still remain to be numerous. They concern primarily the chemistry of elements naturally present in cement (Fe, S, C) and interfering with the radiolysis, as well as the chemistry of elements deliberately added to the wastes (N, Cl). The temperature influence studies are of a great extent since this parameter affects practically the whole of the physical and chemical phenomena proceeding during radiolysis : modification of the kinetics rate constants, primary yields, equilibrium constants, transfer coefficients, etc. With this point of view, the temperature-dependence of the solubility product of $CaO_2.8H_2O$ appears as an important data to be acquired.

Mainly motivated by the safety problem in relation with the "hydrogen" risk, studies in progress about radiolysis find their application in the evaluation of gas term-source for cemented radwastes in storage stage (*i.e.* at the beginning of the exploitation, when dose rates are the highest). They also concern the packaging concepts integrating high activity primary waste forms (nuclear glasses, burned fuel) where the coupling of radiolysis with temperature or corrosion is particularly involved.

References

[1] Zuili D., Couvreur F., Bouniol P., Gorse D., Electrochemical study of the effects of gamma-ray irradiation and temperature on the corrosion behaviour of reinforced concrete, séminaire CEFRACOR Corrosion Acier, January 2001, Paris.

[2] Allen A.O., The radiation chemistry of water and aqueous solutions, Van Nostrand, New York, 1961.

[3] Bouniol P., Aspart A., Disappearance of oxygen in concrete under irradiation: the role of peroxides in radiolysis, Cement and Concrete Research, 1998, **28**, 1669-1681.

[4] Rotureau P., Renault J.P., Lebeau B., Patarin J., Mialocq J.C., Radiolysis of confined water: molecular hydrogen formation, Chem Phys Chem, 2005, **6**, 1316-1323.

[5] Bouniol P., État des connaissances sur la radiolyse de l'eau dans les colis de déchets cimentés et son approche par simulation, Report CEA-R-6069, Direction des systèmes d'information CEA/ Saclay, 2004.

Chapter 9

Obtaining high performance polymeric materials by irradiation

Xavier COQUERET

Introduction

The interaction of high-energy radiation with organic monomers and polymers induces various chemical transformations of increasing importance from both scientific and technological standpoints. Cross-linking and scission in polymers and networks, polymerization of pure and blended monomers, grafting onto synthetic and natural polymers as well as chemical activation of organic materials by oxidation can be induced by irradiation under soft conditions. After more than fifty years of complementary basic and of application-oriented research, the radiation chemistry of polymers has been found to lead to many applications of current technological, economical and societal interest [1,2]. Better understanding of the primary events and subsequent chemical reactions mediated by ionic or free radical intermediates leads to an increasing number of industrial processes that include: food, drug and medical device sterilization, cross-linking of temperature-resistant thermoplastics and elastomers, surface patterning for micro- or nanotechnologies, curing of coatings and composites.

High energy radiation exists under many forms, but the most important types for the potential applications they have in polymer science and technology are X-rays and gamma rays on the one hand, and accelerated electrons on the other. The recent development of swift ion beam accelerators that induce reactions exhibiting some common features with widespread radiation treatments cited at first will be also accounted for.

The designation *ionizing* is sometimes improperly used to qualify high-energy radiation since it is in fact too restrictive because the process of ionisation refers specifically to the ejection of an electron from an atom or molecule thereby creating an ion-pair. Rigorously speaking, high-energy particles or photons interacting with matter may activate a variety of physical and chemical effects, including heat generation, atomic displacements, electronic excitation of atoms and molecules, breaking of chemical bonds, and nuclear reactions (Chapters 1-3). The specific effects depend on the type of radiation, the target, and the irradiation conditions. In many situations, the chemistry that results is actually mediated by free radical species produced by a number of pathways that depend strongly on the chemical constitution, on the purity, and on the physical state of the irradiated matter [3].

The capability of high-energy radiation to bring about chemical changes with a high degree of control over the spatial and temporal features of the exposure has led to the development of a technology known as *radiation processing*. In the particular field of polymers, but not only in that field, the technique has led to products exhibiting improved thermal, chemical or wear resistance without the use of additives. For example, radiation processing has been used to produce rubber latex gloves and other similar items free from sulfur and from toxic nitrosamines, with less allergenic effects to wearers and with less emission of sulfur oxides during final elimination by burning [4]. Many other applications will illustrate the large variety of benefits that have been allowed by the gradual improvement of basic knowledge and by the availability of well-adapted and reliable radiation sources.

Energy deposition resulting from radiation-matter interaction

The complete description of the interaction of some radiation with a substrate is not limited to the detailed inventory of the induced chemical effects. The spatial distribution of the deposited energy, and hence of the resulting chemical effects, is another feature of prime importance (Chapter 1). The location of effects and of the side-reactions can indeed be mastered to a large degree by the appropriate selection of the radiation type and of beam characteristics.The energy loss per unit length of particle track is termed the stopping power in nuclear physics and Linear Energy Transfer (LET) in radiation chemistry. Heavy charged particles are referred to as high LET radiation, whereas X- or gamma rays, and fast electrons are radiation with low and intermediate LET, respectively. LET depends on the nature of the medium, and of the energy and electric charge of the interacting particle [5]. The data reported in **Table 1**, given for water, illustrate this essential feature of polymer processing.

The kinetic energy of incident radiation is dissipated by various cascade processes involving a myriad of elementary interactions that impart the affected atoms with enough

Table 1. *Comparison of linear energy transfers (LET) for various elementary particles in water.*

Particle	Rest Mass (a.u.)	Charge	LET ($nJ\ m^{-1}$)	Observation
Electron	0.00055	−e	0.032	1 MeV (EB or secondary to ^{60}Co gamma)
Alpha	4.004	+2e	14.4	from ^{210}Po
$^{127}I^{n+}$	127	+ne	1120	65 MeV

energy to expel outer shell electrons, or at least give rise to electronic excitation. In a simplified description, heavy charged particles (*e.g.*, proton, deuteron, alpha) lose kinetic energy *via* small energy transfers to atomic electrons in the medium. The more energetic interactions eject electrons from their parent atoms and generate primary ion-pairs. The more energetic secondary electrons can initiate additional ionizations, whereas less energetic secondary electrons induce electronic excitations. At the end of the secondary electron tracks, several ion pairs and excited species are concentrated in a small volume called spur. Only a small fraction of the initial energy is transferred at each event. The cascade of events along the path of the moving particle thus produces a track consisting of clusters of ions or spurs (Chapter 1) (**Scheme 1**).

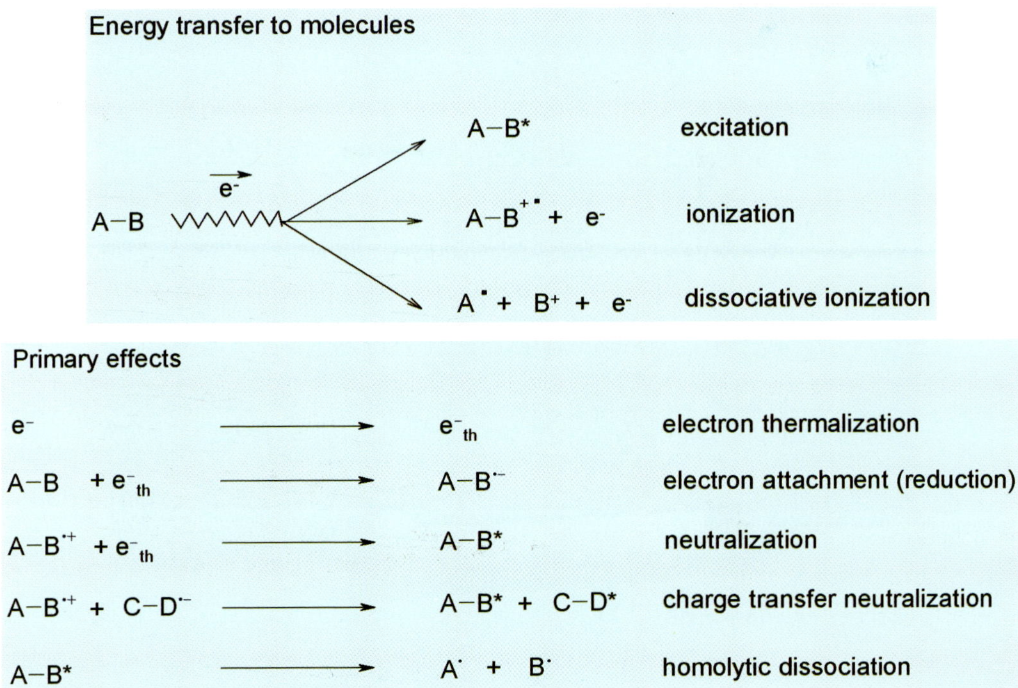

Scheme 1 : *Cascade of events along the path of a high energy particle (electron) interacting with a covalent molecule AB.*

Electrons tracks are less dense than the tracks of heavy charged particles and the spurs are more widely spaced (Chapter 4). The induced physical processes are comparable in many regards to the effect of exposure to heavy particles or ions. X- and gamma rays actually are indirect methods of producing fast electrons in matter. As a consequence, chemical effects can be considered very similar in nature. Some specific features may nevertheless arise from the significantly different dose rates (*i.e.* amount of absorbed energy per unit time) as a consequence of different LET values.

Versatility of radiation processing applied to polymers

A few selected examples illustrate the appropriate choice of these different radiations. Highly penetrating gamma or X- rays are well suited, due to their deep penetration, for radiation processing, such as cross-linking of plastic parts, the items being packaged in cardboxes (**Fig. 1**). On the other hand, medical supplies such as polyolefin syringes, tubing, gloves, blood bags and medical prostheses packaged in sealed films, all made of radiation-

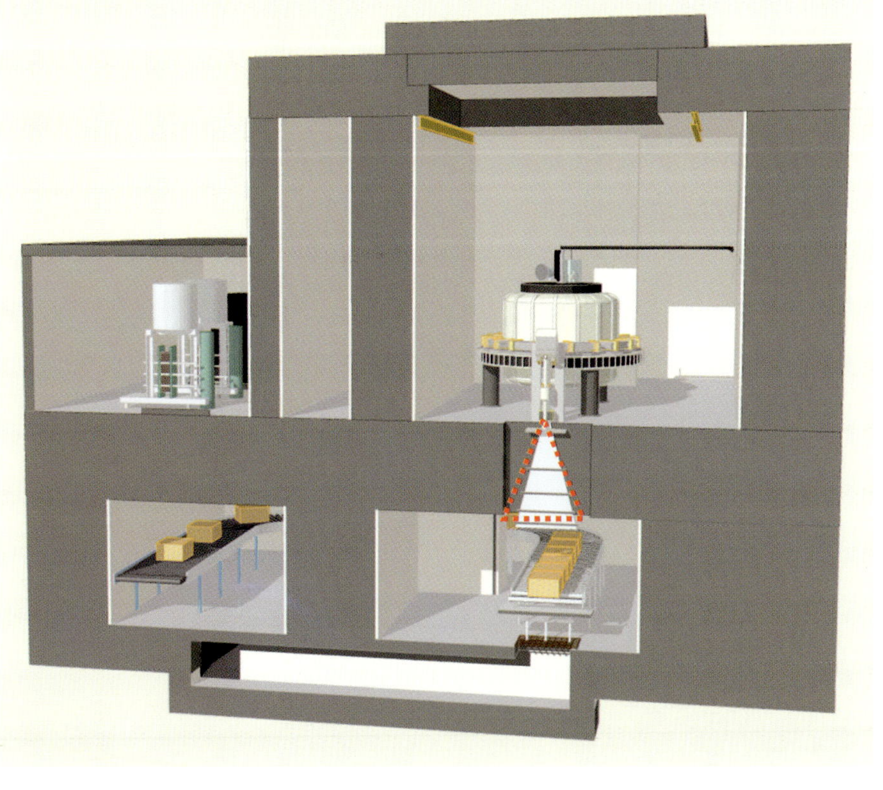

Figure 1 : *10 MeV electron beam accelerator (Rhodotron with its scanned beam circled with the red line) in an EB/X-ray sterilization facility (courtesy of IBA, Belgium).*

resistant polymers, can be sterilized without degradation at the relatively low doses which are sufficient to kill bacteria and viruses. Thanks to such a treatment under high energy radiation, biomedical items can be prevented from contamination during their storage until they are used.

Owing to the low apparent density of the boxes, palletised products can be treated in a single pass in the irradiation chamber. Heat-shrinkable films that find increasing applications in automated packaging processes are produced by cross-linking extruded polyolefin materials with electron beam (EB) accelerators. Treating the thermoplastic wire cable insulation with low to medium energy beams together with high intensity gives a high through-put and an adapted penetration depth of some hundreds micrometers (**Fig. 2**).

Figure 2 : ***700 keV EB accelerators for cross-linking cable insulators.*** *By design of the system, the cable is driven several times through the beam of the electron gun circled by a red line (courtesy of Acome, France).*

Swift heavy ion beams with extremely high LETs are used to degrade strongly polymer chains within tracks over a range of a few tens of micrometers. After irradiation, the top surface of bulk thermoplastics such as polyethyleneterephtalate, polycarbonate, polyvinylidene fluoride or polyimide materials as well as thin films made of the same material are subsequently etched by a wet chemical treatment that reveals pores of which the shape, surface density and dimensions can be controlled by choosing appropriate conditions. High aspect ratio cylindrical or conical traces of diameter ranging from a few nanometers to some micrometers can be produced to form functional porous membranes or to supply templates for the fabrication of nanowires, nanosensors and many other types of mico- or nanodevices (**Fig. 3**).

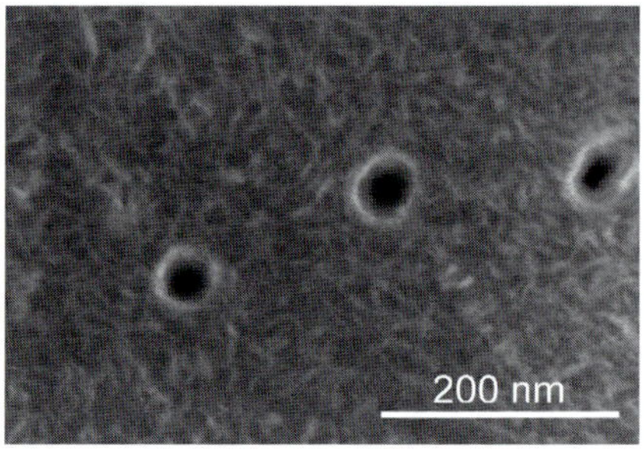

Figure 3 : *Isolated nanometric size pores in a polyvinylidenefluoride (PVDF) membrane obtained by Kr ion irradiation at GANIL and subsequent oxidative etching of a 15 μm-thick PVDF film (courtesy of LSI/CEA-CNRS).*

These few examples illustrate the versatility and the broad applicability of radiation processing to polymers, from simple performance improvement for commodity plastics to advanced developments of radiation chemistry for nanotechnologies.

Radiation sources

At present, the two most common high-energy radiation dedicated to industrial polymer processing are gamma rays and electron beams (EB) (Chapter 2). There are about 200 gamma industrial facilities in operation worldwide, chiefly exploiting the spontaneous emission of highly penetrating 1.17 and 1.33 MeV gamma rays from ^{60}Co radioelement [6]. Electron accelerators cover a broad range of particle energy, from 50 keV for machines dedicated to surface treatments and coating processing, to very high power 10 MeV accelerators for bulk processing. Some facilities in the high energy range can be equipped with a tungsten target intercepting the electron beam to produce X-rays by *bremsstrahlung*, the resulting radiation being very similar in nature and features to the ^{60}Co gamma rays.

The total number of industrial electron accelerators in the world amounts to approximately 1 500 units [6]. Compared to gamma sources, such EB machines have a high-dose rate and therefore short processing times. Because they produce radiation only by means of physical processes, EB facilities are installed and run with easier public acceptance. A limited number of large facilities held by national agencies or related institutions implement swift heavy ions as well as synchrotron radiation lines and develop their applications through large national and international programs.

In addition to these industrial or large facilities, there exist a large number of laboratory accelerators producing various types of radiation for research studies and for discrete production purposes. As an example, lithographic EB maskers are increasingly used by micro- and nanoelectronics R&D laboratories and by related production sites.

Chemical effects induced by high-energy irradiation

The chemical and biological effects of ionizing radiation can be referred to as direct effects when the radiation energy is deposited in molecular targets where chemical transformations are induced. If energy absorption occurs in the external medium (for example water in aqueous solution or biological systems), leading to the formation of radical intermediates which can diffuse to come to react with the molecules, the observed chemical effect is said indirect (Chapter 1).

Various chemical steps generally take place between the end of purely physical processes of energy deposition in polymers and the resulting chemistry. When the energy imparted to a molecular electron is lower than its lowest ionization potential, the resulting excitation (**Scheme 1**) proceeds according to optical selection rules.

The subsequent chemical response is expected to be similar to that obtained by conventional photochemistry. However, ionizing radiation beams have higher penetration. If the interaction involves larger energy deposition, ionization takes place and produces an unstable radical cation and a slow electron. Simultaneous or subsequent dissociation of the radical cation occurs. Thermalized electrons can then induce more conventional chemistry by combination with a cationic species or by electron capture that may occur also with dissociation of the molecular assembly. Negative ions produced by electron attachment can in turn combine with a positive ion by a charge transfer mechanism. Primary and secondary events thus produce excited molecules that can dissociate into free radicals or into molecular products. The former homolytic process may induce chain reactions that can turn into a chemical amplification of the primary effect. Most of the chemical processes caused by ionizing radiations are interpreted in terms of free radical mechanisms [7]. Long-lived cationic chain reactions involving carbenium or oxonium intermediates are unlikely because of the high reactivity of active centers that make transfers and deactivation reactions very fast unless special conditions (purity of involved chemicals, low temperature) are maintained throughout the process. The efficiency of a chemical pathway, expressed by its G-value (number of events produced per 100 eV) or by its radiochemical yield in standard units (moles J^{-1}), quantifies the number of events corresponding to a given chemical transformation *per* absorbed energy unit. The current unit of absorbed dose is the gray (Gy), corresponding to 1 J *per* kg of irradiated material (Chapter 1).

The response of a medium submitted to high-energy irradiation is dependent on the chemical constitution. When irradiated, molecules with carbon-carbon double or triple bonds may produce free radicals which are able to add on a second molecule. As a consequence of the generation of free radicals in these organic compounds, acrylated monomers undergo rapid polymerisation by a chain reaction (**Scheme 2**). Charged radicals (radical anions or radical cations) may as well initiate chain reactions. When the target molecules have previously formed polymers, various types of chemical reactions can then take place, as a function of polymer chemical structure and of irradiation conditions (radiation type and intensity, temperature, ambiance, presence of other chemicals). These reactions essentially include chain scission, cross-linking, oxidation and grafting (**Scheme 3**).

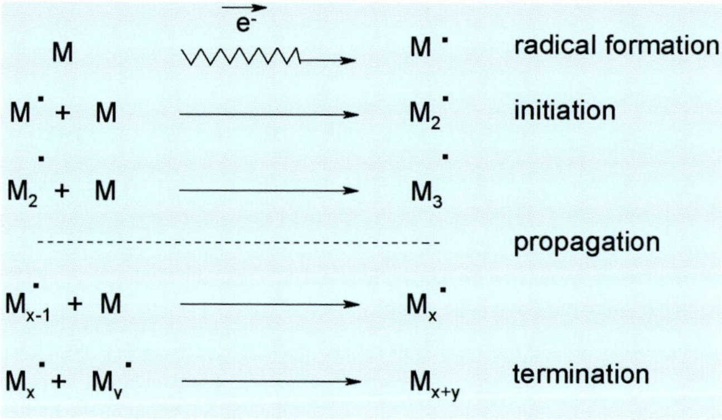

Scheme 2 : *Sketch representing the chain reaction of polymerization induced by exposing monomers to high-energy radiation.*

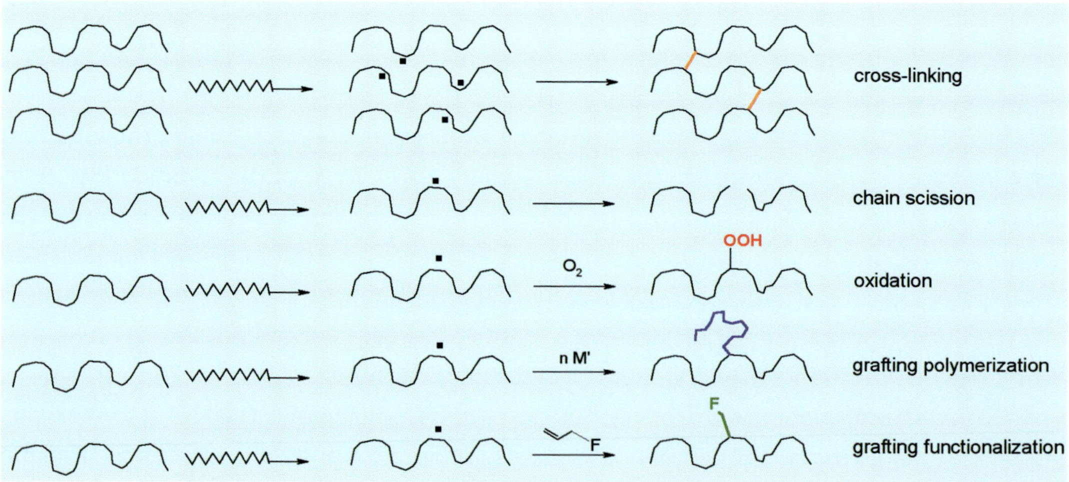

Scheme 3 : *Sketch representing chemical events that can be induced by high energy radiation in polymers.*

Depending on the ultimate use, the range of radiation dose to apply to materials covers several orders of magnitude (**Fig. 4**) [8]. However, at low doses, polymers are generally not significantly degraded, in contrast with bacteria and viruses which can be killed. This difference makes it possible to apply widely radiation-induced sterilization processes to medical supplies, which are often made of and packaged in polymers. In order to avoid confusion, it is worth mentioning that dose, an energetic parameter of the treatment, is independent to a certain extent of the energy of the single particle or photon that interacts with the target material. The latter governs the penetration depth, also affected by the LET in the target material, whereas the first one corresponds to the cumulated energy deposition.

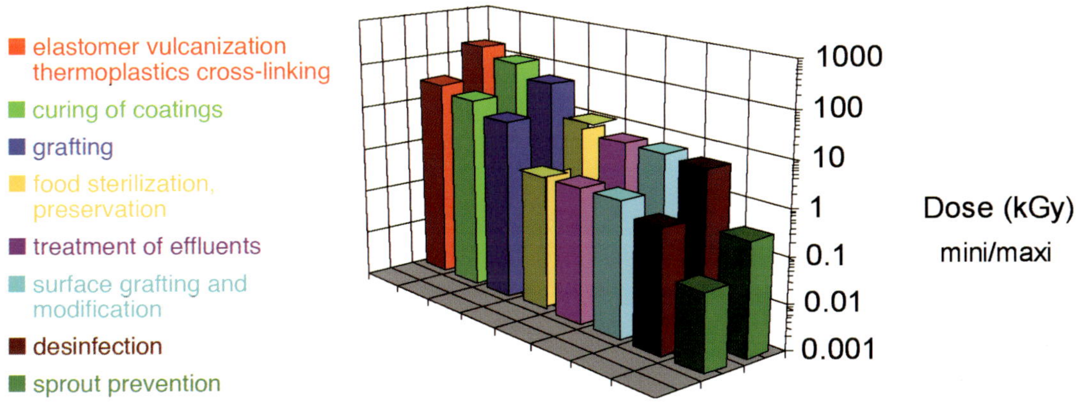

Figure 4 : *Typical radiation dose ranges for the different utilizations of high-energy radiation in polymer processing. Doses for desinfection, sterilization and sprout prevention (biological effects) are given for comparison.*

Chain reactions, essentially polymerizations, can be achieved with medium doses, as a result of the chemical amplification by purely thermal processes of radiation-induced initiation (**Scheme 2**). Processes involving single steps or short kinetic chain length reactions require much higher doses. This is generally the case for the radiation cross-linking of rubbers and thermoplastics.

Cross-linking

When polymers are submitted to high-energy radiation, contrasting behaviours are observed with a dominant effect of cross-linking or of chain scission, depending on the nature of the repetition units (**Scheme 3**).

Table 2. *Radiolytic yield values of cross-linking G(X) and of scission G(S) values for some common polymers.*

Polymer[a)]	Chemical structure	G(X) (10^{-9} mol J^{-1})	G(S) (10^{-9} mol J^{-1})
polyethylene[b)]	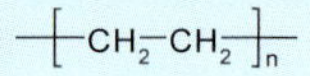	300	90
polypropylene[b)]	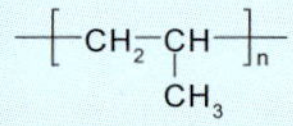	250	110
polybutadiene	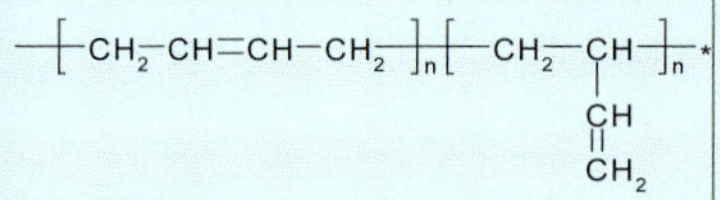	380	-
poly(methyl acrylate)	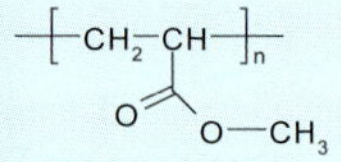	55	18
poly(n-butyl acrylate)	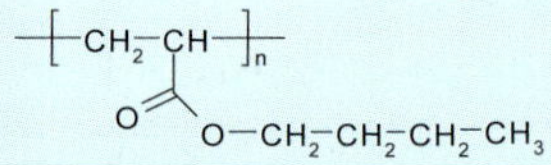	63	18
poly(t-butyl acrylate)	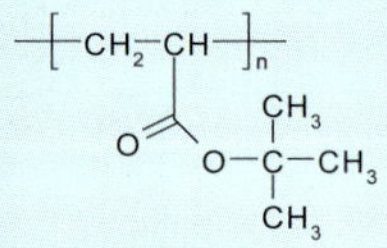	16	18
poly(methyl methacrylate)	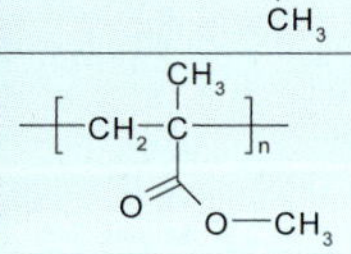	-	120 - 350
poly(vinyl chloride)[c)]	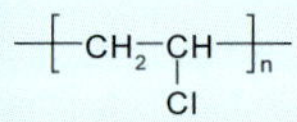	33	23
polystyrene	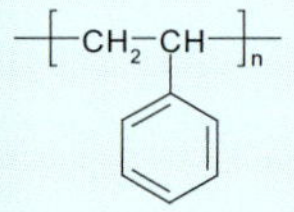	5	<2
poly(ethylene terephtalate)	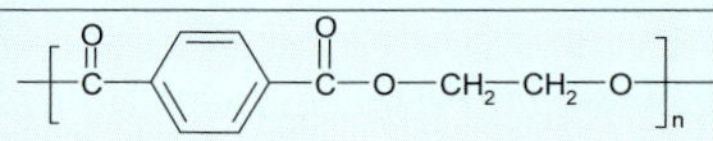	3 - 20	7 - 20

a) *Ionization at 25 °C in the absence of oxygen.* b) *Dependent on microstructure and of crystallinity.* c) *Unplasticized.*

The general trends of the competition between cross-link formation (X) and chain scission (S) can be discussed for some common polymers on the basis of the values of G(S) and G(X) gathered in **Table 2**. Polyethylene (PE) and polyvinylchloride (PVC) form networks, whereas polymethylmethacrylate (PMMA) only degrades strongly with evolution of low molecular weight fragments. EB-lithography utilizes this degradative property to engrave nanometric lines in PMMA films for electronic applications (**Fig. 5**). Tailoring copolymers including repeating units with intrinsically opposite behavior allows controlling the overall behavior [9].

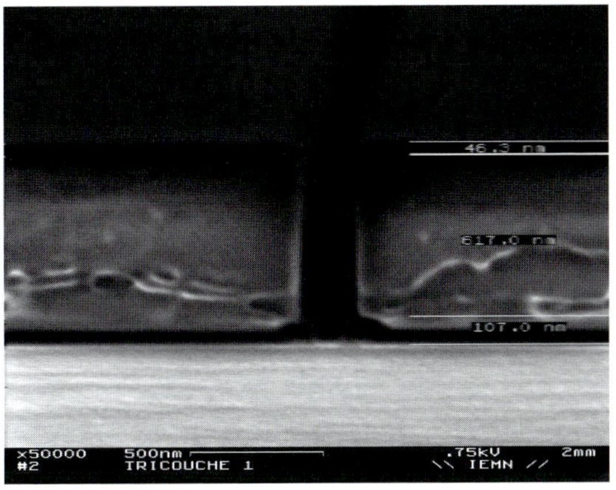

Figure 5 : *Nanometric size transistor grid obtained by localized degradation (central engraved line) of a PMMA-based multilayer coating with EB lithographic equipment (courtesy of IEMN, Villeneuve d'Ascq, France).*

Obviously, the degree of substitution of the main chain has a determining influence on the behavior under irradiation. Polymers including aromatic nuclei as side-groups (polystyrene-PS) or in the main chain (polyethyleneterephtalate-PETP) give a strongly attenuated response to high energy radiation. It must be added – to avoid oversimplification – that morphological and ambiance parameters (microstructure, crystallinity, glassy or rubbery state, degree of plasticization, presence of oxygen) may strongly alter the reported data.

Cross-linking is the largest commercial application of radiation processing applied to polymers. Polymers with cross-linked chains can maintain their shape and useful properties at a higher temperature without flowing (**Fig. 6**). Generally, the materials exhibit after cross-linking better mechanical and chemical resistance. Processing is performed at high speed under the electron beam and neither requires heating to high temperature nor the use of the added cross-linkers necessary to the non-radiative processes.

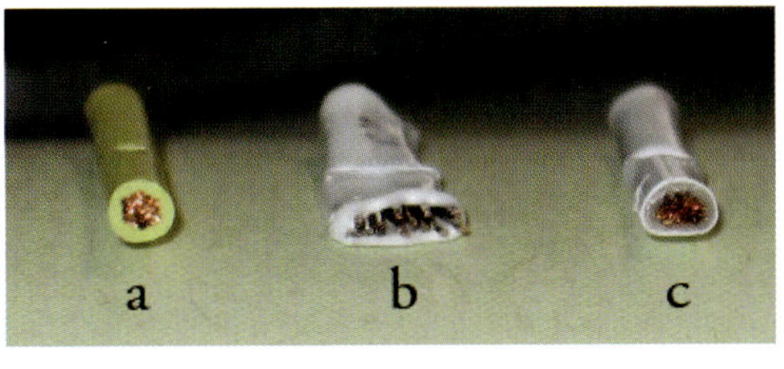

Figure 6 : *Pictures illustrating the compression resistance (after heating at 250 °C for 1 min. under stress.* a) *Electric cable with a radiation cross-linked PE insulating material.* b) *Deformation of cables with uncross-linked PP.* c) *Thermoplastic PE materials submitted to the same test (courtesy of Acome, France).*

A number of manufactured products comprising thermoplastics (polyethylene, polyvinylchloride, polyamides, ...) or elastomers (natural rubber, nitrile rubber, ethylene propylene rubber, silicones) are processed by cross-linking to improve their performances in insulated electric wires and cables, multilayered films for cooking pouches, shape memory tubes, pressure resistant water pipes, expandable foams, automotive parts exposed to motor heat, etc. Unsaturated additives are sometimes used to promote radiation cross-linking, in order to improve productivity or to influence favourably the inherent trend to competing cross-linking and scissioning.

Irradiation can be very useful to stabilize polymer blends, avoiding physical phase-separation of the components or making it possible to block a desired but unstable morphology by sudden cross-linking. Various other applications exploit the shape memory effect that results from radiation cross-linking of semicrystalline thermoplastics: resettable fuses for electric devices, heat-shrinkable insulating tubings and food packages including multilayered metal-polymer complexes are some examples worth mentioning.

Degradation and post-irradiation oxidation

Some important commercial applications take advantage of various chemical effects induced by radiation and resulting in scissioning, branching and partial oxidation. Surface oxidation is achieved in nanolithographic processes using X-rays or EB in order to enhance resolution. If they are well-controlled, the cumulated degradation effects can improve the processability or the compatibility of the plastic in blends with other materials. For example, fluorinated polymers can receive a treatment facilitating melt processing. Radiation-treated polypropylene yields fiber-reinforced composite materials with enhanced performances. Oxidation appears of interest in this particular use. However, for many other applications, the very efficient reaction of generated macroradicals with dissolved and gradually diffusing oxygen forms peroxy derivatives that will induce long-term degradation of polymer properties. Chain reactions further initiated by thermal or photochemical peroxide decomposition can ruin the structural and functional properties of the materials. Anti-oxidants are thus added to minimize the detrimental effect of radio-oxidation. This generic phenomenon represents a key issue in the development of radiation-sterilizable materials for biomedical devices and prostheses.

Radiation grafting

Radiation activation offers a rather unique possibility to generate free radical centers in a reactive or in a latent form within polymer materials for initiating grafting polymerization. In the pre-irradiation technique, the material is treated first, preferably in the presence of oxygen, to induce the formation of radicals or peroxy species that are subsequently activated thermally or chemically in the presence of the monomer to be grafted. In semi-crystalline polymers, the long-lived free radicals formed in the ordered domains can also be responsible of the grafting upon appropriate thermal activation and monomer diffusion (In **Scheme 4**

step b, the monomer is glycidyl methacrylate GMA) [10]. An alternative process consists in treating the polymer-reactive additive mixture by simultaneous irradiation grafting.

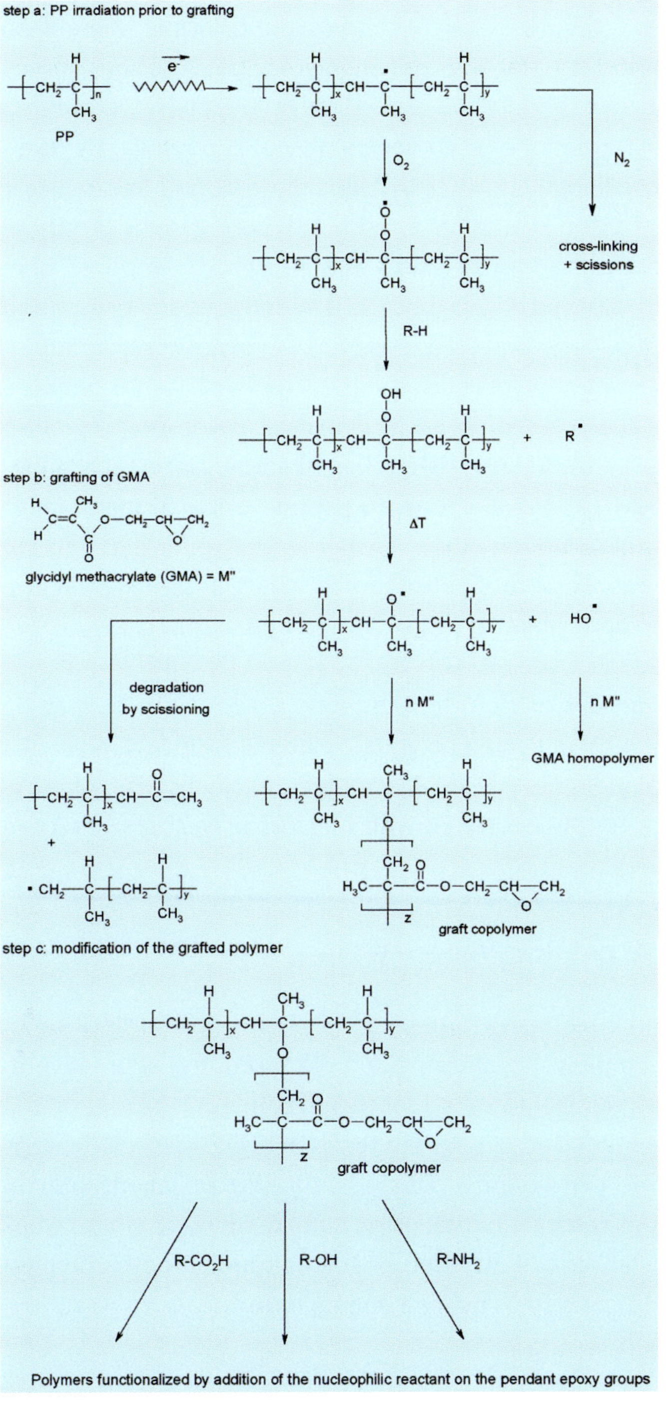

Scheme 4 : *Simplified representation of the pre-irradiation method* (step a) *applied to the grafting reaction of glycidyl methacrylate (GMA) onto polypropylene* (step b) *for the functionalization of the material by reaction with nucleophiles* (step c).

The technique is applicable to polymer substrates under the form of films, fibers, powders or bulk objects. Free radicals are produced in relatively high concentration during the ionizing treatment of synthetic (polyolefins, polystyrene, polyamides, fluorinated polymers) as well as natural polymers (cellulose, starch, chitosan). In principle, there is no restriction regarding the type of monomer to form the graft polymer, but sufficient affinity of the monomer for the polymer substrate and for the grafted polymer favours the grafting yield. Suppression of homopolymer formation (*i.e.* without grafting) is possible, by using additives such as Mohr's salt ($FeSO_4$, $(NH_4)_2SO_4$, $6H_2O$), that quench diffusing free radicals, but convert hydroperoxy species into free radical by a redox reaction. Various industrial applications have been developed. Textile industry uses this process to improve fiber dyeability (grafting of polar polymer), to reduce hydrophilicity and soil stainability (grafting of fluorinated polymers) or to keep sterile medical dressings (grafting of antiseptics). Super-absorbents have been produced by grafting hydrophilic synthetic polymers onto cellulose or starch substrates. Plastic-based biomedical devices are modified with hydrophilic layers to enhance biocompatibility (prostheses), to reduce the risks of haemostasis (catheters, artificial kidney, extracorporeal circulation devices). Novel chromatographic supports, membranes with ion-binding properties for decontamination of wastewater as well as for fuel cells are produced by grafting and are under intense study. Control of graft length and microstructure is under investigation to improve the performances of the grafted materials.

Grafting of functional side-groups without forming long polymer chains may be achieved in a similar way by the reaction of activated polymeric materials with low molecular weight compounds carrying functional groups of appropriate reactivity. The physical stabilization of unstable blends of amorphized starch with reactive plasticizers has been achieved by EB-irradiation [11].

Radiation-induced cross-linking polymerization

Radiation curing of inks and coatings by cross-linking polymerization is another important application that expends rapidly. It represents a most promising alternative to solvent-based processes that are progressively banned for their detrimental impact on the environment. This technology offers several advantages in terms of ecoconception (energy saving, reduction of volatile organic compounds VOCs) [12]. *In-situ* polymerization of restorative resins infused into damaged archaeological objects made of wood or into weak artistic pieces of porous structure is another illustration of the unique in-depth chemical effects that can be induced under soft conditions by high energy radiation [13].

High-energy radiation treatment is an incomparably more suited alternative to intense UV light for initiating in depth the cross-linking polymerization of solvent-free compositions including monomers and reactive prepolymers. The rapid polymerization process is triggered by exposure to the radiation. Owing to the presence of multifunctional monomers, a 3-dimensional covalent network is formed, causing the drying of the initial formulation and yielding generally mechanically and chemically resistant materials (**Scheme 5**).

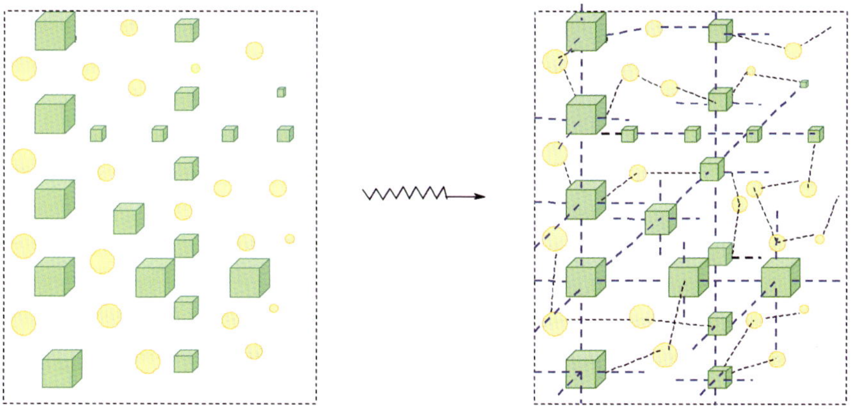

Scheme 5 : *Radiation-induced cross-linking polymerization of monofunctional (●) and multifunctional (■) monomer blends.*

The liquid or semi-solid formulations can be handled and applied to substrates by various processes (sprayer, roller, blade coater) as solvent-based coatings. Acrylate derivatives and some vinylaromatics are well-suited to give fast curing by free radical mechanism. Epoxy or vinyl ether functional analogues can be cured by a cationic process requiring the incorporation of onium salts initiators. Pigments and additives are added to the formulation in order to facilitate the coating application, to enhance curing efficiency and to adjust performance and properties of the coating. UV radiation is extensively used for labels, cartons and containers in the packaging industry. High energy radiation curing provides additional advantages compared to UV light. No initiator is needed for free radical polymerization, and the penetrating power of electrons is greater than UV light at usual concentration in photo-initiator and/or in formulations with high pigment or filler concentration. Electrons are more efficient than UV rays in opaque coatings, and composites and they activate monomers even after vitrification of the network undergoing gradual curing. This results in lower amounts of unreacted monomers and other extractables. Grafting reactions can also take place at the substrate-coating interface, thus improving adhesion. Recent advances in formulation and in process control give insight into major industrial developments in the food packaging industry and into industrial coatings (flooring, outdoor building parts).

Curing of fiber-reinforced polymer composites using high-energy electron beams offers similarly significant advantages for fabricating a variety of aerospace, ship and ground vehicle components (**Fig. 7**) [14]. The EB curing process greatly reduces the time required to cross-link the polymer matrix compared to conventional heat curing. The process allows curing without external thermal activation for high throughput. It yields materials with reduced residual stresses for good part fit-up. Curing times are shorter, and the curable formulations have longer storage stability. The absence of highly volatile monomers renders the industrial process safer and more environment friendly.

Figure 7 : *Glass/polyester 10 meter-long ship hull cured by EB and X-ray induced polymerization. EB gun is circled by red line. (Courtesy of Intermarine and Astrium Space Transportation.)*

Emulsion Polymerization

Emulsion polymerization is a type of radical polymerization that usually starts with a dispersion incorporating water, one or some hydrophobic monomers, and a surfactant [15]. Polymerization actually takes place not in the large droplets of dispersed monomers but in the latex particles that form spontaneously in the first instants of the process. These latex particles are typically 100 nm in size, and comprise many individual monomer swollen polymer chains surrounded by the surfactant, forming the so-called emulsion. Their number is essentially stable along the reaction, but their size increases gradually since monomer continues to diffuse and to swell the polymer particle. Because initiation takes place at a slow rate in the continuous aqueous phase, the kinetic chain length inside a single particle is extremely long (polymer chains of high molecular weight) and, at a given instant, half of the population of latex particles is undergoing polymerization, whereas the other half has been visited by an even number of free radicals, thus having been terminated by bimolecular termination. The inactive state lasts until polymerisation is resumed by a newly entered free radical.

Emulsion polymerization is used to manufacture several commercially important polymers. These emulsions increasingly find applications in adhesives, paints, paper coating and textile coatings, for they contain little or no amount of volatile organic compounds. Most of them are currently obtained by conventional thermally initiated polymerisation process, but gamma radiation from ^{60}Co sources can be used to generate at a low and well controlled rate the hydroxy radicals in the water phase, so as to sustain the alternating switching on/off of the polymerization in the heterogeneous medium [16]. The use of radiation makes it possible to decouple the thermal activation of the polymerization process desired for increasing the production rate from the other effect of temperature influencing molecular weight. Indeed, when a thermal initiator is used, it also decomposes faster at higher temperatures, producing more polymers for the same reaction time, but with shorter polymer chains. This feature of interest associated with radiation-induced initiation is likely to receive increasing industrial development in the coming years.

Conclusion

The main features and the early achievements of radiation processing in the domain of polymer materials have been briefly presented, and are summarized in **Scheme 6**. Efficient processes for the improvement of thermoplastics and rubbers properties have been made available for decades and continue to be improved [17]. The current trends show the enormous potential of radiation-based processes to address environmental concerns by proposing performing and societally acceptable alternatives to outdated technologies based on unsuitable chemistries and unreasonable energy consumption. Both commodity and technological products will increasingly rely on radiation treatment of polymers. Use of radiation in recycling technologies and for new utilization of biomass is expected to play an increasing role in a near future. This trend is obvious in speciality fields prone to intense innovation, as the sector of biomedical materials and devices where nanotechnologies and the fine use of radiation chemistry have lead to considerable advances (**Scheme 6**).

Acknowledgements

The author wishes to express his thanks to Dr. C. Lagreve (Acome), Dr. M.C. Clochard (CEA-DRECAM), Dr. B. Defoort (EADS-Astrium), Mr. G. Massaro (IBA), Dr. M. François (IEMN), for supplying original illustrations appearing in this article. Sponsoring of research programs on electron-beam initiated cross-linking polymerization by the EADS Foundation as well as support of the Polynano project by MENESR (ACI Nanosciences n°120) are grateftully acknowledged.

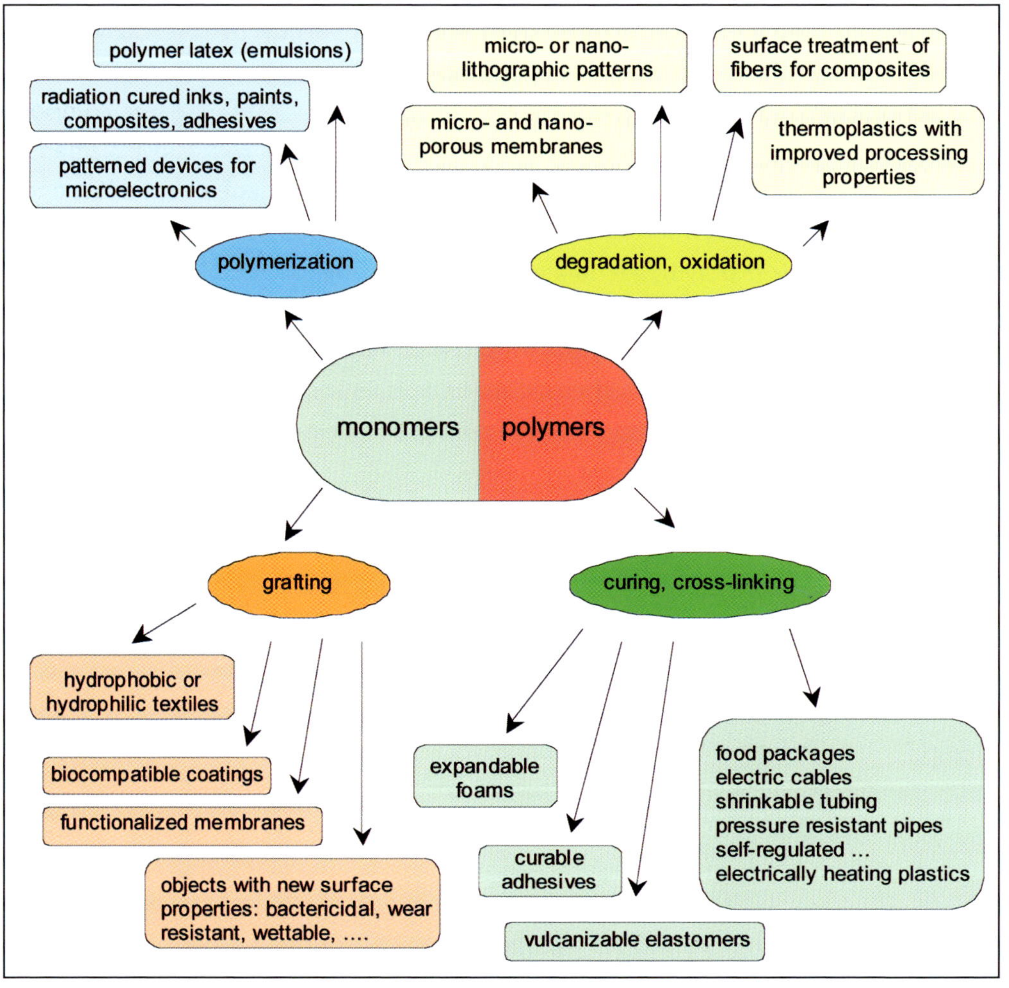

Scheme 6: *Industrial applications of monomer and polymer irradiation.*

References

[1] Chapiro A., "Radiation chemistry of polymeric systems", Wiley, New-York, 1962.

[2] Ivanov V.S., "Radiation chemistry of polymers", VSP Publishers, Utrecht, 1992.

[3] Turner J., "Atoms, Radiation, and Radiation Protection", Wiley, New-York, 1995.

[4] Sarma K.S.S., Majli A.B., Electron Accelerators and their Applications, IANCAS Bulletin, 1999, **15** (4).

[5] Swallow A.J., "Radiation Chemistry - An introduction", Longman, London, 1973.

[6] Machi S., Prospects for the application of radiation processing and the activities of the IAEA, Radiat. Phys. Chem., 1988, **52**, 591-597.

[7] Lund A., Shiotani M., "Radical Ionic Systems Properties in Condensed Phases", Kluwer, Dordrecht, 1991.

[8] Bly J.H., "Electron Beam processing", International Information Associates, Yardley, 1988.

[9] Turgis J.D., Vergé C., Coqueret X., Composition effects on the EB-induced cross-linking of some acrylate and methacrylate copolymers, Radiat. Phys. Chem., 2003, **67**, 409-413.

[10] Le Thuaut Ph., Martel B., Crini G., Maschke U., Martel B., Coqueret X., Morcellet M., Grafting of cyclodextrins onto polypropylene nonwoven fabrics for the manufacture of reactive filters - Part 1 : Synthesis parameters, J. Appl. Polym. Sci., 2000, **77**, 2118-2125.

[11] Olivier A., Cazaux F., Coqueret X., Compatibilization of starch-allylurea blends by electron beam irradiation: Spectroscopic monitoring and assessment of grafting efficiency, Biomacromolecules, 2001, 1260 -1266.

[12] Drobny J.G., "Radiation Technology for Polymers", CHIPS, Weimar, 2002.

[13] Cœuré P., Khoi Tran Q., Procédés radiochimiques pour la conservation des biens culturels, Actualité Chimique, 1999, **225**, 141-143.

[14] Defoort B., Lopitaux G., Dupillier J.M., Larnac G., Coqueret X., Electron-beam initiated polymerization of acrylate compositions, 6: Influence of processing parameters on the curing kinetics of an epoxy acrylate blend, Macromol. Chem. Phys., 2001, **202**, 3149-3156.

[15] Odian G, "Principles of Polymerization", 3rd edition, Wiley-Interscience, New York, 1991.

[16] Wang X., Zhang Z., Preparation of polystyrene latex particles by gamma-rays-induced emulsifier-free emulsion polymerization, Radiat. Phys. Chem., 2006, **75**, 1001-1005.

[17] Clough R.L., High energy radiation and polymers: a review of commercial processes and emerging applications, Nucl. Instrum. Methods, B, 2001, **185**, 8-33.

Chapter 10

Radiosterilization of drugs

Bernard TILQUIN

Introduction

In the early 1990s, whilst radio-sterilization of medical devices was flourishing, the radiolysis mechanisms in irradiated solid-state drugs were not or poorly described [1]. The existing analytical methods failed actually to detect final radiolytic products since they were various and present only at the level of traces. Moreover, the European Pharmacopoeia (Ph. Eur.) tests are not adapted yet for the detection of chemical changes in irradiated drugs. The improvement of the sensitivity of to-day analytical methods enables us now to study the behaviour of a drug under ionizing radiations and to determine the conditions of the radiosterilization treatment.

For a better understanding of the context of this work, it seems fundamental to remind first some notions of sterility and sterilization. Then we will dwell on radiosterilization by describing the process and its advantages, without forgetting its limitations. These latter lead us to a review of the investigations on the radio-induced degradation of drugs. *In fine* we propose the latest regulations on radiosterilization.

The effects of irradiation on many different types of substances are now studied at CHAM (Analytical Chemistry Department at the University of Louvain, School of Pharmacy) [2] and in other groups [3]. All the most advanced analytical tools are explored to detect the traces and to study radiosterilization: Electron Spin Resonance (ESR) is the most sensitive technique and the spectra were found to be specific to free radicals produced in irradiated solid molecules. HPLC (High Performance Liquid Chromatography) coupled with various detection systems, such as mass spectrometry (MS) – Diode Array Detector (DAD), is the most promising technique for analyzing traces of final products.

Notion of sterility

Sterility can be defined as a complete absence of viable organisms. In order to reach sterility, all the living germs are supposed to be eradicated. This concerns pathogen as well as non-pathogen micro-organisms. The latter could cause diseases under optimal growth conditions.

The means of attaining sterility is called sterilization. In fact, this killing process must be rather considered as an inactivation. The death of a cell is indeed effective when it cannot reproduce itself. For this reason, the effect of a sterilization process can be followed by survival curves.

As the notion of absolute sterility, as it was defined, is not applicable in practice, the notion of probability is used, where the likelihood of the presence of a contaminated unit or article should be very low. The likelihood of the presence of contaminated unit is used to define the efficiency of the sterilization (Security Assurance Level, SAL).

It is noteworthy that, at least presently, sterilization is not required for all drugs. It depends on the type of administration. Hence, it concerns mostly ophthalmic preparations, sterile topical products and injectable solutions, including intramuscular, intravenous and sub-cutaneous ways.

Sterilization methods

The techniques which are described in the US and the European Pharmacopoeias [4,5] in order to reach nowadays sterility are the following.

Steam or dry heating

First of all, thermal sterilization, which includes the use of steam or dry heat, is the earliest method used. Steam sterilization is up to now the most widely employed sterilization process. It is carried out in a chamber called autoclave. A typical autoclave cycle corresponds to a period of 15 minutes at 121 °C. In the case of dry-heat, the process is undertaken in an oven with heated and filtered air, distributed uniformly. The cycle is of two hours at 160 °C.

Both methods enable reaching the recommended SAL value of at least 10^{-6}. They are considered as the techniques of reference. However, a major limitation is that they are unapplicable to heat-labile substances (and packaging when used as a terminal process). The number of drugs, which cannot withstand high temperatures, is increasing especially with the new generation of treatments based on the use of proteins. Alternatives to thermal sterilization are therefore necessary.

Filtration

Among them, sterilization by filtration is frequently used. The filter consists in a membrane, which removes physically the micro-organisms. The fragility of this filter membrane is one of its most serious disadvantages. Moreover, it is not terminal and the operations that are carried out after the sterilization by filtration must be done under aseptic conditions. Finally, a SAL of 10^{-6} cannot be reached.

Gas sterilization

Gas sterilization is a process in which the active agent is a gas. The agent generally used is ethylene oxide, which is an alkylator. It has major drawbacks. On one hand, it is highly flammable and on the other hand, it is carcinogenic, toxic and mutagenic. Thus the residual ethylene oxide left inside the drug is highly hazardous. Lastly, as for all the sterilizing gases, its efficiency is restricted by a limited penetration into the innermost system areas.

Aseptic processes

Another option is the aseptic processing, which involves a series of aseptic steps. Each of them must be highly controlled since the introduction of micro-organisms must be avoided all along the process. Aseptic filtration or aseptic processes are time-consuming, expensive, difficult to monitor and the SAL value is limited to 10^{-3}-10^{-4}, so alternative methods for sterilization of thermosensitive solid drugs are needed.

Radiosterilization

Though Minck reported already in 1916 that X-rays reduced the microbial population, the application of radio-sterilization started only in 1950. The ability of ionizing radiation to inactivate microorganisms has been well documented [6]. In fact, until the 1990s, the analytical techniques used to measure the drug degradation failed to detect final products present in traces, and no systematic study could thus performed on parameters such as absorbed dose, dose rate, oxygen content or temperature of irradiation. For that reason, the radio-sterilization was very restricted in pharmacy, whilst it was flourishing for medical devices, particularly for thermo-labile polymeric materials (Chapter 9). Then, due to the development of highly sensitive analytical techniques, the radiation chemistry research could be oriented towards the study of the radiolysis mechanisms in irradiated solid-state drugs or in aqueous solutions and the analysis of the final products in terms of their potential toxicity . The aim of the research is to determine the conditions under which sterilization is achieved at the required SAL value for micro-organisms eradication with a safe and minimized production of non-toxic radiolytic products.

Since the inactivation of bacteria follows an exponential decay process with a limiting value tending towards zero, the absolute sterility can never be obtained. The D_{10} value is the absorbed dose required to reduce a microbial population to 10% of its initial value, so that in industrial applications, a SAL value of 10^{-3} is reached after $3 \times D_{10}$ value. For drugs, a dose of 25 kGy (or kJ kg^{-1}) is generally higher than $6 \times D_{10}$ value and achieves the minimum SAL required of 10^{-6}. Many reviews demonstrated that for drugs with low bioburdens (initial contamination by microorganisms), sterilization was achieved with doses even lower than 15 kGy.

The radio-sterilization process is based on the exposure of a product to ionizing radiation. In Pharmacopoeias, two types of radiation are considered : gamma rays or electron beams. Generally, the gamma sterilization is more documented [7]. The irradiation types are characterized by their dose rate (much higher for electron beams) and their penetrating power (much higher for gamma rays) (**Tab. 1**).

Table 1. *Properties of irradiation types.*

Specificity	Gamma rays	High energy electrons
Dose rate	1 kGy/h	10^7 kGy/h
Penetration depth	1-2 m	6-5 cm for 10 MeV
Irradiation Facility	Radio-active source	Electron accelerator

Advantages

By comparison with earlier methods, numerous advantages are admitted for the radio-sterilization. The process :

- can reach a very high efficiency (SAL = 10^{-6}) due to the high living cell sensitivity to radiation.

- is a cold process applicable to thermo-sensitive molecules and thermosensitive packaging.

- permits to sterilize the drug inside its packaging (terminal method).

- enables to sterilize homogeneously large containers (due to the penetration of gamma rays).

- enables to sterilize large batches within a short pulse train (electron beam).

- produces related compounds (radiolytic products) in lower concentration that initial impurities.

- is easily monitored and controlled, the variable being the absorbed dose.

- is an economical process.

- does not induce – it would be needless to repeat again – any radioactivity.

Disadvantages

- The major disadvantage of radio-sterilization is the peculiarity of the radiation chemistry induced simultaneously to sterilization, which requires a special study: the irradiated drug is thus considered as a "new" drug. While micro organisms are killed, various products of degradation of the main compounds may appear in traces amounts though they are often the same as in thermal sterilization.

- In the solid state, the drugs are radio-resistant until the Standard Absorbed Dose (SAD) (25 kGy) but some changes in the organoleptic properties may be reported at this dose; by decreasing the doses, some discoloration and/or smell may be reduced.

The SAD [4] may be lowered if it is validated provided a SAL of 10^{-6} is achieved . The doses required to achieve this SAL are available in format when ANSI/AAMI/ISO methodology is used [8].

The chemical changes are much more important in aqueous solution than in solid state; as a rule, a solute in diluted aqueous solutions is completely degraded after 1.5 kGy. In pharmacy, excipients (isotonic solutions) are needed ; new approaches are improved with these additives and are now able to inhibit the phenomenon

Dose distribution

The dose mapping inside the irradiated object is important. In the dose range 25-32 kGy, regulatory agencies require a ratio of Dmax/Dmin of 1.28 maximum value. The distribution of absorbed dose depends on the radiation used and is affected by the geometry of the vials (containers) as well as the nature (electron density) of packaging material.

Penetration of gamma-rays is more homogeneous. For E-beams, the dose depth distribution is related to the energy levels. If possible, double E-beams are recommended to flatten the dose distribution curve. Another way is to use scattering foils and reflection plates. The packaging material (vials and stoppers) must be suitable.

Solid-state Sterilization

In the solid state, the drug is directly ionized and excited by the ionizing radiation. The reactive species formed may not diffuse (except for a few electrons and mobile hydrogen atoms) and radical fragments are trapped in the solid matrix, what is called the cage effect. The cage effect favours recombination reactions of fragments into the initial molecule and thus the drug transformation yield is usually very low.

The capacity of the European Pharmacopoeia purity tests to detect chemical changes induced in irradiated drugs was explored. The pharmacopoeia tests failed to detect products from radiolysis of chloramphenicol even at extremely high doses (> 400 kGy), and the only noticeable changes detected were in the color of the drug [9].

The techniques of ESR were used [10] to study radicals formed in solid state. ESR proved to be a very sensitive technique to detect radicals (10^{-10} mol l^{-1}) trapped for days, weeks and even years and discriminate between irradiated and non-irradiated drugs. Radical amounts increased linearly with absorbed doses up to 10 kGy, the radiolytic yield was constant, and ESR post-dosimetry was proposed as a new method to identify radiosterilized drugs [11]. Thermoluminescent signals are present prior to any irradiation in all the β-lactam drugs tested [12]. The irradiated drugs showed a modification in the quality and quantity of the thermoluminescent temperature peaks.

Gas chromatography (GC) with an infrared (IR) detector was introduced as a method to detect volatile radiolytic products, some of which were hypothesized to be responsible for the bad smells emanating from irradiated drugs. Thiocyanic acid was held responsible, for example, for the sulfurous smell in irradiated ampicillin. The head-space (HS) injection technique for GC and the on-line MS detection allowed new approaches to detect radiosterilization [12]. Many volatile radiolytic products were identified from the mass spectral libraries. Some of the compounds identified such as aldehydes, esters and sulfides were quite malodorous. A few of the volatile radiolytic products came from the degradation of drug molecules by the ionizing radiation, whereas residual solvents played a key role in the formation of other volatile radiolytic products.

The HPLC-UV-DAD was used for the analysis of final products; results showed that many related compounds of cefotaxime were present prior to irradiation. Some of them had maxima around 320-380 nm and could be responsible for the change in color of the dissolved irradiated drug. The radiolytic products (25 kGy) were present in traces and were all below the qualification limit of 0.1%. The HPLC-UV-MS results showed that some radiolytic products detected were impurities already present in the non-irradiated cefotaxime and increased after irradiation. The other products were unique to radio-sterilization [14].

For some β-blockers, the radical yields were found to be very low ($<10^{-9}$ mol J^{-1}) and they were considered to be radioresistant [11]. The radiolytic related compounds were also analysed by HPLC-UV, in order to test for a correlation between radical and product yields. For the β-blockers, no correlation was established between the yields of radicals and of final products [15].

Aqueous solution sterilization

In aqueous solutions, the water is ionized and excited by the ionizing radiation, and the reactive species (e_{aq}^-, $^{\bullet}H$ and $^{\bullet}OH$) (Chapter 1) formed may diffuse and react with the solute. The amount of reactive species formed by water radiolysis is proportional to the absorbed dose multiplied by their respective radiation chemical yields.

The solute, which is in rather low concentration is indirectly transformed by the numerous reactive species of the water radiolysis, so the yield of solute transformation is much higher than in the pure solid state.

There is a consensus that in aqueous solutions the degradation of the drug is proportional to the absorbed dose, because the yields are constant in a certain domain of concentration (solute dilution curve, Chapter 1). The greater the absorbed dose the greater the solute degradation. To a given absorbed dose corresponds a given amount of reactive radicals formed by the water radiolysis. Thus, if the concentration of the drug solute is increased in diluted solution, the relative percentage of degradation of the solute is decreased [16]. In pharmacy, typical concentration is 10^{-3} mol l^{-1}.

Moreover, the absorbed doses should be low, since at higher doses, radiolytic products may interfere [17].

In order to inhibit degradation by the radical scavenging, two approaches were studied: one is to let an excipient to play the role of a radical scavenger, so protecting the drug from the radical attack, the second is to freeze the solution into a solid state where the cage effect favours the recombination of primary radicals of the solvent prior to the reaction with the drug.

Radioprotection of drugs

The degradation of the drug solute results from the indirect effect through reactions with products of water radiolysis. Thus, addition of scavengers of these reactive species, able to react faster than the drug, will protect the drug solute. In order to optimize the radioprotection of the drug solute, the radical mechanisms of its degradation should be studied. The choice of the scavenger and its concentration will depend on the solute concentration and reactivity with the radical(s) responsible for the solute degradation. This could be further complicated by additional direct effects.

At the radio-sterilization doses, simulation predicts a greater loss than in the experimental results, possibly because some radiolytic products react with the water radiolysis radicals thus protecting the drug solute [16]. The simulations also predict similar solute concentrations without dose rate effects for E-beam or gamma irradiations whereas the opposite is found in experimental results [17]. The complexity of the radiolysis mechanisms at sterilization doses appears with the increase of the analytical efficiency.

The use of radio-protecting excipients was shown to be more promising to lower the loss of potency of drugs in aqueous solution without oxygen than an increase in the dose rate [16,17]. Type and concentration of the chosen radioprotector are important parameters. Metoprolol tartrate [18] is commercialized as an injectable drug with 0.9% sodium chloride as an isotonicity agent, but sodium chloride does not protect metoprolol tartrate from degradation by both E-beam and gamma irradiations. 1,2-propanediol (2%) significantly reduces the loss in metoprolol tartrate, because it reacts rapidly with hydroxyl radicals; mannitol (5%) also reacts quite well with hydroxyl radicals and protects the drug. Pyridoxine or nicotinamide (vitamins) in metoclopramide aqueous solutions give recoveries far above 90% up to a 15 kGy (E-beam or gamma rays), 90% meet the pharmacopoeial specifications concerning the chemical potency. Both excipients react with the hydrated electrons and the hydroxyl radicals [19].

Frozen aqueous solutions

Lowering the irradiation temperature in order to freeze the drug aqueous solution limits the diffusion of the radicals which are the reactive species (indirect effect) and favours their recombination in the cage.

In a few recent studies, the degradation of the drug solute was lowered by irradiating frozen aqueous solutions [20,21]. Their results show that:

- Frozen solutions of vitamins and glucose are stable to irradiation.
- The degradation of cyanocobalamine (vitamin B12) is reduced by freezing the solution.
- Calcium pantothenate solutions (vitamin B5) are successfully sterilized at – 196 °C.
- Insulin may be radiosterilized by freezing the solutions.
- Injectable drugs containing 1 % morphine hydrochloride and 100 UI / ml heparin are sterile and not degraded when irradiated with 25 kGy at – 80 °C and – 196 °C, respectively.

Cryo-irradiation is thus a promising method to sterilize protein solutions, as well as preparations derived from human blood.

At 10 kGy and without excipients, the loss of human insulin in aqueous solutions is almost complete. Addition of radio-protecting excipients (free radicals scavengers) and cryo-irradiation allowed to decrease insulin degradation. The best radio-protector used was ascorbic acid in aqueous solution and oxidized glutathione in the frozen solutions.

Only the combination of these two approaches (addition of scavenger and freezing) enables the irradiated human insulin in aqueous solution to meet the European Pharmacopoeia requirements for chemical potency [22].

Guidelines

ICH (standing for International Conference on Harmonization of Technical Requirements for Registration of Pharmaceuticals for Human Use) is a project that brings together the regulatory authorities of Europe, Japan and the United States and experts from the pharmaceutical industry in the three regions to discuss scientific aspects of product registration. For the European region, the organism that publishes and distributes the guidelines adopted by ICH is the European Agency for the Evaluation of Medicinal Products (EMEA) [23].

In the Note for guidance and development of pharmaceuticals, the section on the manufacturing process stipulates clearly that, wherever possible, products intended to be sterile should be terminally sterilized in their final container. The methods of sterilization considered as terminal by the European Pharmacopoeia [5] are steam, dry-heat, gas or gamma-radiation sterilization. If terminal sterilization is not possible, sterilization filtration or aseptic processing are authorized but the choice must be fully justified.

One last consideration, which is given in the Note for guidance on limitations to the use of ethylene oxide in the manufacture of medicinal products, must be taken into account. First of all, as described previously, ethylene oxide is a cytotoxic molecule, a carcinogen and a mutagen agent. Epidemiological data have clearly shown that the incidence of leukaemia and other tumours is larger for workers exposed to this substance at their work place. In view of these toxicological backgrounds, the use of ethylene oxide is only acceptable when two conditions are fulfilled : it must be "*pharmaceutically absolutely necessary*" and the residual in the product cannot exceed a limit of 1 ppm. Obviously workers should be also protected.

All these recommendations are summarized in the Decision trees for the selection of sterilization methods, edited by the EMEA [23] (**Fig. 1**). Two cases are considered : on one hand, the aqueous products and on the other hand, the non-aqueous liquid, semi-solid and dry powder products. **Figure 1** shows the order of preference of the sterilization methods for the second group. The terminal ones are ranked in the first place. Among them, thermal sterilization is still referred as the best choice, radio-sterilization ranking right after. Since gas sterilization is excluded and non-terminal methods are listed as the last choice, radio-sterilization now precedes all these methods. It is deemed as the recommended alternative method to thermal sterilization.

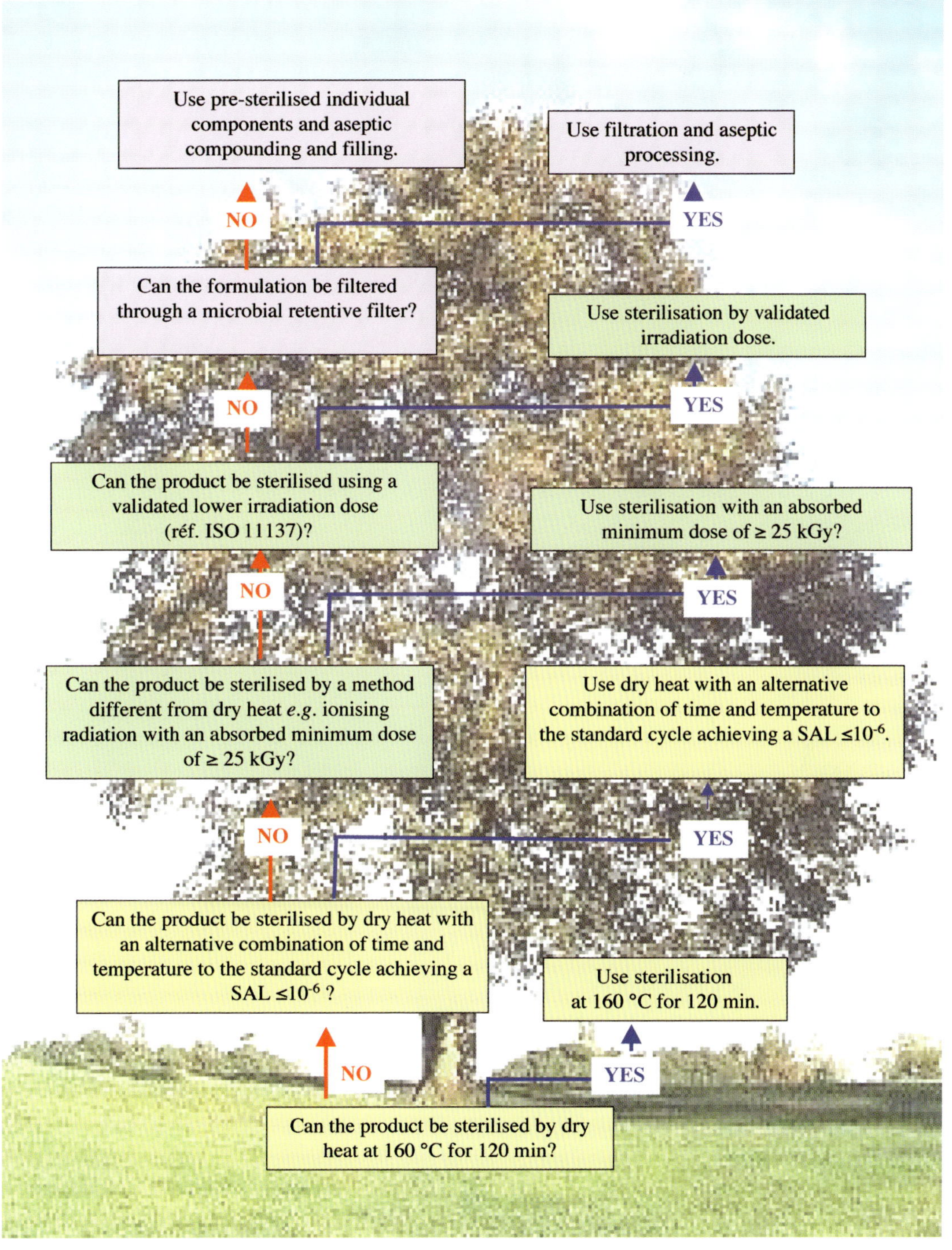

Figure 1 : *Decision tree for sterilisation choices for non-aqueous liquid,semi-solid or dry powder products* [23].

For aqueous solutions, some delay is necessary for research results to be applied in industry and so a new selection tree for aqueous solutions taking into account radio-sterilization is expected.

To conclude, radio-sterilization of drugs appears as a promising terminal process, which is recommended in Guidelines and, when applied under controlled conditions, ensures the required Security Assurance Level for eradicating the dangerous micro-organisms without producing toxic radiolytic products. For a new generation of drugs, to which heat-sterilization cannot be applied due to their thermo-lability, radio-sterilization is the only alternative treatment.

Acknowledgements

The author wish to thank all the Ph.D pharmacists of the CHAM laboratory with special thanks to A. Engalytcheff, A. Maquille and C. Slegers for their contribution to this manuscript. Thanks to F. Gelders for assistance in the preparation.

References

[1] Jacobs G.P., A review: radiation sterilization of pharmaceuticals, Radiat. Phys. Chem., 1985, **26**, 133-142.

[2] Tilquin B., Comprendre la radiostérilisation. J. Pharm. Belg., 2007, **62**,10-20.

[3] Radiosterilization and decontamination of pharmaceutical and pharmaceutical raw materials. IAEA, CRP Report, Vienna (A), 2005.

[4] USP, "Sterilization and sterility assurance of compendial articles", *in* The United States Pharmacopoeia 24, The National Formulary, United States Pharmacopeial Convention INC., Rockville, MD, 2000, **19**, p. 2143-2148.

[5] Ph. Eur., "Textes Généraux sur la Stérilité", *in* European Pharmacopoeia 5th Edition, p. 475-477, Council of Europe, Strasbourg, 2005.

[6] Nordhauser F.M., Olson W.P. (eds) Sterilization of Drugs and Devices, Technologies for the 21st Century Interpharm Press Inc., Buffalo Grove, IL, 1998.

[7] Reid D., Gamma processing technology: an alternative technology for terminal sterilization parenterals, J. Parenteral Sci. Technol., 1995, **49**, 83-89.

[8] ISO 11137. Sterilization of health care products - Requirements for validation and routine control-Radiation sterilization, International Organization for Standardization, Geneva, Switzerland, 1995.

[9] Zeegers F., Gibella M., Tilquin B., Analysis of some products from the radiosterilization of solid chloramphenicol, Radiat. Phys. Chem., 1997, **50**, 149-153.

[10] Zeegers F., Tilquin B., Étude par R.P.E. de la décroissance radicalaire dans les medicaments irradiés, ESR applications in organic and biorganic material. Catoire E. (eds), Springer-Verlag Berlin, 1992, 292-305.

[11] Engalytcheff A., Deridder V., Debuyst R., Tilquin B., Determination of radical yields in solid-state drugs as one technique to identify drugs that will withstand radiosterilization : radioresistance of beta blockers, Radiat. Res., 2003, **160**, 103-109.

[12] Stocker P., Gibella M., Tilquin B., Lesgards G., Thermoluninescence of some ionized pharmaceuticals, J. Chim. Phys – Chim. B.D., 1999, **96**, 174-177.

[13] Barbarin N., Rollmann B., Tilquin B., Role of residual solvents in the formation of volatile compounds after radiosterilization of cefotaxime, Int. J. Pharm., 1999, **178**, 203-212.

[14] Barbarin N., Tilquin B., Study of non-volatile degradation compounds produced by radiosterilization of cefotaxime, Radiat. Phys. Chem., 2001, **60**, 359-367.

[15] Engalytcheff A., Vanhelleputte J.P., Tilquin B., HPLC detection and quantification of radiolytic products of eight β-blockers irradiated in the solid state and hypotheses on their origins, Pharmaceut. Res., 2004, **21**, 1103-1108.

[16] Slegers C., Tilquin B., Theoretical approach to the destruction or sterilization of drugs in aqueous solution, Radiat. Phys. Chem., 2005, **72**, 363-365.

[17] Slegers C., Tilquin B., Final product analysis in the e-beam and gamma radiolysis of aqueous solutions of metoprolol tartrate, Radiat. Phys. Chem., 2006, **75**, 1006–1017.

[18] Slegers C., Maquille A., Deridder V., Sonveaux E., Jiwan J.L.H., Tilquin B., LC-MS analysis in the e-beam and gamma radiolysis of metoprolol tartrate in aqueous solution: Structure elucidation and formation mechanism of radiolytic products, Radiat. Phys. Chem., 2006, **75**, 977–989.

[19] Maquille A., Private Communication, 2007.

[20] Juanchi X., Albarran G., Negron-Mendoza A., Radiolysis of cyanocobalamin (vitamin B12), Radiat. Phys. Chem., 2000, **57**, 337-339.

[21] Moyne P., Botella A., Peyrouset A., Rey L., Sterilization of injectable drugs solutions by irradiation, Radiat. Phys. Chem., 2002, **63**, 703-704.

[22] Terryn H., Maquille A., Houée-Levin C., Tilquin B., Irradiation of human insulin in aqueous solution : first step towards radio-sterilization, Int. J. Pharm., 2007, **343**, 4-11.

[23] EMEA (European Agency for the Evaluation of Medicinal Products), Decision trees for the selection of sterilization methods, CPMP (Committee for Proprietary Medicinal Products) / QWP / 054 / 98 Corr), 2000.

Chapter 11

Food irradiation: wholesomeness and treatment control

Jacques RAFFI et Jacky KISTER

Food irradiation by X- or gamma-rays and by electron beam has been studied for more than four decades, mainly to reduce spoilage losses and to improve hygienic quality [1-3]. This treatment is legally accepted in more than 50 countries, but still prohibited in other ones by lack of public acceptance, despite numerous researches sponsored by the International Food Irradiation Project (IFIP), the International Atomic Energy Agency (IAEA) (in particular *via* the program for Analytical Detection Methods for Irradiation Trade or ADMIT), the Food and Agriculture Organisation (FAO), the World Health Organisation (WHO) and the European Union (*via* two concerted actions and the working group of the European Committee of Normalisation or ECN).

Principles and wholesomeness of food irradiation

Action of ionising treatments

In principle X-rays, as well as electron and gamma rays, can be used; the main effect of gamma and X-rays is Compton scattering, *i.e.* ionisation of atoms giving rise to "primary electrons"; these electrons, like those generated in an accelerator, lead to a number of other "secondary electrons" and ions. This cascade of secondary electrons (about 4000 to 6000 ionisations per initial secondary electron) loses energy in ionising the foodstuff molecules, with consequent production of free radicals and, thereby, of "radiolytic products" (Chapter 1).

The physical, chemical and biological effects are linked to the irradiation dose which is the quantity of energy absorbed by the matter, measured in gray: 1 Gy = 1 J/kg. The positive effects (increase of hygienic quality) are mainly linked to radiolytic effects on DNA molecules of viable cells from micro-organisms or living foodstuffs. The doses applied may vary from 0.05 to 50 kGy, but usually between 3 and 10 kGy. The possible applications are put in **Table 1** but the true choice was made, comparing the cost of the process with the obtained advantage. Thus it depends on the foodstuff, on the irradiation goal, on the chosen technique (gamma or X-rays, electron beam) and, also, on the country involved.

Table 1. *Doses recommended for possible applications of food irradiation.*

Dose (kGy)	Purpose
0.05 to 0.15	Sprouting inhibition
0.15 to 0.5	Insect desinfestation and parasite disinfection
0.5 to 1	Delay of physiological process (*e.g.* ripening) (***a***)
1 to 3	Spoilage of pathogenic micro-organisms - Shelf-life extension
1 to 7	Elimination of pathogenic micro-organisms (***b***)
7 to 25	Decontamination of certain food additives and ingredients (***b***)

a) Mainly for developing and non temperate countries. b) The main application in developed and temperate countries such as France.

Irradiation facilities

Up to now, only electron beam accelerators and gamma rays cells (**Fig. 1**) were used for commercial applications. The power of these facilities is measured in watts for electron beam accelerators, in becquerels for gamma-rays irradiators; in this last case, we generally speak of activity, instead of power: 1 becquerel (Bq) corresponds to a source where one disintegration happens per second (the old unit, 1 Ci = 3.7×10^{10} Bq).

National authorities control the security of irradiation facilities; the maximal energy of these radiations is fixed, according to WHO recommendations, by construction of the accelerator or by choice of the radioactive isotope (mainly cobalt 60) in order to avoid any induction of additional radioactivity in the foodstuff.

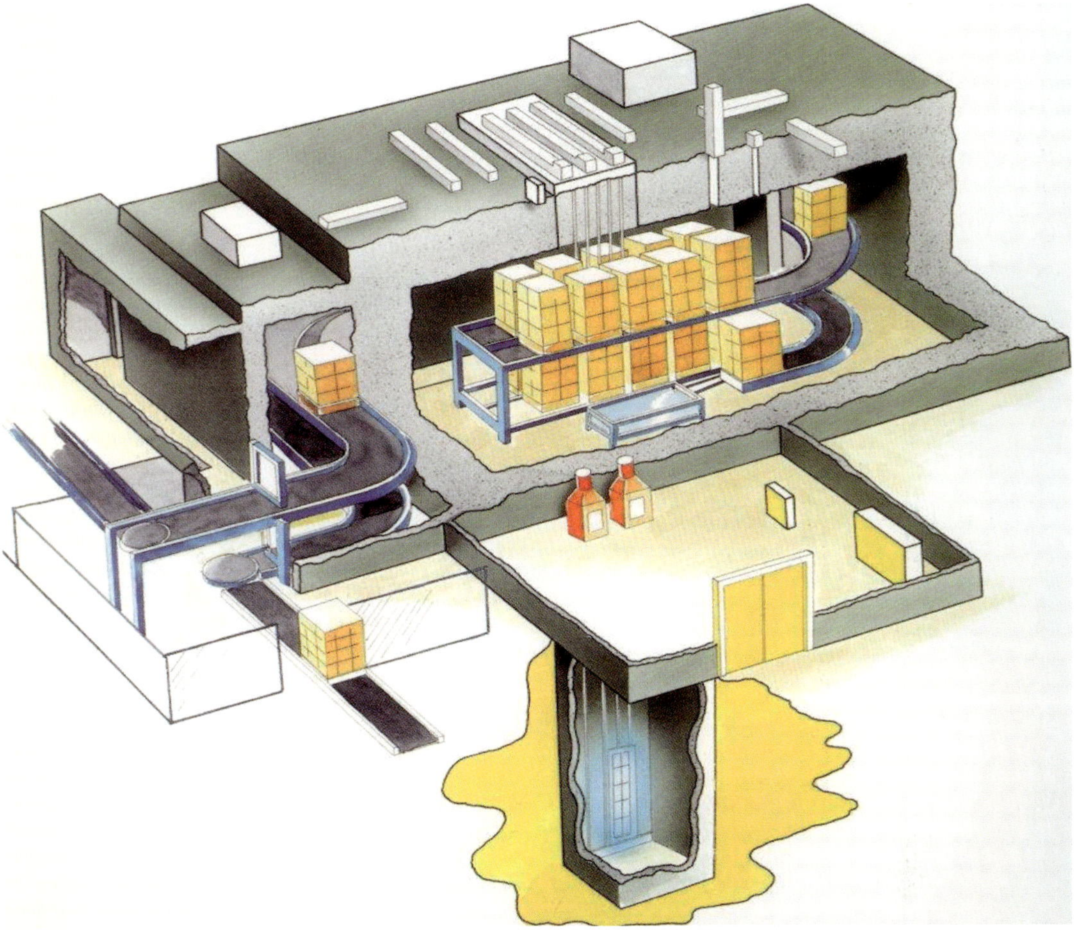

Figure 1 : *Scheme of a gamma irradiation facility (usually 1 to 3 x 10^8 Ci,* i.e. *about 10^{17} Bq). When not used, the ^{60}Co radiactive source is stored down in the pool. During irradiation, the source is raised to the upper position and parcels are moved in front of and around the source. A thick concrete shielding surrounds the irradiation chamber.*

Nutritional effects

The consequences of the radiolysis depend on the dose, on the temperature during the treatment, on the food constituents [4,5], and on the mobility of the foodstuff. However, some general rules can be drawn:

- Irradiation has practically no effect upon the proteins.
- The only important radiolytic effect upon carbohydrates is a depolymerization of polysaccharides, which can lead to positive consequences (better digestibility, decrease

of heating time, irradiated prunes judged as softer by the consumers...) or negative ones (strawberries irradiated above 3 kGy judged as too soft by the consumers...).

- Irradiation of lipids, especially unsaturated ones, may lead to unwanted flavour, which is the main technological limit of the treatment. Even if lowering the irradiation temperature can always reduce the quantities of these undesirable radiolytic products, this generally increases the treatment cost.

- Like thermal treatment, radiation processing causes some loss of vitamins, but they are only important at sterilisation doses, level which is only used for special applications such as diets for immunologically-incompetent patients.

- Consequently, irradiation *"will not lead to nutrient losses to an extent that would have an adverse effect on the nutritional status of individuals or populations"* [6-8].

Wholesomeness of irradiated foodstuffs

First at all, let us point out two main ideas:

- To ascertain that a chemical compound is dangerous is a scientific absurdity; curare can be use in medicine, *i.e.* it is not always a poison. On the opposite, sugar may be dangerous (for every one, not only for diabetic patients)... but surely at high concentrations. What is dangerous, in fact, is "a couple" linking the chemical structure of a compound with the quantities ingested.

- The absolute wholesomeness of a product (or of a treatment) is impossible to prove (if it exists); what we can only prove is the toxicity, depending of the use of the product, using assays on animals fed with irradiated diets.

Extensive animal feeding studies designed to detect the potential presence of toxic substances in various irradiated foods have been carried out since the 1950s, mostly in USA and Europe. None of the studies carried out under the auspices of the IFIP (24 involved countries, 67 technical reports published in 12 years) showed any indication that the irradiated foods contained radiation-produced carcinogens or other toxic substances. In a French study, for example, nine chemical compounds that had been identified in irradiated starch were fed daily to rats in amounts calculated to be 800 times the amounts the animals might be expected to consume from a normal daily intake of irradiated starch. No toxic effect was found even at this exaggerated rate of intake (chapter 8 of [1]).

Radiation chemistry studies had shown that the radiolytic products of major food components are present in low quantities (a maximum of about 10 ppm at 10 kGy) and, moreover, do not depend on the food from which they were derived. Moreover, for major food components [4,5], these products had also been identified in foods subjected to other accepted types of food processing such as appertisation, or under natural auto-oxidation. This understanding is very important for considering general clearance of irradiated foods:

- when foods of similar composition are similarly irradiated, their chemical responses are similar and they are, accordingly, toxicologically equivalent (for instance: beef, chicken and pork meats);

- when an irradiated food in a class of similar foods is cleared as safe, then other members of that class are, correspondingly, wholesome (for instance strawberries and raspberries).

Many investigations have been devoted to possible health hazards: in 1980, a Joint FAO/IAEA/WHO Expert Committee meeting [6] concluded that *"the irradiation of any food commodity up to an overall average dose of 10 kGy presents no toxicological hazard; hence, toxicological testing of foods so treated is no longer required"*. Moreover, in September 1997, an expert study group from WHO, FAO and IAEA, came to the conclusion that *"foods treated with doses greater than 10 kGy can be considered safe and nutritionally adequate when produced under established Good Manufacturing Practice"* [7, 8].

Since then no scientific result has raised doubts about these conclusions.

Methods of detection of irradiated foods

These methods may be helpful to label the foodstuffs, to control trade and for information of the consumers. However, the most difficult problem has been that the chemical changes that occur in irradiated foodstuffs [4,5] are very small and generally similar to those produced by classic food treatment processes (heating, freezing) or natural spoilage (auto-oxidation).

Many detection methods have been discussed in the literature [3,9,10], but considerable progress has been made, due to the IAEA ADMIT program and, mainly, to the European concerted actions [11-14]. We briefly describe here the published European protocols leading to a proof of irradiation or a proof of no irradiation (**Tab. 2**).

Table 2. *Detection of irradiated foodstuffs (DIF) with ECN protocols.*

Protocol	Date	Proof Methods	Application scope
EN 1784	1996 2003 (1)	DIF containing fat by GC analysis of hydrocarbons	crude meat, cheese, avocado, papaya, mango
EN 1785	1996 2003	DIF containing fat by GC/MS analysis of 2-alkylcyclobutanones	crude meat whole liquid egg
EN 1786	1996	DIF containing bone by ESR spectroscopy	meat and fish
EN 1787	1996 2000	DIF containing cellulose by ESR spectroscopy	pistachio nuts, paprika powder
EN 1788	1996 2001	DIF containing silicate studied by TL	aromatic herbs, spices, shrimps
EN 13708	2001	DIF containing crystalline sugars by ESR	dried figs, mangoes papayas & raisins
Protocol	**Date**	**Screening Methods**	**Application scope**
EN 13751	2002	DIF using photostimulated luminescence	herbs, spices, seasonings, shellfish
EN 13783	2001	DIF using Direct Epifluorescence Filter Technique/ Aerobic Plate Count (DEFT/APC)	Poultry meat, spices
EN 13784	2001	DNA Comet Assay for the DIF	Meats, spices, dried fruits
EN 14569	2004	DIF using the *limulus* amoebocyte lysate test in conjunction with a gram negative bacterial count (LAL/GNB)	Poultry meat

(1) The second date corresponds to a first revision of the protocol, which must be done, at least, every five years.
***GC**: gas chromatography, **MS**: Mass Spectrometry, **ESR**: electron spin resonance, **TL**: Thermoluminescence.*

Two "chemical" protocols voted by the ECN are relative to food containing lipids; the percentage of radiolytic products from lipids such as volatile hydrocarbons, aldehydes or butanones is directly linked to the chemical composition of lipids. In the case of another treatment such as heating there is no direct correlation between the different percentages.

For instance, the two main hydrocarbons radio-induced from each fatty acid of structure CN:M (*i.e.* having chains with N carbon atoms and M double bonds), have structures in CN-1 : M (with one carbon less than the parent fatty acid), and CN-2 : M+1 (with two carbons less and one extra double bond more), leading to protocol EN 1784 (**Fig. 2**).

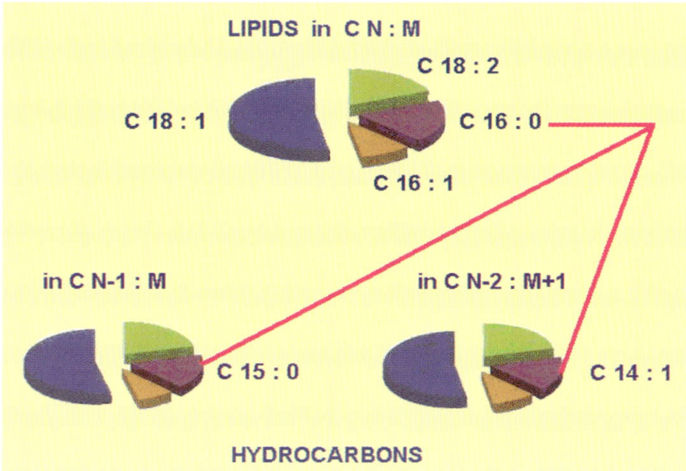

Figure 2 : *When avocado samples have been irradiated, the lipids in CN ; M (N carbon atoms and M double bonds) lead to two main hydrocarbons in C N–1 : M and C N–2 : M+1: the three pie-charts are quite identical (in contrast with any of other food treatments).* *This constitutes a specific test of the irradiation.*

The natural amount of aldehydes in foodstuffs is too high, with regard to the radio-induced one, to allow the establishment of a protocol. But the radio-induced cyclobutanones lead to the protocol EN 1785 and it must be noted that, up to now, the butanones seem to be characteristic of the irradiation treatment.

Three protocols used the Electron Spin Resonance (ESR) technique; ESR allows the observation of unpaired electrons, especially free radicals induced by irradiation, if they are stable during commercial storage of the food. This only occurs in the solid and dry components of the food, where the reactivity of the radicals with each other or with water is low. The first protocol (EN 1786) is relative to meat and fish bones (**Fig. 3**), the second one (EN 1787) to food containing cellulose such as berries, and the third one (EN 13708) to food containing crystallized sugars.

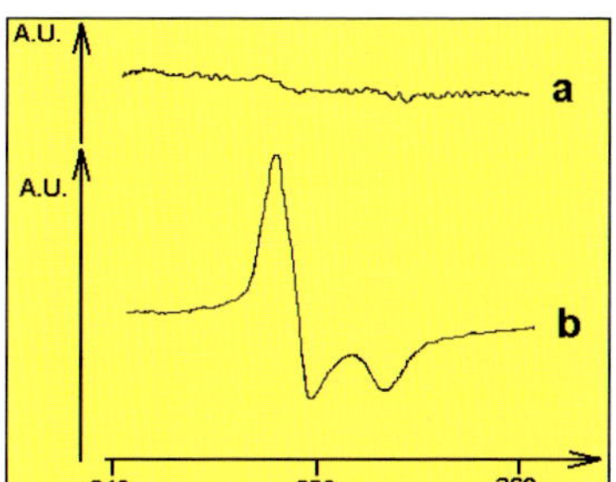

Figure 3 : *ESR spectra of unirradiated* (**a**) *and 3 kGy irradiated* (**b**) *bones. The magnetic field in the abscissae is in mT (milli tesla) and the intensity is in a.u. (arbitrary unit).*

A last European protocol uses thermoluminescence (TL): irradiation results in positive ions and free electrons, which can be relatively stable in solid and dry foodstuffs. When the food is fastly heated, the recombination of ions and electrons leads to light emission (luminescence). It was later pointed out that the seat of the phenomenon is not in the vegetable material itself but in contaminating mineral particles adhering to the products. On this basis, an European protocol (EN 1788) has been approved for herbs, spices and shrimps.

Some other methods were also approved but only lead to an irradiation presumption (*i.e.*, in fact, a proof of treatment) and can be used as screening protocols (**Tab. 2**) as they are easy to use in food control laboratories.

A possible future of food irradiation

We must notice that each country has its own law [15], a complexity that hampers the development of the process. In Europe, the Brussels Commission wrote a first directive (European "law") in 1986 but the European Parliament rejected it; one reason for that was the impossibility, in 1988, to bring the proof whether or not a foodstuff has been irradiated. Since the publication of the first European protocols, a new directive was written and then voted by the European Parliament in 1999 [16]. This directive was now adapted in each law of European countries but the final positive list (*i.e.* included authorized irradiated foodstuffs) is still under discussion. Moreover the European Union is also blocked by:

- the attitude of certain countries (such as Germany, despite the fact this country was one main contributor to this technology and to the proof of its wholesomeness);

- the lacking enthusiasm of other members (a consequence of a bad use of the "precautionary principle"?);

- the lobbying of certain associations of consumers, chemical industrials and food marketing networks.

But Europe is not alone in the world: the Food and drug administration (FDA) approved for instance, on December 1997, the use in the US of irradiation to kill harmful bacteria such as *Escherichia coli* in beef, lamb and pork. The FDA was acting on a three years old petition since 1994, but gave the allowance after the recall last August of 11,000 tons of hamburger meat feared tainted with *E. coli*. If applied in France, the accident, which occurred in South-West of France in late 2005, could not have happened.

Some developing countries which export into Europe products disinfested or debacterised by chemical fumigation well know that, once, this will be completely forbidden (Toronto agreement, December 1995); but a working group of the French Academy of Medicine found that, even in France, at present time, some industrialists prefer to use this forbidden treatment (without labelling) instead of irradiation (which requires a labelling). WHO also point out the interest of irradiation to fight against problems of salmonella, lysteria etc. [17-20].

The next years will probably be a turning point with regard to the development of food irradiation:

- When the directive will be applied, it will be easy to inform the consumers (who at present have no choice) with an "official development" of the treatment, including good labelling of the irradiated foodstuffs, or

- Whether or not there will be official development in Europe, there will be in all cases introduction in Europe of foreign irradiated foodstuffs, due to the need to replace some chemical treatments and to fight problems of salmonellae, but more or less visible for the consumers which means no opportunity for a good information. This is also why, despite a lot of existing texts on food irradiation (see for example the *Codex Alimentarus*, [21]), there is a demand for an ISO protocol for "*Good processing practices for the irradiation of foods for human consumption*" [22]. Moreover, last year, the USA asked the World Trade Organization to recognize a technical rule on processing and handling of irradiated food.

The choice is clear. The issue of consumer protection must not be understood as a means to prevent the consumer to be informed and to make an informed choice [2]: "*Shall we let Sound Science decide the safety of irradiated food?*" [23]. Now the ball for acceptance by the public is in the court of policy and health authorities but also in the court of consumers associations…

References

[1] Vasseur J.-P., Ionisation des produits alimentaires, Tec & Doc - Lavoisier - APRIA - Teknéa, Paris, 1991.

[2] Ehlermann D., Four decades in food irradiation, Rad. Phys. Chem., 2005, **73**, 346-647.

[3] Raffi J., The state of food irradiation and of detection of irradiated foodstuffs, Res. Adv. Food Sci., 2002, **3**, 11-19.

[4] Elias P., Cohen A., Radiation chemistry of major food components, Elsevier, Amsterdam, 1977.

[5] Elias P., Cohen A., Recent advances in food irradiation, Elsevier, Amsterdam, 1983.

[6] Joint FAO/IAEA/WHO Expert Committee, Wholesomeness of Irradiated Food, WHO, N°659, Geneva, 1981.

[7] WHO, Food irradiation – Sky's the limit, Press Release WHO/68, 1997, 19 September.

[8] Joint FAO/IAEA/WHO Study Group, High-dose irradiation: Wholesomeness of food irradiated with doses above 10 kGy, WHO Techn. Rep. Ser., 1999, **890**, Genève.

[9] Delincée H., Analytical detection methods for irradiated foods. A review of the current literature, IAEA-TECDOC-587, Vienna, IAEA,1992.

[10] Raffi J., Methods of identification of irradiated foodstuffs and relative products, in "Handbook of food analysis", Nollet L. (ed), 2nd edition, Marcel Dekker/CRC Press, 2004, 1919-1940.

[11] Raffi J., Stevenson M.H., Kent M., Thiéry J.M., Belliardo J.-J., A European intercomparison on Electron Spin Resonance identification of irradiated foodstuffs, Int. J. Food Sci. Technol., 1992, **27**, 111-124.

[12] Raffi J., Delincée H., Marchioni E., Hasselmann C., Sjöberg A.-M., Léonardi M., Kent M., Bögl K.-W., Schreiber G., Stevenson M.H., Meier W., New methods for the detection of irradiated food, CEC, BCR, Luxembourg, 1993, EUR 15261 EN.

[13] McMurray C., Stewart E., Gray R., Pearce J., Detection methods for irradiated foods, current status, Royal Chem. Soc, 1996.

[14] Raffi J., Stachowicz W., Migdal W., Barabassy S., Kalman M., Yordanov N., Andrade E., Prost M., Callens F., Final report of Copernicus contract CIPA-CT94-0134 on "Establishment of an eastern network of laboratories for identification of irradiated foodstuffs", CEC, 1998.

[15] Database on approvals for irradiated foods, Food & Environmental Newsletter, 2006, **9**, 21-59.

[16] Directive 1999/2/CE du Parlement Européen et du Conseil du 22/2/99, J. Off. Comm. Eur., 13/3/99.

[17] WHO/FAO, Food Irradiation, A technique for preserving and improving the safety of food, Geneva, 1988.

[18] WHO, Study on the safety and nutritional adequacy of irradiated food, Press Release WHO/68, 19 September 1994.

[19] WHO, Multi-drug resistant salmonella typhimutium, fact sheet N°139, January 1997.

[20] WHO, Food safety – Joint FAO/IAEA/WHO Study Group on high Dose Irradiation, Weekly Epidemiological record, 1998, **73**, 9-11.

[21] Codex Alimentarus, CAC/RCP 19-1979, Rev. 2-2003, recommended international code of practice for irradiation processing of food, 2003.

[22] ISO/CD 22810, Food irradiation – Good processing practices for the irradiation of foods intended for human consumption, 2006.

[23] Loaharanu P., Shall we let Sound Science decide the safety of irradiated food? American Council of Science and Health, Facts & Fears, 15 December 2006, http://www.acsh.org/factsfears/newsID.895/news_detail.asp

Part III

Radiation damage to biomolecules, radioprotection and radiotherapy

Chapter 12

Radiation-induced damage to DNA : from model compounds to cell

Thierry DOUKI and Jean CADET

Deoxyribonucleic acid (DNA) is a key biomolecule involved in the storage of genetic information and its transmission from one cell generation to the other at each cellular division. These biological functions are made possible by the complementarity between nucleic bases on the two opposite strands: thymine specifically pairs with adenine and guanine with cytosine. In the DNA molecule, bases are linked to a sugar (2-deoxyribose), and the resulting nucleosides are bound to a phosphate group. The obtained nucleotides are the basic monomeric units of DNA and form a chain by linkage between two consecutive deoxyribose rings through a phosphodiester bond. A wide array of chemical and physical agents, including ionizing radiations, is able to damage the chemical structure of DNA. A first result of such an event is the killing of the cell. Another major consequence of the formation of DNA damage is the modification of the genetic information (a mutation) at the damaged sites in the newly synthesized strand upon replication of DNA (**Inset**). This process not only involves the irradiated cell but all the following generations and may represent an initiating event in the formation of tumors. In the present review, emphasis will be placed on the chemical pathways involved in the radical degradation of the DNA components (**Fig. 1**). These processes are mostly accounted for by the attack of the highly reactive $^{\bullet}OH$ radical produced upon radiolysis of the water molecules (Chapter 1) surrounding DNA and by the direct ionization of nucleic components by the incident radiation. The relative contribution of the indirect and direct effects to radiation-induced DNA damage has been and still is an issue of debate and a subject of intense investigations (Chapter 13). Another specific feature

Inset : Induction of mutation by DNA damage

The figure below illustrates how a damaged base (a thymine in this case) leads to a change in the genetic information, namely a mutation. The two DNA strands are symbolized by the vertical lines linked by the nucleobases. The modified thymine appears in grey. Following apparition of the damage, the hydrogen bonding properties are lost or modified. Upon the 1st replication, a guanine instead of an adenine may be incorporated in front of the damage. In the following generations of cells, the guanine is normally paired with cytosine, leading to a T:A to C:G mutation.

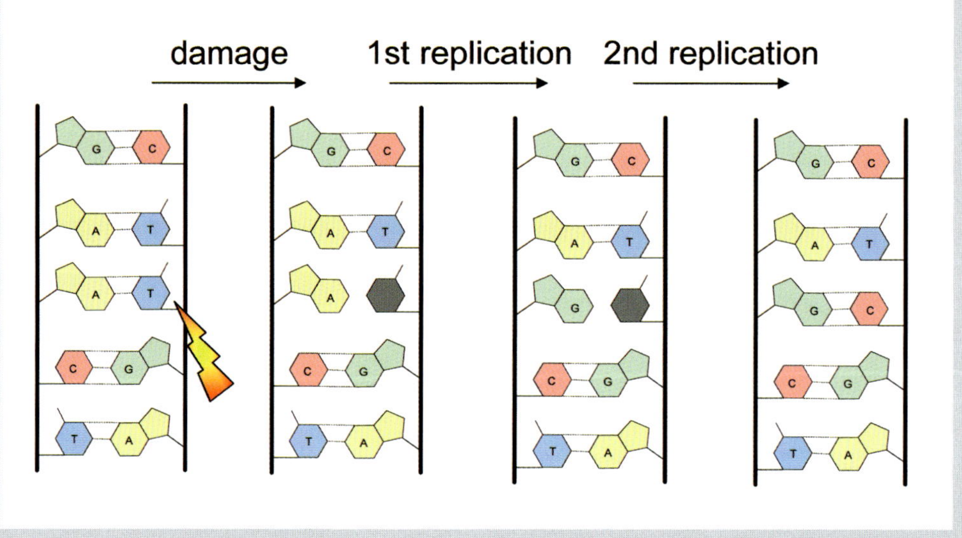

of the interaction between ionizing radiation and the genome is the co-localization of several lesions along the track of the incident particle, resulting in highly deleterious clustered damage. Comprehensive mechanisms of degradation are available on most of the DNA nucleobases and the 2-deoxyribose moiety as inferred from studies on model systems and isolated DNA. In contrast, much less information is available on cellular DNA despite recent development of biochemical and chemical analytical tools aimed at measuring base lesions.

Radiation-induced degradation pathways of DNA monomers

Extensive studies on monomeric compounds, such as bases and nucleosides, have led to the isolation and the characterization of the final radiation-induced degradation products [1]. In addition, electron spin resonance, pulse-radiolysis studies and comparison

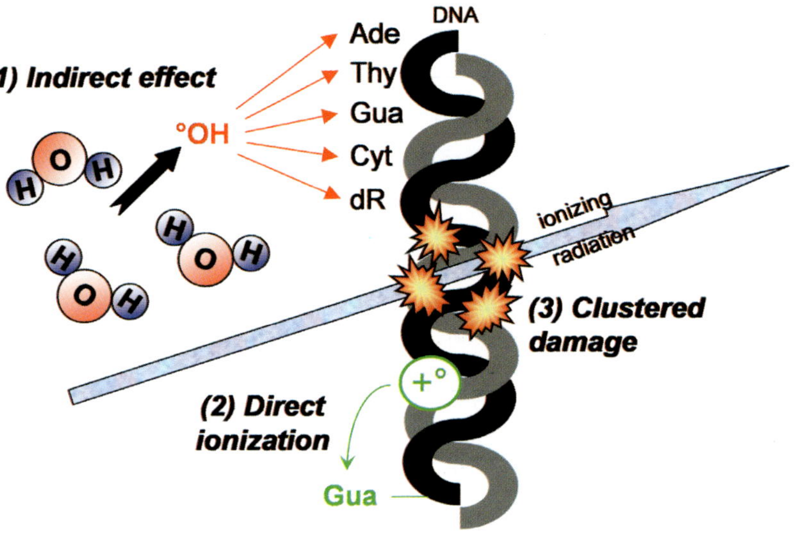

Figure 1 : *The different pathways involved in radiation-induced damage to DNA.* 1*) Indirect effect corresponds to the degradation of DNA by the reactive species produced upon water radiolysis. The most damaging radical is* $^{\bullet}OH$ *that may react with deoxyribose and the four bases.* 2*) The direct effect is the ionization of DNA component by the incident radiation. The resulting positive charges migrate within the double-helix towards guanine.* 3*) Both mechanisms take place along the track of the incident radiation, leading to the localization of several damages in a small portion of DNA. (Ade: Adenine, Thy: thymine, Gua: guanine, Cyt: cytosine, dR: 2-deoxyribose.)*

with photosensitization experiments have provided information on the structure and the reactivity of the main radical intermediates produced in the radical reactions [2]. Altogether this has permitted to propose detailed mechanisms for the main radiation-induced degradation pathways. The emphasis is placed here on oxidation reactions in the presence of oxygen that are predominant in cells. It should though be kept in mind that reductive pathways, that mostly involve initial addition of electron or hydrogen atom to pyrimidine bases, may also give rise to stable final products under anaerobic conditions.

$^{\bullet}OH$ radical-induced damage to nucleobases

Among the species produced upon radiolysis of water, hydroxyl radical ($^{\bullet}OH$) is the most reactive. Indeed, its reaction rate with the four bases and related nucleosides is diffusion-controlled. The main reactive sites of hydroxyl radicals on nucleobases are the double-bonds of the heterocycles. Accordingly, addition of $^{\bullet}OH$ at the C8 position of adenine and guanine yields the corresponding reducing 8-hydroxy-7,8-dihydropurin-7-yl radical (**Fig. 2**). Oxidation of this intermediate leads to the formation of related 8-hydroxypurines that are in dynamic equilibrium with their more stable 8-oxo-7,8-dihydropurine tautomeric form. Competitive reduction of the latter purine radical gives rise to imidazole ring opened compounds: the formamidopyrimidine (Fapy) derivatives. Addition of ˙OH to the C4 position of the moiety, that is in fact the predominant radical reaction, yields an oxidizing oxyl type radical subsequent to a fast dehydration process. Addition of superoxide radical to the latter intermediate gives rise to an imidazolone product that is quantitatively hydrolyzed into its oxazolone derivative.

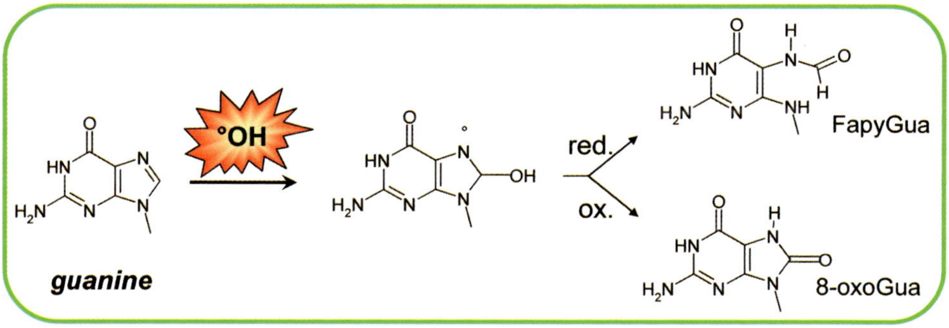

Figure 2 : ***Hydroxyl radical-induced degradation of guanine.*** *Addition of ˙OH to the C8 position leads to the reducing 8-hydroxy-7,8-dihydroguan-yl radical that is then either oxidized into 8-oxo-7,8-dihydroguanine (8-oxoGua) or reduced into formamidopyrimidine lesion (FapyGua).*

In the case of pyrimidine bases, hydroxyl radical preferentially adds to the C5 and, to a lesser extent, to the C6 positions of thymine and cytosine (**Fig. 3**). The obtained hydroxyl radical adducts may undergo a fast reaction with molecular oxygen. Reduction of the resulting peroxyl radicals yields related hydroperoxides, that then give rise to final stable products such as 5,6-hydroxy-5,6-dihydropyrimines (glycols; ThdGly for thymidine) by reduction, together with ring rearrangement and fragmentation products. ˙OH radicals may

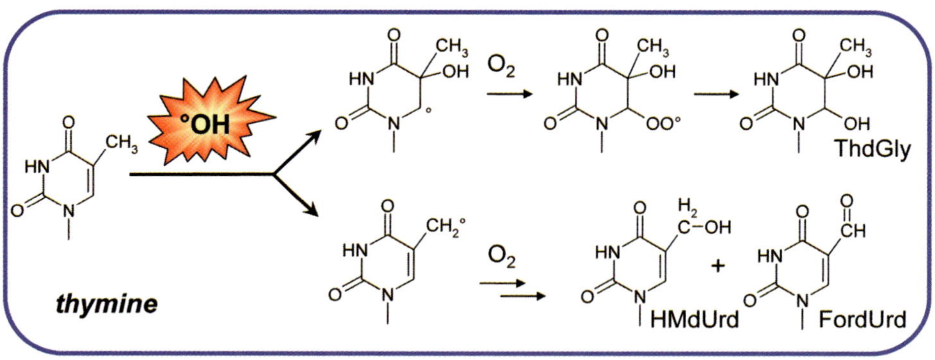

Figure 3 : ***˙OH radical-induced degradation of thymine. Oxidation involves either ˙OH addition to the C5-C6 double bond or ˙OH-mediated hydrogen abstraction from the methyl group.*** *Molecular oxygen then adds to the resulting carbon-centered radicals. This reaction yields hydroperoxide that further evolves into stable products, such as thymidine glycol (ThdGly), 5-hydroxymethyl-uracil (HMdUrd) and 5-formyl-uracil (5-FordUrd). N- represents the bond to the DNA backbone.*

also abstract a hydrogen atom from the methyl group of thymine. In the presence of oxygen, the radical produced leads to the formation of a hydroperoxide, 5-(hydroperoxymethyl) -2′-deoxyuridine. The two final stable products of this degradative pathway are the corresponding alcohol and the aldehyde, namely 5-(hydroxymethyl)-2′-deoxyuridine (HMdUrd) and 5-formyl-2′-deoxyuridine (FordUrd), respectively.

One-electron oxidation reactions

Lesions induced by the direct ionization of DNA by ionizing radiation are difficult to study in aqueous solution because of the predominant effect of the species produced upon water radiolysis (Chapter 13). However, the chemical pathways triggered upon ionization of bases and nucleosides have been unraveled by one-electron oxidation reactions provided by chemical and photochemical systems. It was thus shown that hydration of the purine radical cations gives rise to the reducing 8-hydroxy-7-yl radical also observed upon $^{\bullet}OH$ addition at position 8 (**Fig. 4**). Competitive deprotonation of the radical cations leads to the formation of related oxidizing neutral radical that gives rise to imidazolone and oxazolone from guanine and to hypoxanthine from adenine. Study of the one-electron oxidation of 2′-deoxyguanosine has been hampered by the consumption of 8-oxo-7,8-dihydroguanine as soon as it is formed due to its low ionization potential by the oxidizing guanine radical.

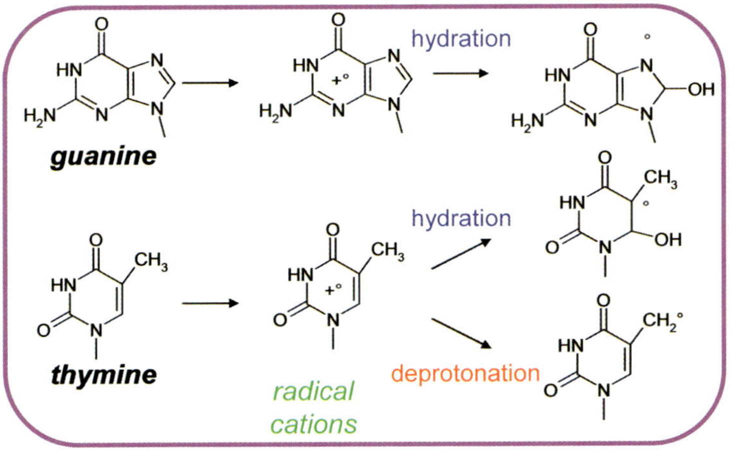

Figure 4 : *Reactivity of radical cations of guanine (upper panel) and thymine (lower panel). In both cases, the radical cations produced by ionization deprotonate or are hydrated to yield the same radicals than those produced upon reaction with $^{\bullet}OH$ radicals.*

One-electron oxidation reactions of pyrimidine bases are also well established, mostly through the study of the menadione-mediated type I photosensitization of 2′-deoxycytidine and thymidine [3]. The radical cations once formed may undergo two competitive pathways.

Hydration of the thymidine radical cation leads to the predominant formation of the oxidizing 6-hydroxy-5,6-dihydrothymid-5-yl radical whereas deprotonation mostly generates the methyl-centered radical (**Fig. 4**). As already pointed out, these two pyrimidine radicals are also produced upon reaction with $^{\cdot}OH$. However, the methyl-centered radical represented only 5% of base radicals when produced by $^{\cdot}OH$ while it corresponds to 30% of the thymidine radicals produced upon one-electron oxidation. Type I photosensitization of 2′-deoxy-cytidine leads to the formation of the 6-hydroxy-5,6-dihydro-2′-deoxycytidil-5-yl radical also produced upon reaction with $^{\cdot}OH$. Additional oxidative pathways involve the formation of 2-deoxyribonolactone and free cytosine as well as production of 2′-deoxyuridine as the result of deprotonation of the pyrimidine radical cation at C1′ and NH_2 group, respectively.

Radiation-induced damage to oligonucleotides and double-stranded DNA

Studies on isolated DNA and sugar model compounds have allowed the determination of the main $^{\cdot}OH$-mediated degradation pathways to the 2-deoxyribose moiety. In addition, data are also available on the influence of the double-helix structure on the fate of purine and pyrimidine base radicals.

Sugar damage

Studies involving gas chromatography measurements, isotopic labeling experiments and synthesis of photolabile precursors of sugar radicals have provided relevant information on the $^{\cdot}OH$ induced hydrogen abstraction to 2-deoxyribose that in most cases leads to DNA strand breaks [4] (**Fig. 5**). The most reactive site of the sugar ring of DNA is C4′. In the presence of molecular oxygen, the resulting carbon centered radical gives rise to a peroxyl intermediate that may decompose predominantly into phosphoglycolate with the release of base propenal. Hydrogen abstraction at the 5′ position (**Fig. 5**) is also a common reaction that yields among other oxidation products the 5′-aldehyde derivative under aerobic conditions. These reactions, as well as other minor ones involving H2′ and H3′, lead to frank strand breaks. In contrast, abstraction of H1′ in the presence of oxygen gives rise to 2-deoxyribonolactone without cleavage of the phosphate-sugar backbone. This oxidized abasic site is unstable under even mild alkaline conditions leading to the formation of a strand break.

Radiation-induced base damage within isolated DNA

Some of the final radical-degradation products of nucleobases have been detected and quantified within isolated DNA. Interesting notable changes in the product distribution

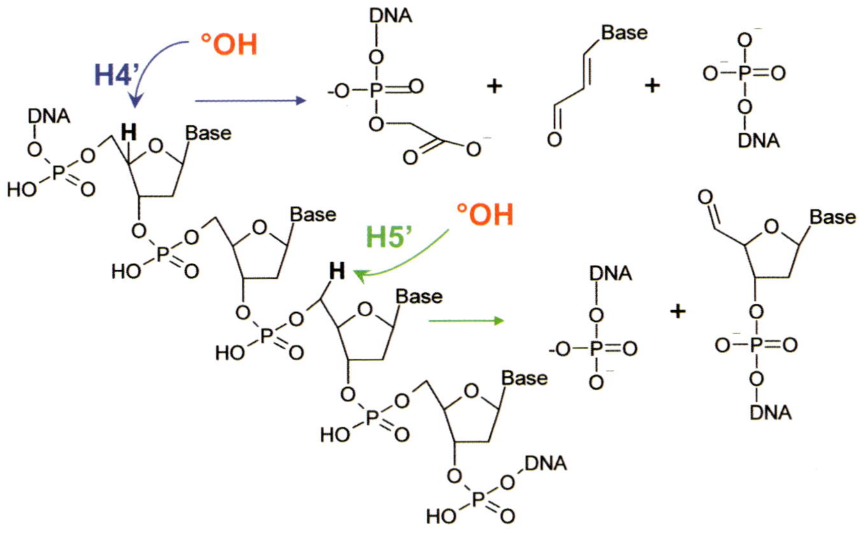

Figure 5 : *Examples of DNA single strand breaks produced upon ˙OH-mediated hydrogen abstraction to the 2-deoxyribose moiety.* *Attack at the H4' position induces the release of base propenal and the formation of phosphoglycolate, while abstraction of H5' leads to the formation of a 5'-aldehyde.*

have been observed within DNA duplex with respect to that obtained with free bases or nucleosides. The most striking one is the predominant formation of 8-oxo-7,8-dihydroguanine in DNA under aerated conditions whereas the related nucleoside is barely detected upon gamma irradiation of aerated aqueous solutions of dGuo. Indeed, 8-oxoGua bond to the phosphodiester backbone can no longer diffuse and react with oxidizing species like in solutions of free nucleosides. Moreover, formation of FapyGua and FapyAde at the expense of the 8-oxo-purine lesions was observed under reducing conditions, as observed with the free nucleosides. These results confirm the common origin of these two classes of modified purines that arise from the reducing 8-hydroxy-7,8-dihydropurin-7-yl radicals (**Fig. 2**).

Data have also been obtained on the formation of oxidized pyrimidines within isolated DNA exposed to ionizing radiation in aerated aqueous solutions, conditions where ˙OH is the main reactive species. Thymidine glycols and the methyl oxidation products HMdUrd and FordUrd were found to be produced in significant yields. Interestingly, the ratio between the yields of these two classes of damage is much lower than for the free nucleoside. This may be explained by a higher accessibility to ˙OH of the methyl group than the C5-C6 double bond in double-stranded DNA. The available information on oxidatively generated damage to the cytosine base within double-stranded DNA is much more limited than for the other bases. Evidence was however gained for the formation of 5-hydroxy-2'-deoxycytidine together with low amounts of 5-hydroxy-2'-deoxyuridine.

Tandem lesions

Extension of the study of base damage from free bases or nucleosides to oligonucleotides and isolated DNA has permitted the identification of a new class of base lesions that involves the reaction of a DNA radical with an adjacent nucleotide on the same strand [5]. Thus, exposure to X-rays of dinucleotides carrying thymine and guanine in aerated aqueous solution gave rise to lesions exhibiting both an 8-oxoGua moiety and a formylamine residue (dF), produced by the oxidation of the guanine and thymine bases, respectively. These 8-oxoGua/dF tandem lesions, which are produced by a single initial radical event, were also found to be generated in a significant yield within isolated DNA (**Fig. 6**) [6]. Mechanistic studies showed that the initial step in the formation of 8-oxoGua/dF was the addition of $^{\bullet}OH$ to the C5-C6 double bond of thymine, followed by reaction of molecular oxygen and addition of the resulting thymine peroxyl radical to the C8 position of guanine. Other types of tandem lesions arising from a single radical event involve crosslinking of thymine to an adjacent purine *via* its methyl group and cyclonucleosides that arise from the intramolecular addition of the 5′-carbon centered radical to the C8 position of a purine base of the same nucleotide.

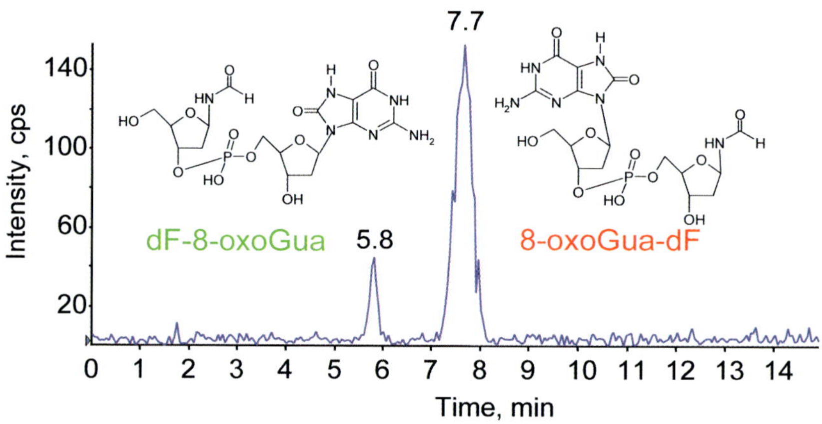

Figure 6 *: Detection of the 8-oxoGua/formylamine tandem lesions in isolated DNA exposed to gamma radiation in aerated aqueous solution. DNA was digested by a mild enzymatic treatment in order to release the lesions without cleavage of the phosphodiester bond. The sample was then injected onto a reverse phase HPLC column. The detection was provided by a mass spectrometer monitoring the main fragmentation of the two tandem base lesions.*

Charge transfer

The double-stranded structure of DNA not only affects the chemistry of $^{\bullet}OH$-induced damage but also may modulate oxidation reactions that take place upon initial direct ionization of a given nucleotide. Indeed, the positive holes created in the latter process have been shown to migrate through DNA double-helix, as extensively discussed in Chapter 14. They are eventually trapped at sites exhibiting the lowest ionization potential, namely the guanine bases and especially those located on the 5′-end of GG doublets. This was inferred from electron spin resonance studies performed at low temperature on DNA samples exposed to ionizing radiation and from photoionization experiments of oligonucleotides under more physiological conditions. In the latter works [7-9], damage was mapped at the sequence level by electrophoretic techniques following conversion of the base damage into strand breaks by either alkaline treatment or incubation with specific *N*-glycosylase repair enzymes. Although the range of charge migration and the underlying mechanism(s) are still a matter of debate, the predominance of base damage over direct strand breaks and the overwhelming degradation of guanine are common features reported by all the groups. It may be added that direct measurement of base damage by a more specific chromatographic approach confirmed these observations [10].

Radiation-induced damage to cellular DNA

Measurement techniques

A real gap exists between the amounts of mechanistic information gathered in model systems and isolated DNA and those available at the cellular level. The measurements of oxidatively generated lesions have been hampered by the low level of modifications (in the range of 1 per million normal nucleotides) and usually the low amounts of the available DNA samples. The bulk of the quantitative data on radiation-induced damage to cellular DNA are related to single (SSB) and double-strand breaks (DSB). Indeed, a number of electrophoretic methods allow their detection either following DNA extraction such as pulsed-gel electrophoresis for monitoring DSB or in the whole cell such as the alkaline comet assay or alkaline elution for measuring both SSB and DSB. The latter methods may also be used to quantify base damage after conversion into strand breaks by using DNA repair glycosylases that cleave DNA at the site of the lesion. These biochemical techniques are sensitive and are not subjected to artifactual oxidation. However, they do not provide specific information on the chemical nature of the damage since repair enzymes excise a broad spectrum of lesions. More specific data on modified bases can be obtained by applying chromatographic methods that require DNA hydrolysis into monomeric units prior the measurements. Liquid chromatography associated with electrochemical detection has been used extensively for the quantification of 8-oxo-7,8-dihydro-2′-deoxyguanosine. A more versatile approach involves tandem mass spectrometry that makes possible the quantification of several modified nucleosides in one analysis. Gas chromatography coupled to mass spectrometry has also been proposed. However, flaws in the sample treatment that involved a silylation step at high temperature have led to the overestimation of the level of lesions. The limitations of chromatographic methods include a need for rather large amount of DNA and occurrence of spurious oxidation upon extraction of DNA. Recent efforts have been devoted to minimize the latter risk. Altogether, less than 10 modified bases can be accurately quantified within cellular DNA while more than 50 different types of radiation-induced lesions have been identified in nucleosides.

Radiation-induced DNA damage in cells

Evidence is now available in a wide variety of cell types for the formation of single and double-strand breaks in a *ca.* 25:1 ratio within DNA after exposure to X- and γ-rays. Biochemical approaches showed that modified bases are produced in amounts similar to those of single strand breaks. It was also found, by using both enzymatic and HPLC-MS/MS assays that purine and pyrimidines bases are damaged to a similar extent (**Fig. 5**). Thymine glycols were found to be produced with the highest frequency. Methyl oxidation products of thymine including HMdUrd and FordUrd were also shown to be efficiently generated. 8-OxoGua was found to be another major lesion but FapyGua was at least equally frequent. The observation of an efficient radiation-induced formation of FapyGua strongly suggests that cellular DNA is in a reducing context likely due to a relatively low intracellular oxygen concentration and the presence of reducing species. Formation of FapyAde and 8-oxodAdo was also observed but the overall yield of these two adenine lesions was about one order of magnitude lower than that of related guanine damage, suggesting a much lower

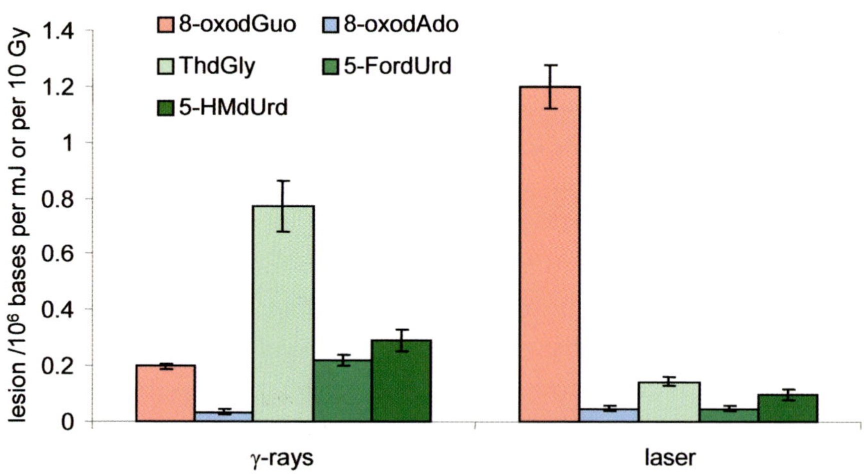

Figure 7 : *Yield of formation of oxidized bases in human THP1 monocytes exposed to γ-rays or 266 nm high intensity laser beam. Yields are expressed in lesion/10^6 bases per 10 Gy and per mJ, respectively. Red, blue and green bars correspond to guanine, adenine and thymine damage, respectively.*

susceptibility of the adenine moiety to radical oxidations in DNA. Altogether, the available data show the predominance of the hydroxyl radical-induced damage within the DNA of cells exposed to low linear energy transfer (LET) radiation. Interestingly, ionization of cellular DNA by exposure to high intensity 266 nm laser pulses leads to the overwhelming formation of 8-oxodGuo (**Fig. 7**), in agreement with occurrence of charge transfer, guanine bases being the sink for this process.

A distribution of oxidized bases similar to that obtained with γ-rays is observed upon exposure of human cells to high LET particles such as heavy ions [11]. However, with both $^{12}C^{6+}$ and $^{36}Ar^{18+}$ ions beams, a decrease in the yield of all lesions was observed [12]. This observation is reminiscent of the effect of increasing LET on the radiolytic yield of $^{\bullet}OH$ that decreases as the result of efficient radical recombination in the particle track. A same trend has been observed for both single and double-strand breaks [13]. These results strongly suggest a major role of the hydroxyl radical chemistry in the formation of radiation-induced DNA damage in cells.

Clustered damage

The decrease in the yield of lesions upon increasing LET seems contradictory with the well established increase in cell lethality for a same dose of radiation (relative biological effectiveness). Another striking result is the relatively low yield of damage induced within DNA when a lethal dose is applied to cells. Indeed, 4 Gy of γ-rays which is a lethal dose is needed to double the basal level of 8-oxodGuo as inferred from comet assay measurements. Both observations can be explained by the implication of other types of radiation-induced damage that are likely to have strong biological impact. Indeed, in contrast to other types of oxidative stresses which involve the production of oxidizing species homogeneously within the cell, ionizing radiation triggers induction of chemical reactions mostly in localized area along the incident track. In the recent years, modelization studies have shown that even low LET particles are able to induce a significant fraction of damage that consists in two individual modifications located only a few nucleotides apart on the same strand or on the opposite strands [14]. Interestingly, the complexity of the clustered damage increases with increasing LET. Such a process explains why DSBs are considered as specific radiation-induced DNA lesions. However, DSBs represent only a fraction of the complex, so-called "clustered", lesions that may consist in two oxidized bases or a modified base and a strand break, or any larger combination of these individual lesions (**Fig. 1**). Observation of an increase in the ratio between the yields of double- and single-strand breaks with increasing LET strongly supports this hypothesis [13]. More direct evidence for the formation of clustered damage has been provided by the use of repair enzymes in order to convert closely located modified bases (or a modified base and as SSB) into double-strand breaks [15]. Recent studies involving synthetic oligonucleotides have shown that base damage and/or strand break located within a cluster are less efficiently repaired by comparison with isolated lesions [16,17]. Therefore, clustered lesions may persist inside the genome and interfere with DNA transcription and replication, leading ultimately to mutagenesis and cell death.

Conclusion and perspectives

Radiation-induced damage to DNA arise from either the reaction of highly reactive $^{\bullet}OH$ radicals produced by radiolysis of water or, to a lesser extent, of the direct ionization of DNA components. In addition, the clustering of the lesions along the radiation track is expected to be highly damaging to the cell. Recent data, mostly based on the use of microbeam of charged particles, have revealed other levels of complexity. Indeed, deleterious effects such as mutagenesis are observed in cells that are not directly hit, a process so-called the "by-stander" effect. Another interesting effect is the induction of mutations in cells that were hit only in the cytoplasm and not in the nucleus where DNA is located. In addition, other radiation-induced reactions, such as dissociative capture of low energy electrons and K-shell ionization events, have also been proposed to be involved in the induction of DNA lesion. However, evidence for the implication of the two latter processes at the cellular level is still required. It is clear that major efforts are still needed to gain further insights in the complex mechanism of formation of radiation-induced damage to DNA and to better assess the biological role of these modifications.

References

[1] Cadet J., Berger M., Douki T., Ravanat J.-L., Oxidative damage to DNA: Formation, measurement and biological significance, Reviews Physiol. Biochem. Pharmacol., 1997, **131**, 1-87.

[2] Steenken S., Purine bases, nucleosides, and nucleotides: aqueous solution redox chemistry and transformation reactions of their radical cations and e- and OH adducts, Chem. Rev., 1989, **89**, 503-520.

[3] Wagner J.R., van Lier J.E., Johnston L.J., Quinone sensitized electron transfer photooxidation of nucleic acids: Chemistry of thymine and thymidine radical cation in aqueous solution, Photochem. Photobiol., 1990, **52**, 333-343.

[4] Knapp Pogozelski W.K., Tullius T.D., Oxidative strand scission of nucleic acids: route initiated by hydrogen abstraction from the sugar moiety, Chem. Rev., 1998, **98**, 1089-1107.

[5] Box H.C., Dawidzik J.B., Budzinski E.E., Free radical-induced double lesions in DNA, Free Radic. Biol. Med., 2001, **31**, 856-868.

[6] Bourdat A.-G., Douki T., Frelon S., Gasparutto D., Cadet J., Tandem base lesions are generated by hydroxyl radicals within isolated DNA in aerated aqueous solution, J. Am. Chem. Soc., 2000, **122**, 4549-4556.

[7] Schuster G.B., Long-range charge transfer in DNA: transient structural distortions control the distance dependence, Acc. Chem. Res., 2000, **33**, 253-260.

[8] Treadway C.R., Hill M.G., Barton J.K., Charge transport through a molecular π-stack: double helical DNA, Chem. Phys., 2002, **281**, 409-428.

[9] Giese B., Long-distance charge transport in DNA: the hopping mechanism, Acc. Chem. Res., 2000, **33**, 631-636.

[10] Douki T., Angelov D., Cadet J., UV Laser photolysis of DNA: effect of duplex stability on charge-transfer efficiency, J. Am. Chem. Soc., 2001, **123**, 11360-11366.

[11] Pouget J.-P., Frelon S., Ravanat J.-L., Testard I., Odin F., Cadet J., Formation of modified bases in cells exposed either to gamma radiation or high-LET particles, Radiat. Res., 2002, **157**, 589-595.

[12] Douki T., Ravanat J.-L., Pouget J.-P., Testart I., Cadet J., Minor contribution of direct ionization to DNA base damage induced by heavy ions, Int. J. Radiat. Biol., 2006, **82**, 119-127.

[13] Roots R., Holley W., Chatterjee A., Irizarry M., Kraft G., The formation of strand breaks in DNA after high-LET irradiations: a comparison of data from in vitro and cellular systems, Int. J. Radiat. Biol., 1990, **58**, 55-69.

[14] Nikjoo H., O'Neill P., Terrissol M., Goodhead D.T., Quantitative modelling of DNA damage using Monte Carlo track structure method, Radiat. Environ. Biophys., 1999, **38**, 31-38.

[15] Sutherland B.M., Bennet P.V., Sidorkina O., Laval J., Clustered damage and total lesions induced in DNA by ionizing radiation: oxidized bases and strand breaks, Biochemistry, 2000, **39**, 8026-8031.

[16] Blaisdell J.O., Wallace S.S., Abortive base-excision repair of radiation-induced clustered lesions in Escherichia coli, Proc. Natl. Acad. Sci. USA, 2001, **98**, 7426-7430.

[17] David-Cordonnier M.-H., Boiteux S., O'Neill P., Efficiency of excision of 8-oxo-guanine within DNA clustered damage by XRS5 nuclear extracts and purified human OGG1 protein, Biochemistry, 2001, **40**, 11811-11818.

Chapter 13

Mechanisms of direct radiation damage to DNA

Michael D. SEVILLA and William A. BERNHARD

Introduction

It has been estimated that approximately 50% of the DNA damage from γ-irradiation is due to direct-type effects, which encompass two types of events: those stemming from energy deposited in DNA itself and those starting with ionization of the DNA solvation shell and rapidly followed by transfer of radical cations and ejected electrons to DNA (quasi-direct effects) [1-3]. The 50% remaining damage is from water radicals via the indirect effect (Chapter 12).The large contribution of direct-type effects stems from the fact that the DNA of higher organisms is highly condensed in the form of chromatin, effectively excluding water near DNA and, thereby, reducing the importance of the indirect effect [1]. For these reasons a full understanding of the direct-type effects is required for a comprehensive model of the effects of radiation on DNA *in vivo* [3]. Calculations indicated that about 90% of the energy deposited in liquid water is *via* ionization; presumably, a similar fraction would apply in the case of DNA [4]. Ionization creates sites of electron-loss (a hole or radical cation) and sites of electron-gain (radical anion). These radical species account for the majority of direct-type damage observed in DNA. The other portion of the energy is deposited as excitations, which at least for low LET (linear energy transfer) radiation do not appear to give rise to significant amounts of chemical damage. Thus, one should be able to account for nearly all the stable direct-type damage produced in DNA by tracking the fate of the initially formed sites of electron-loss and electron-gain. Not all ejected electrons upon reaching low energies, <20 eV, terminate by one-electron reduction of a particular DNA site. Some low energy electrons (LEE) terminate by dissociated electron attachment (DEA) [5], (Chapter 5). It appears, however, that damage due to DEA for low LET radiation is relatively minor in comparison to the inelastic process that results in radicals directly formed by one-electron gain and one-electron loss.

Trapped radicals

Given in **Figures 1a** and **1b** are the structures and labels for free radicals and some stable end products which are most important to direct radiation damage to DNA and are discussed here. Because all of these radicals are inherently reactive, it is advantageous to irradiate DNA samples at low temperatures, 4 K or 77 K, where the radiation produced radicals are trapped and stable for long time periods and then perform measurements [6]. The relative yield of each specific primary electron-loss and electron-gain radical trapped in γ-irradiated DNA depends on the temperature and other variables such as DNA conformation, DNA hydration state, DNA packing, total dose and the LET of the radiation. **Table 1** shows the relative initial yields of the radicals trapped in DNA hydrated to 14 mol H_2O/mol nucleotide [7,8]). The samples were irradiated and measured at 77 K. The first row shows that for double stranded (ds) DNA exposed to γ-rays (low LET radiation) four types of radicals predominate. Two are due to electron loss, Gua(N1-H)$^{\bullet}$ and dRib$^{\bullet}$, and two are due to electron gain, Thy$^{\bullet-}$ and Cyt(N3+H)$^{\bullet}$. The formation reactions for these radicals are shown in **Figure 2.** These results suggest that, at 77 K, the neutral sugar-phosphate radicals originating from the deoxyribose-phosphate backbone contribute a larger fraction of damage at high LET than at low LET. Although not shown in the **Table 1**, $^{\bullet}$OH radical located in the outer hydration layer of DNA is also found in small quantities, 0-5%, depending on the degree of hydration [9]).

Table 1. *Approximate initial relative radical[a] yields in Irradiated double stranded DNA at 77 K* [7,8].

Radiation type	Initial LET (keV/µm)	Gua$^{\bullet+}$ ^{a}Gua(N1-H)$^{\bullet}$	Ade$^{\bullet+}$ ^{a}Ade(N-H)$^{\bullet}$	^{a}Cyt$^{\bullet-}$ Cyt(N3+H)$^{\bullet}$	Thy$^{\bullet-}$	^{b}dRib(Cn-H)$^{\bullet}$	RO(O)$_2$P$^{\bullet-}$
γ	≈ 0.3	35%	< 5%	29%	27%	ca. 10%	—
$^{36}Ar^{18+}$	≈ 400	22%	< 5%	24%	28%	ca. 27%	ca. 0.1%

a) *Note that Gua$^{\bullet+}$ in DNA undergoes reversible deprotonation from N1, Ade$^{\bullet+}$ rapidly deprotonates from its NH_2 group (pK_a=1) and Cyt$^{\bullet-}$ undergoes rapid protonation at N3.* b) *May include a small but unknown fraction of neutral base radicals.*

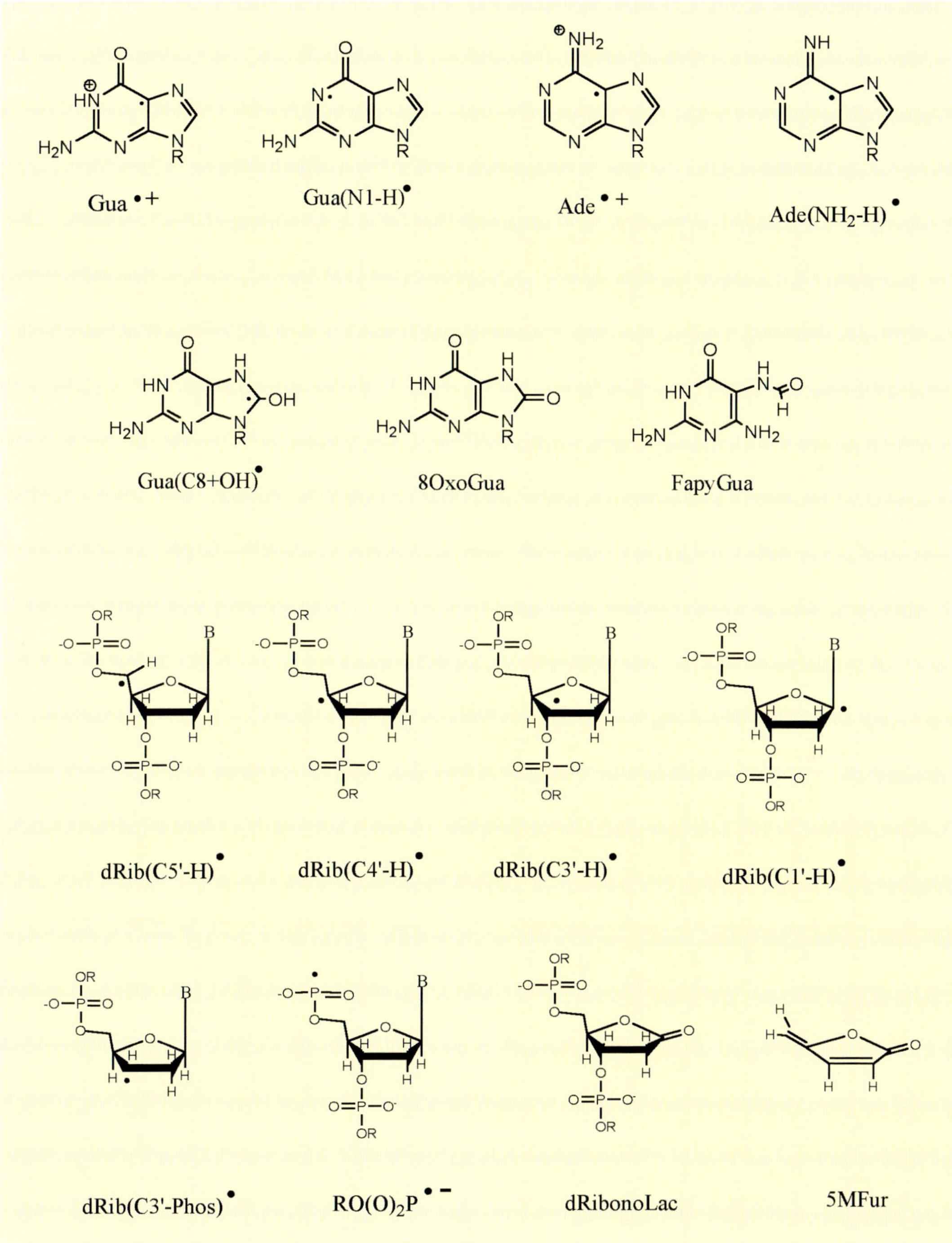

Figure 1a : *Nomenclature used in text for free radical structures and stable end products derived from oxidation and excitation pathways.*

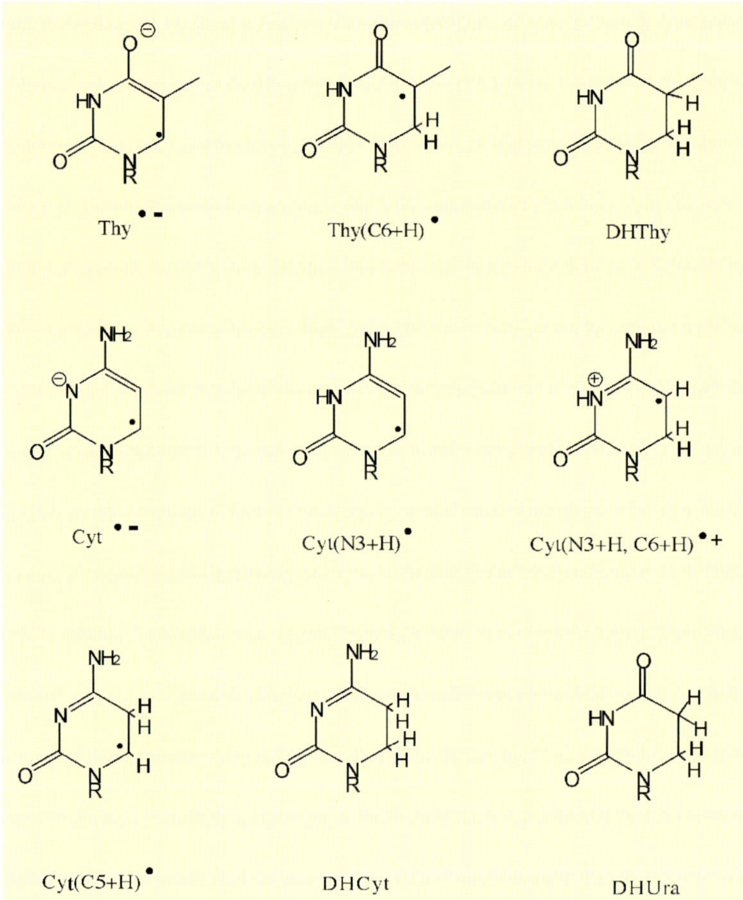

Figure 1b : *Nomenclature used in text for free radical structures and stable end products derived from oxidation and excitation pathways.*

Base radicals

Electrons are selectively captured by the pyrimidines and the holes formed on the base stack, or transferred to the base stack, are nearly all trapped by Gua, giving $Gua^{\cdot+}$. This species is in a reversible equilibrium with its deprotonated form $Gua(N1\text{-}H)^{\cdot}$ (**Fig. 2**). While initial distributions between the pyrimidines are nearly equal upon annealing, or irradiation to high dose, the excess electron transfers to cytosine as the most stable site as a result of protonation of $Cyt^{\cdot-}$ at N3 forming $Cyt(N3{+}H)^{\cdot}$ [10]. Molecular orbital calculations show that the major factor that affects the relative yields of DNA base ion radicals is the difference in both the ionization potential and electron affinity of the DNA bases [11]. In addition, experimental and theoretical work shows that variations in relative radical yields are also affected by the

DNA strandedness. For example, the transfer of both the hole and the excess electron is more facile through stacked bases in dsDNA than in ssDNA in which the base stacking is disrupted.

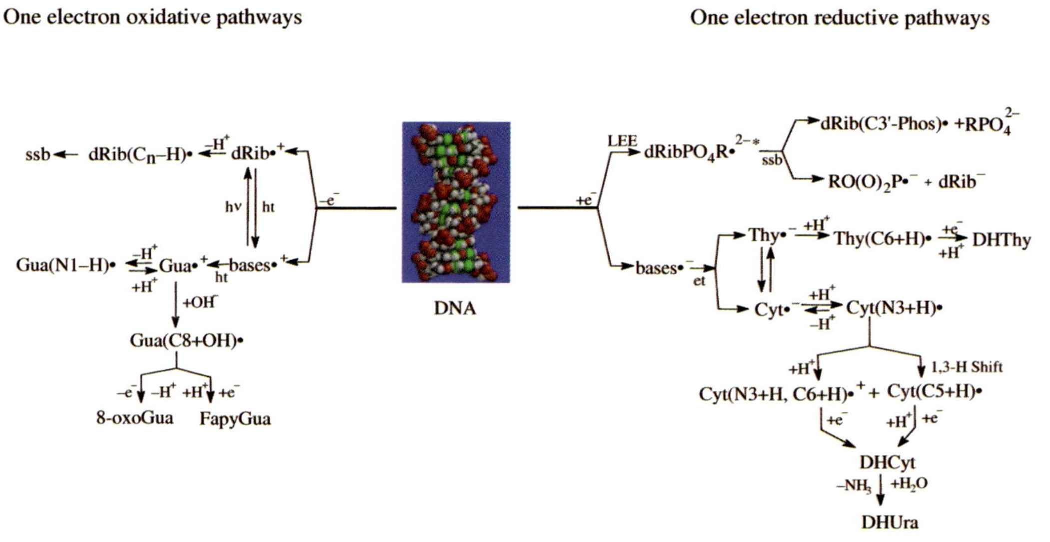

Figure 2 : *Important reaction pathways stemming from the ionization of DNA (with a model of DNA in the common B-form configuration). Electron loss pathways are on the left and electron gain pathways on the right. Notation: LEE (low energy electron), ssb (single strand break), ht (hole transfer), and et (electron transfer).*

While the deprotonation reactions from nitrogen sites described above are reversible, ion radicals of the DNA bases may also react irreversibly in characteristic ways to form neutral base radicals as shown in **Figure 2** [1,3]. DNA base cation radicals may deprotonate from carbon sites and/or undergo hydroxide ion addition. For the thymine radical cation, Thy$^{\bullet+}$ can form the Thy(Me-H)$^{\bullet}$ radical by loss of a proton from the methyl group before transfer of the hole to the more stable guanine site [1]. For the guanine radical cation, Gua$^{\bullet+}$ reaction with water yields the 8-hydroxyguanine radical, Gua(C8+OH)$^{\bullet}$ on annealing to temperatures over 200 K. This species undergoes subsequent oxidation to form 8-oxoguanine or reduction to form FapyGua (2,6-diamino-4-hydroxy-5-formamidopyrimidine) (**Fig. 2**) [12]. Pyrimidine anion radicals formed initially protonate irreversibly at the C5-C6 position. Thy$^{\bullet-}$ protonates at C-6 to form the well known 5,6-dihydrothymin-5-yl radical, Thy(C6+H)$^{\bullet}$. Cyt$^{\bullet-}$, and its reversibly protonated form Cyt(N3+H)$^{\bullet}$, protonate irreversibly at C5 or C6 to form Cyt(C5+H)$^{\bullet}$ and Cyt(N3+H+,C6+H)$^{\bullet+}$ (**Fig. 2**) [10]. Subsequent one-electron reduction of Thy(C6+H)$^{\bullet}$ followed by protonation leads to the stable end-product of dihydrothymine DHThy (**Fig. 2**). In a similar reaction sequence, DHCyt is formed. In water, deamination of DHCyt gives DHUra, a stable lesion found in DNA [3,10]. The fate of the radiation-induced hole is then chiefly 8-oxoGua and FapyGua, whereas, the excess electron results in the 5,6-dihydropyrimidines, DHThy and

DHUra. While these are the major products, if they are spatially isolated from other lesions, they are readily repaired enzymatically and consequently have no deleterious effect; but when formed in clusters, the likelihood of cellular injury increases dramatically [2].

In the past it has been often suggested that neutral DNA base radicals may abstract a hydrogen atom from the sugar phosphate backbone to form a strand break; however, theoretical work clearly indicates that abstraction from the DNA sugar backbone by DNA base neutral radicals is thermodynamically unfavorable in most cases [11]. It is likely that strand breaks are largely accounted for by direct ionizations of the sugars with additional strand fragmentations induced by low energy electrons (LEE) (discussed below).

DNA sugar radicals

There are long-standing questions regarding the identity and the yields of the sugar radicals in irradiated DNA [13,14]. The DNA sugar-phosphate backbone makes up about 50% of the electron density of the DNA and, therefore, about 50% of all holes and electrons are initially generated on the DNA backbone. Nearly all ejected electrons are eventually trapped on the pyrimidine bases (Cyt and Thy) (**Tab. 1**). However, of the holes generated in the DNA backbone, both base release studies [15] and ESR (electron spin resonance) evidence on sugar radicals [14] indicate that there is a partial transfer of radiation-induced holes to the bases. Without hole transfer from the sugar backbone to the DNA bases, about 25% of all radicals resulting from holes and electrons would then be expected on the sugar phosphate backbone. However, ESR work suggests about 10% of the total radicals are on the sugar phosphate backbone [14,16]. In agreement with this later value, product analysis also suggests that hole transfer to the bases reduces the sugar damage from the calculated maximum of 25% to *ca.* 13% of the total [15]. (Note that based on initial ionizations 50% of the holes correspond to 25% of all radicals.) There is a clear correlation between most sugar radicals and the subsequent formation of a strand break. However, studies on plasmid DNA indicate that the yield of trapped sugar radicals fall short of the yield of strand breaks by about 0.07 μmol/J [16]. This shortfall is yet to be explained and it has been suggested to involve double oxidations at a single site. One additional known contributor to strand break formation, which does not involve a radical cation precursor, is the mechanism of DEA by LEE [19]. The actual yield of strand breaks due to LEE is not known and is presumed to be relatively small.

Two of the sugar radicals stabilized at 77 K in γ-irradiated DNA have been tentatively identified; these are the dRib(C1-H)' and dRib(C3-H)' radicals [14]. These are most likely formed by deprotonation of sugar cation radicals. In recent work [8], with both gamma and ion beam irradiated DNA, evidence was reported for two radical species that result from fragmentation of the DNA sugar phosphate backbone. They are a C3′ dephosphorylation radical, dRib(C3′–Phos)$^•$, and a phosphoryl radical, $RO(O)_2P^{•-}$. Both these species are associated with immediate single strand breaks (SSB). Deprotonation of a deoxyribose-phosphate radical cation cannot explain these fragmentation reactions. Based on the well known radiation chemistry of alkyl phosphates, it is likely that dRib(C3′–Phos)$^•$, and $RO(O)_2P^{•-}$, are formed by dissociative electron attachment likely *via* LEE (**Fig. 2**).

Recent work has also found another mechanism for production of sugar radicals, *i.e.* deprotonation from electronic excited states of DNA base cation radicals [17]. This work suggests that combination of an excited state with a DNA base cation such as $Gua^{•+}$ results in hole transfer to the sugar followed by deprotonation to form a sugar radical. This mechanism is considered to be important in high LET radiation damage in which excitations are copious. Thus, three mechanisms for sugar radical formation in DNA are now known:

1. The major pathway is hole deprotonation of the directly ionized sugar phosphate backbone.

2. Secondary pathways likely to be more important in high LET radiation are:
   - dissociative electron attachment caused by LEE (low energy electrons) and
   - electronic excitation of existing DNA base cation radicals.

Three of the deoxyribose carbon centered radicals, dRib(C3′-H)$^•$, dRib(C4′-H)$^•$, and dRib(C5′-H)$^•$, give rise to prompt SSB via a phosphate elimination reaction. The strand break mechanism for the dRib(C1′-H)$^•$ radical differs substantially. Like the three radicals above, it will release a free unaltered base but instead of forming a strand break, a subsequent one-electron oxidation results in deoxyribonolactone (dRibonoLac) formation (**Fig. 1a**) [1,18]. This abasic site is quite stable. Heat and a catalyst, such as spermine, turn this lesion into a strand break plus 5-methylenefuranone (5MFur) [18]. The later is a signature end product of C1′ sugar damage.

Low Energy Electrons

As describe above DNA strand breaks can be induced by LEEs. Recent work has shown LEEs, with energy as low as 1 to 5 eV [19], are effective in this regard. The decay of localized transient anion states (resonances) within DNA is the *principal* mechanism leading to SSB and double strand breaks (DSB) by electrons with energies below 15 eV [19]. Theoretical calculations support the fragility of the DNA backbone to LEEs and suggest that even 1 eV electrons can cause SSB, perhaps *via* a *"resonant capture mechanism"* [20]. From these reports, it is clear that there may be a very low energy threshold for strand break induction by LEE which act directly on DNA. As mentioned above, ESR studies do find evidence for specific radical species (dRib(C3'–Phos)$^{\bullet}$ and $RO(O)_2P^{\bullet-}$) in low abundance that likely result from LEE induced strand cleavage.

Transfer of electrons and holes, induced by irradiation, through duplex DNA

As pointed out above ESR studies show that after low temperature γ-irradiation, electrons are trapped on cytosine and thymine and holes on guanine (**Tab. 1**). However, the distances that excess electrons and holes travel before being irreversibly trapped have been in active dispute in the literature until recently [21]. Earlier results were contradictory and gave electron migration distances from just a few DNA bases pairs up to hundreds of base pairs. The structure of DNA with overlapping π-orbitals of the stacked bases even gave rise to the suggestion that DNA is a "molecular wire." However, recent work [21] shows that DNA is in fact a poor semi-conductor, which is by definition an insulator. Since hole and electron migration within DNA and its hydration layer are processes important to the radiation chemistry of DNA, a careful and systematic investigation of these processes has been performed over the last several years in several laboratories. These investigations included consideration of rate of transfer through the stacked bases and the role of the hydration sheath, base sequence, temperature, inter duplex as well as intraduplex transfer; these efforts have helped to elucidate charge migration processes in DNA [21].

With regard to hole/electron transfer distances, these recent ESR studies clearly show that after the initial migration and trapping, electron and hole transfer distances at 77 K are limited to *ca.* 10 base pairs and that transfer occurs predominantly *via* tunneling [21]. However, at ambient temperatures, holes can actually range up to 100 base pairs *via* an activated hopping mechanism, from Gua to Gua. A point of controversy in the literature was the value of β (the tunneling decay constant), which governs the reduction in the rate of electron transfer, k, as the tunneling distance, D, increases [$k = k_o \exp(-\beta D)$]. Early work had suggested values of β as low as 0.1 Å^{-1}; these results are now known to be a result of activated processes, "hopping" (Chapter 14). ESR work has clearly shown that electron transfer, by tunneling, through DNA, occurs with a β of 0.7-0.9 Å^{-1} [21]). For inter-duplex transfer, a higher beta value (near 1.3 Å^{-1}) is suggested.

High LET irradiation studies

The effect of LET on a variety of radiation endpoints, including mutations, double strand breaks, single strand breaks, transformations, cell inactivation, cell death, chromosome aberrations and chromatin breaks have been extensively studied. However, for high LET radiation, an understanding of the chemical processes that lead to strand breaks after the direct deposition of energy in DNA (and its solvation shell) is in its infancy. Modern track structure theory commonly posits two distinct zones of energy deposition for high LET ions. In one, commonly referred to as a "core," glancing collisions generate continuous ionizations and excitations; knock-on collisions, with small impact parameters and large depositions of energy, also occur in the core. The second zone of energy deposition results from a cascade of high energy electrons (δ-rays) which are emitted at larger angles from the core; these result in a "halo" or "penumbra" consisting of widely spaced spurs (Chapter 4). The processes that occur in this region are thought to mimic those found in low LET radiation, in which radiation damage results largely from scattered high-energy electrons. The chemical processes that occur in the penumbra are expected to be similar to those of low LET irradiation such as gamma irradiation. However, the processes in the core are clearly of a different character and one expects different mechanisms. Indeed, ion beam studies show that sugar radical formation is greatly augmented at high LET (second row of **Tab. 1**). As mentioned above for DNA samples subject to ion beam irradiation at 77 K, two immediate strand break radicals, dRib(C3′–Phos)$^{\bullet}$ and $RO(O)_2P^{\bullet-}$, have been identified in surprisingly high yields compared to low LET radiation such as γ-radiation [8]. The high energy density along the track core of the ion results in ionizations and excitations in close proximity creating reaction profiles not found at low LETs. Thus, these radicals have been suggested to result from LEE electrons, perhaps by interacting with an already vibrationally excited sugar ring. Lower levels of $Gua^{\bullet+}$ found in high LET samples (see **Tab. 1**) with correspondingly higher levels of sugar radicals are in accord with the now well established mechanism that excitations of $G^{\bullet+}$ directly result in sugar radicals [17].

Acknowledgments

We appreciate the careful reading and helpful suggestions provided by David Becker. The authors thank the National Cancer Institute of the NIH for support, grant R01-CA32546 to WAB and grant R01-CA045424 to MDS.

References

[1] Becker D., Sevilla M.D., The chemical consequences of radiation damage to DNA, *in* "Advances in Radiation Biology", Lett J.T., Sinclair W.K. (eds), Academic Press, San Diego, 1993, p. 121-180.

[2] Ward J.F., The Complexity of DNA Damage: Relevance to Biological Consequences, Int. J. Radiat. Biol., 1994, **66**, 427-432.

[3] Bernhard W.A., Close D.M., DNA damage dictates the biological consequence of ionizing radiation: the chemical pathways, *in* "Charged Particle and Photon Interactions with Matter", Hatano Y., Mozumder A. (eds), Marcel Dekker, New York, 2004, p. 431-470.

[4] Uehara S., Nikjoo H., Goodhead D.T., Comparison and Assessment of Electron Cross Sections for Monte Carlo Track Structure Codes, Radiat. Res., 1999, **152**, 202-213.

[5] Huels M.A., Handorf I., Illenberger D., Sanche L., Resonant dissociation of DNA bases by subionziation electrons, J. Chem. Phys.,1998, **108**, 1309-1312.

[6] Sevilla M.D., Becker D., ESR Studies of Radiation Damage to DNA and Related Biomolecules, Royal Society of Chemistry Specialist Periodical Report: Electron Spin Resonance, 2004, **19**, 243-278. (See previous reviews referenced within.)

[7] Wang W., Yan M., Becker D., Sevilla M.D., The influence of hydration on the absolute yields of primary free radicals in gamma-irradiated DNA at 77 K. II. Individual radical yields, Radiat. Res.,1994, **137**, 2-10.

[8] Becker D., Razskazovskii Y., Callaghan C., Sevilla M.D., ESR of DNA Irradiated with a Heavy Ion Beam ($^{16}O^{8+}$): Evidence for Damage to the Deoxyribose Phosphate Backbone, Radiat. Res., 1996, **146**, 361-368. (See references within.)

[9] ***a.*** La Vere T., Becker D., Sevilla M.D., Yields of OH• in Gamma-Irradiated DNA as a function of DNA Hydration: Hole Transfer in Competition with OH• Formation, Radiat. Res., 1996, **145**, 673-680.

b. Debije M., Strickler M., Bernhard W.A., On the efficiency of hole and electron transfer from the hydration layer to DNA: An EPR study of crystalline DNA X-irradiated at 4 K, Radiat. Res., 2000, **154**, 163-170.

[10] Debije M.G., Bernhard W.A., Thermally stable sites for electron capture in directly ionized DNA: free radicals produced by the net gain of hydrogen a C5/C6 of cytosine and thymine in crystalline oligodeoxynucleotides. J. Phys. Chem. A, 2002, **106**, 4608-4615.

[11] Colson A-O., Sevilla M., Application of molecular orbital theory to the elucidation of radical processes induced by radiation damage to DNA, *in* "Theoretical and Computational Chemistry: Computational Molecular Biology", Lesczycyski J. (ed), 1999, **8**, 245-277.

[12] Shukla L.I., Adhikarya A., Pazdro R., Becker D., Sevilla M.D., Formation of 8-oxo-7,8-dihydroguanine-radicals in gamma-irradiated DNA by multiple one-electron oxidations, Nuc. Acid Res., 2004, **32**, 6565-6574.

[13] Close D.M., Where are the sugar radicals in irradiated DNA?, Radiat. Res., 1997, **147**, 663-673.

[14] Shukla L.I., Pazdro R., Becker D., Sevilla M.D., Sugar Radicals in DNA: Isolation of neutral radicals in Gamma Irradiated DNA via hole and electron scavenging, Radiat. Res., 2005, **163**, 59-602.

[15] Razskazovskiy Y., Debije M.G., Bernhard W.A., Direct Radiation Damage to Crystalline DNA: What is the Source of Unaltered Base Release?, Radiat. Res., 2000, **153**, 436-441.

[16] Purkayastha S., Milligan J.R., Bernhard W.A., Correlation of Free Radical Yields with Strand Break

Yields Produced in Plasmid DNA by the Direct Effect of Ionizing Radiation, J. Phys. Chem. B, 2005, **109**, 16967-16973.

[17] Adhikary A., Malkhasian A.Y. S., Collins S., Koppen J., Becker D., Sevilla M.D., UVA-Visible photo-excitation of guanine cation radicals produces sugar radicals in DNA and model structures, Nuc. Acid Res., 2005, **33**, 5553-5564.

[18] Roginskaya M., Bernhard W.A., Marion R.T., Razskazovskiy Y., The Release of 5-Methylene-2-Furanone from Irradiated DNA Catalyzed by Cationic Polyamines and Divalent Metal Cations, Radiat. Res., 2005, **163**, 79-84.

[19] Huels M.A., Boudaiffa B., Cloutier P., Hunting D., Sanche L., Single, Double, and Multiple Double Strand Breaks Induced in DNA by 3-100 eV Electrons, J. Am. Chem. Soc., 2003, **125**, 4467-4477.

[20] Barrios R., Skurski P., Simons J., Mechanism for Damage to DNA by Low-Energy Electrons, J. Phys. Chem. B, 2002, **106**, 7991-7994.

[21] Cai Z., Sevilla M.D., Studies of excess electron and hole transfer in DNA at low temperatures, *in* "Topics in Current Chemistry: Long Range Transfer in DNA, Vol II", Shuster G. (ed), Springer-Verlag, New York, 2004, **237**, 103-128. Also see other Chapters in the volume, especially those by Douki T., Cadets J., and Berlin Y.A. and Siebbeles L.D.A.

Chapter 14

Charge motion in DNA

Yuri A. BERLIN and Laurens D. A. SIEBBELES

Introduction

Double stranded deoxyribonucleic acid (DNA) represents a macromolecular assembly consisting of two intertwined helices with an aromatic π–stack core, where the bases of the pyrimidine deoxynucleotides (thymine, T; cytosine, C) and purine deoxynucleotides (adenine, A; guanine, G) participate in Watson-Crick base pairing (A:T; C:G) (Chapters 12 and 13). This unique structure discovered more than 50 years ago defines such biological properties of DNA as its capability for coding, storage and propagation of genetic information. Therefore it is not surprising that for a long time, DNA has almost exclusively been the province of biologists and life scientists.

During the past fifteen years, however, the situation has gradually changed mainly due to the development of new methods for direct manipulation of single DNA molecules and for their labeling with redoxactive probes. Using these experimental techniques, researchers were able to expand information about molecular characteristics of DNA to include its mechanical, electronic, and transport properties [1]. These properties and new experimental methodology for their studies have attracted attention of scientists working in different fields that are often far apart from biology and genetics. In particular, the ability of DNA to serve as a medium for long-range charge transfer has stimulated interest in the possibility to exploit this molecule in nanoscale electronics, molecular computing, and in electrochemical biosensoric devices. The same property is shown to be important for developing new ways to detect structural changes due to protein binding and base mismatches. For these potential applications, the elucidation of mechanisms responsible for charge transport phenomena in DNA turns out to be crucial.

This challenging problem is also vital for current research on oxidative damage of DNA, which may cause apoptosis, mutations, and cancer.

A key structural element, which determines transport properties of DNA, is the array of π-stacked base pairs. The striking resemblance of the base pair stack to conductive one-dimensional aromatic crystals prompted the proposal that the interior of the double helix can provide a one-dimensional pathway for charge migration due to formation of a π-band across different stacked bases [2]. The validity of this early mechanistic picture was tested in experiments on the photo-induced oxidation of DNA, dated back to the early 1990s (for review see *e.g.* [3]). In typical experiments designed to study the transfer of an electronic hole (*i.e.* a positive charge corresponding to an ionized nucleobase), a donor and an acceptor of these charge carriers are intercalated in the stack of native base pairs or chemically attached to the phosphate-sugar backbone. For properly chosen donor and acceptor species, such chemical modifications enable one to generate holes under irradiation of the sample by light due to the removal of electron from nucleobases to the photoexcited donor. The subsequent hole transfer from the donor to the acceptor bridged by the base pair sequence can be probed by measuring, for instance, the quenching of the fluorescence of the donor for sequences of different lengths or the damage yield at certain sites along the sequence. Later, a similar approach was also utilized in experiments on the photo-induced reduction of DNA aimed to probe transport of excess electrons along the stack of base pairs [4]. In the latter case negative charge carriers were generated using the photoexcited donor to inject electrons into DNA or to transfer them directly to the acceptor. Some additional information about motion of electrons and holes in the interior of the double helix has become available from the series of works on low-temperature γ-radiolysis of crystalline DNA, its ice and glassy aqueous solutions [5].

The experimental studies mentioned above have shown that the ordered π-electron system of the natural DNA bases in duplex B-form DNA (here simply referred to as DNA) indeed provides an appropriate pathway for the motion of excess positive and negative charges once generated on extended and chemically well-defined stacks of base pairs. However the observed dependence of the charge transfer efficiency on the base pair sequence [6] suggests that the actual behavior of generated electrons and holes is more complicated than the band-like picture of charge motion proposed at the initial stage of investigations [2].

Structural disorder in DNA and disordered energy landscape

Qualitatively the plausible scenario of the DNA-mediated charge transfer can be inferred from the consideration of the relevant energy landscape. The latter shows how the energy of a charge carrier changes as a hole or an excess electron is consecutively deposited on each nucleobase involved in the formation of the π–pathway between a donor and an acceptor. To construct the simplest landscape possible, one should take into account the energetics of individual bases and the structural disorder arising from the choice of A, T, G, or C at each substitution base site along the backbone of the DNA helix. These two factors lead to the static energy disorder that determines the compound mechanism of the entire process of charge transfer as well as mechanisms governing its elementary steps.

In the case of holes the static energy disorder arises due to the differences between individual nucleobases in the values of the oxidation potential and the ionization energy. To be more specific, G is known to be the most easily oxidized nucleobase since its *in vitro* reduction potential is about 0.4 eV lower than that of A, and significantly lower than the reduction potentials of C and T. The same conclusion follows from the results obtained for energies of a positive charge located at different native nucleobases using various computational methods of quantum chemistry [7-9]. Therefore the energy of a hole when residing on A, C, or T sites is higher than on G. Hence the G base is a more probable place for the positive charge to be localized than three others. This hierarchy of hole energies (G < A < C,T) also holds when stacking interactions between neighboring nucleobases are taken into account [9], although their ionization potentials become smaller in comparison with the values found for individual bases. In particular, ionization potentials of GG doublets and GGG triplets formed by stacking two or three adjacent Gs on the same strand are shown to be less than the ionization potential of a single G base. As a consequence, the energy of holes on GG and GGG sites is lower than the energy of G^+ by several tenths of eV.

Thus, due to different energetics of nucleobases, structural disorder in the stack of A:T and G:C pairs gives rise to static disorder in energies of holes residing on individual bases. Furthermore, a close examination of the resulting energy landscape reveals three main groups of states. The first group consists of states with the lowest energy. They serve as deep hole traps arising when several adjacent Gs are stacked on the same strand and form a multiple GG…G unit (**Fig. 1**). Hole states associated with individual G bases belong to the second group, which is intermediate in energy between deep traps and holes residing on A, T, and C bases. For this reason, these states will be defined henceforth as "intermediate". Three other native nucleobases A, T, and C are responsible for the formation of the third group comprising of hole states with the highest energies. Since A, T, and C can be considered as a building block of the bridge connecting two neighboring G bases, all states belonging to the third group will be referred to as "bridging".

Three groups of states discussed above are separated by two energy gaps. At room temperature the width of both gaps exceeds a typical thermal energy E_{th} given by the product of the Boltzmann constant k_B and temperature T. Therefore, in the simplest case considered here the groups do not overlap and can be considered as isolated.

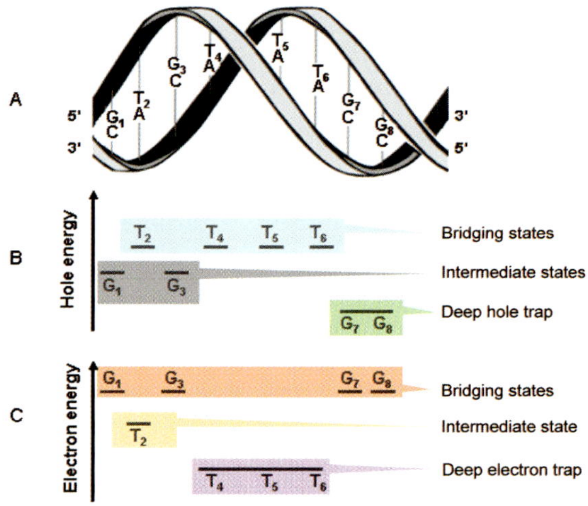

Figure 1 : *A fragment of the DNA double helical structure* (A), *and energy landscapes for hole* (B), *and excess electron motion* (C) *along the stack of base pairs.*

The main elements of the energy landscape for positive charge transfer mentioned above are shown in **Figure 1B**, using the fragment of the DNA duplex schematically depicted in **Figure 1A** as an illustration. In this particular case, the doublet G_7G_8 is an example of the deep hole trap, sites G_1 and G_3 exemplify intermediate states, while sites T_2, T_4, T_5, and T_6 correspond to "bridging" states.

Certainly the energy landscape of **Figure 1B** is oversimplified. In particular, energies of holes on each of single G's are assumed to be equal. Moreover, the hole energies for T_2, T_4, T_5 and T_6 bridge units are also supposed to be identical for simplicity. Meanwhile, quantum mechanical calculations [7-9] suggest that flanking bases can affect the hole energies on A, T, G, and C bases due to stacking interactions, thus further increasing the degree of static energy disorder within the DNA π-stack.

A similar structure of the energy landscape can be expected for the process of excess electron transfer on the basis of the redox potential data available in the literature, see *e.g.* [10]. However, now the traps have to be pyrimidine deoxynucleobases stacked on the same strand (*e.g.* TT dimers or TTT triplets), the intermediate states should be associated with the anions T^- and/or C^-, while bridging states will correspond to G and/or A bases (see **Fig. 1C**). This conclusion directly follows from the hierarchy of the measured reduction potentials, which decrease in the order $C \approx T >> A > G$ [10]. Electron affinities calculated both for individual nucleobases and for their different trimers exhibit the same trend [11].

Mechanisms of charge transfer: the entire process and elementary steps

The energy landscapes discussed suggest two distinct mechanisms of charge transfer between the donor D and the acceptor Ac located at the opposite ends of the nucleobase sequence [12,13].

If bridging (and for certain systems also intermediate) states in the landscape are much higher in energy compared with D and Ac, a charge will be transferred via superexchange mediated tunneling. Two characteristic features of this coherent quantum mechanism should be mentioned in the context of electronic properties of DNA. First, the two-center superexchange charge transfer occurs in a single step and hence does not involve genuine chemical intermediates. Second, the rate of the whole process, k_{CT}, rapidly decreases with the donor-acceptor distance R following the familiar exponential law

$$k_{CT} = k_0 \exp(-\beta R), \qquad (1)$$

where k_0 is the pre-exponential factor and β is the falloff parameter. In the case of charge transfer in DNA the β values are theoretically expected to be of the order of 1 Å^{-1}.

Another situation arises if intermediate states are comparable in energy with D. Now a charge can be injected from the D site to the proximal base (G in the case of holes and T or C in the case of electrons) with the subsequent temporal localization in the corresponding intermediate state of the energy landscape. Thereafter, a hole or an electron is able either to return back to D or to undergo a transition to the adjacent unoccupied intermediate state through the intervening nucleobases associated with the bridging states (*e.g.* A or T for holes and G or C for electrons). The latter transition represents the first step in a series of consecutive incoherent hopping transitions that allows charge carriers to move along the stack of base pairs using intermediate states as stepping stones. As a consequence, a charge is able to reach a remote Ac site separated from D by several hundred angströms, where hopping transport is terminated by trapping.

Thus, unlike the coherent single-step superexchange, incoherent hopping in DNA involves several steps, *i.e.* injection of charge carrier, their transport along π-pathway due to successive transitions between nucleobases with appropriate energetics, and trapping. In addition, there are several other distinctions between hopping and superexchange mechanisms. In particular, the former mechanism implies the formation of reactive chemical intermediates (*e.g.* $G^{\cdot+}$ and pyrimidines anions for hole and electron transfer, respectively), which propagate from the site of their generation to the distant site of the reaction. By contrast, the single-step superexchange does not include any intermediate active species. The distance dependencies of the charge transfer rate for two mechanisms turn out to be also distinct. For unbiased hopping on a long one-dimensional regular lattice with (N+1) sites separated by the distance a, this dependence can be approximated by the algebraic function [14]

$$k_{CT} \infty 1/N \approx 1/(aR) \qquad (2)$$

rather than by the exponential law (1) typical for superexchange.

The above-mentioned features of the superexchange and hopping mechanisms were documented in a number of charge transfer processes observed in DNA. As has been demonstrated in a number of experiments, the single-step superexchange mechanism dominates in DNA oligomers with short base pair sequences (R < 20 Å). Representative examples of such systems and superexchange-driven reactions are given in **Table 1**. By contrast, experiments with longer nucleobase sequences (R > 20 Å) reveal that the multi-step hopping mechanism prevails. According to the current consensus, the latter mechanism governs a number of processes in different duplexes listed in **Table 2**.

Table 1. *Examples of DNA oligomers with short base pair sequences and superexchange driven elementary processes observed in these systems* [a)].

System	Elementary Process	Reaction Scheme
Stilbene (S) capped DNA hairpins	Hole generation on a single G base	$^1S^*$-(AT bridge)-G → S^--(AT bridge)-G^+
	Charge recombination	S^--(AT bridge)-G^+ → S-(AT bridge)-G
Acridine (Acr) modified DNA	Photo-induced hole generation on a single G base	Acr*-(AT bridge)-G → Acr^--(AT bridge)-G^+
Hairpins and double stranded short oligonucleotides	Hole trapping by GGG triplets	G^+-(short AT bridge)-(GGG) → G-(AT short bridge)-$(GGG)^+$
Stilbenediether (Sd) capped DNA hairpins	Electron injection	$^1Sd^*$-(GG bridge)-T → Sd^+-(GG bridge)-T^-
Pyrene(Py)-modified duplexes containing uracil bases (U)	Photo-induced generation of negative charges	… – U – C – … (U attached to Py*) → … – $U^{\bullet-}$ – C – … (U attached to $Py^{\bullet+}$) → … – U – $C^{\bullet-}$ – … (U attached to $Py^{\bullet+}$)

a) Asterisk denotes the excited state of the corresponding structural unit.

Table 2. *Examples of DNA oligomers with long base pair sequences and processes govern by multi-step charge hopping in these systems [a].*

System	Process	Reaction Scheme
4′-modified oligonucleotides with the GGG unit	Hole migration from the site of generation to the GGG trap via intrastrand and interstand $G^+ \to G$ ("zigzagging") transitions	G^+ T C A G C T C A G T C T G C A (G G G) C A G T C G A G T C A G A C G T C C C → G T C A G C T C A G T C T G C A $(G\ G\ G)^+$ C A G T C G A G T C A G A C G T C C C
	Migration of a hole selectively generated at the G site to the triple G trap along one strand of duplex with TT bridges of equal length	G^+ TT G TT G … TT (GGG)→ G TT G TT G … TT $(GGG)^+$
Anthraquinone (AQ)-linked duplex DNA oligomers	Photo-induced propagation of radical cations	AQ*-T(GG)T(GG)T(GG)T(GG)T(GG)T(GG)TATA ↓ AQ^--$T(GG)^+$T(GG)T(GG)T(GG)T(GG)T(GG)TATA ↓ … ↓ AQ^--T(GG)T(GG)T(GG)T(GG)T(GG)$T(GG)^+$TATA
Stilbenedicarboxamide (Sa) capped DNA hairpins	Photo-induced charge separation	^{1}Sa*–AAGAGA(GGG)→Sa^-–AAG^+AGA(GGG)→… …→Sa^-–$AAGAGA(GGG)^+$
DNA oligomers with internally conjugated aromatic amine donor (X) and 5-bromo-2′-deoxyuridine acceptor (Y)	Photo-induced excess electron transfer	X*–(A:T and G:C pairs)–Y → X^+–(A:T and G:C pairs)–Y^-
DNA hairpins with flavin-capped donor (F) and the TT-dimer acceptor	Excess electron-transfer-based repair of cis-sin TT dimer	F*–(A:T and/or G:C pairs)–TT → F^+–(A:T and/or G:C pairs)–TT^-

a) Asterisk denotes the excited state of the corresponding structural unit.

As has been emphasized in earlier publications [15-17], there is no dichotomy between the two mechanisms of charge transfer in DNA considered above. On the contrary, each can contribute to the mechanistic picture of the entire process. The superexchange mediated tunneling controls the rate of short-range (< 20 Å) elementary hops of a charge carrier between neighboring nucleobases that produce intermediate states in the corresponding energy landscape, while multi-step hopping is responsible for the long-range migration of charge along the stack of nucleobases.

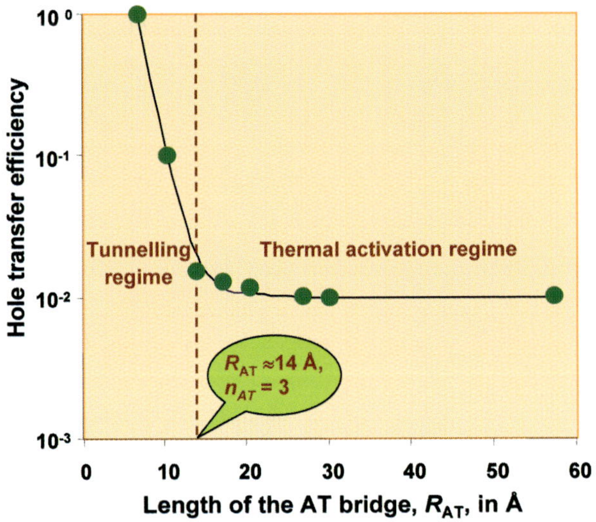

Figure 2 : *Efficiency of hole transfer from site-selectively generated $G^{\cdot+}$ to the GGG triplet across A:T bridges of various length R_{AT}. Points correspond to the experimental data of Giese et al. [18]. The length dependence calculated for the same system [16] is shown by the solid line. The intersection of the dotted line with the horizontal axis gives the length of A:T bridge and the number of A:T pairs, at which the rates of quantum tunneling and classical thermally-induced transitions become equal.*

The superexchange mediated tunneling, however, is not the only plausible mechanism of short-range steps of hopping motion in DNA: at finite temperatures this mechanism is in competition with classical thermally-induced transitions of charge carriers between two neighboring "resting" sites (G for holes, C and/or T for excess electrons). As can be seen from equation (1) and **Figure 2**, the tunneling rate exponentially decreases with the distance separating these two states, while the rate of the thermally-induced transition W_{th} is mainly determined by a thermal population of the bridge. Therefore W_{th} does not vary with distance, but depends on the energy gap between intermediate and bridging states, E_{ib}, in accordance with the Arrhenius law

$$W_{th} = W_0 \exp[-E_{ib}/(k_B T)], \qquad (3)$$

where W_0 is the pre-exponential factor. Due to the distinction in distance dependencies, a changeover from the superexchange mediated tunneling to the thermally activated regime of the elementary hopping step can be expected as the distance between neighboring intermediate states becomes equal to a certain critical value [16]. In the case of hole transfer, the tight-binding model for the elementary step of hopping motion [16] suggests that a positive charge can be transferred between two neighboring G sites *via* superexchange only if the A:T bridge connecting these two sites consists of less than 3-4 base pairs. Since the mean plane-to-plane distance between base pairs in B-DNA is known to be 3.4 Å, this corresponds to the situation where the AT bridge has the length $R_{AT} \leq 14$ Å. Otherwise (*i.e.* for $R_{AT} > 14$ Å), the elementary hopping step includes thermal activation of holes into the tight-binding band followed by their ballistic or hopping motion along the A:T bridge. For holes, the latter process is known in the literature as A-hopping. Recent experiments (for review, see [18]) provide strong evidence for such thermally-induced transitions through long bridges with the number of A:T pairs $n_{AT} = 4\text{-}10$ ($R_{AT} \approx 17\text{-}37$ Å) and $n_{AT} = 4\text{-}16$ ($R_{AT} \approx 17\text{-}58$ Å). Moreover, measurements [18] of the hole transfer efficiency as a function of R_{AT} for the process

$$\begin{matrix} G^+ \\ C \end{matrix} \begin{pmatrix} T \\ A \end{pmatrix}_{n_{AT}} \begin{matrix} (G & G & G) \\ C & C & C \end{matrix} \rightarrow \begin{matrix} G \\ C \end{matrix} \begin{pmatrix} T \\ A \end{pmatrix}_{n_{AT}} \begin{matrix} (G & G & G)^+ \\ C & C & C \end{matrix} \quad (n_{AT} = 1\text{-}16) \qquad (3)$$

and the analogous theoretical dependence [16] were found to be in good agreement (see **Fig. 2**), thus supporting theoretical predictions concerning two competing mechanisms of elementary hopping step.

Thus, a plausible scenario for the entire process of charge transfer from D to Ac along the sequence of nucleobases involves variable-range hopping between intermediate states corresponding to the bases with the appropriate oxidation or reduction potentials. Short steps made by a moving charge in this multi-step transport process occur due to the coherent superexchange mediated tunneling. By contrast, long steps require thermal activation of charge carriers needed to overcome the energy gap between intermediate and bridging states. Once this thermally-induced transition has completed, electrons or holes can reach the next "resting" site undergoing ballistic motion or hopping along the pathway provided by bridging states.

As follows from the detailed kinetic analysis of this scenario [13,15], the model of variable-range hopping allows quite accurate predictions of both sequence and distance dependencies for the efficiency of charge transfer through stacks with various combinations of base pairs. This, in turn, provides reasonable estimations of the distance scale for the propagation of charge in DNA duplexes. Based on these estimations verified by steady-state experiments (see *e.g.* [19] and references therein) one can conclude that typically the upper limit for the distance traveled by charges in DNA is about 200-300 Å. Charge transfer over such large distances can be accomplished because of the weak distance dependence of the reaction rate (*cf.* equation (2)) offered by variable range hopping. In addition, a coexistence of quantum and classical steps of hopping process is also favorable to the long-range migration of electrons and holes in the interior of double helix.

How much time a charge needs to move through DNA

Although steady-state experiments have yielded a wealth of knowledge about the mechanism of charge transfer in DNA, the results obtained with this type of measurements are insufficient to decide how fast a charge can be transferred over a certain distance. Absolute values of the charge transfer rate can be obtained from time-resolved pump-probe laser experiments, such as those performed by Lewis and coworkers [20]. These experiments were carried out on DNA sequences containing a stilbenedicarboxamide (Sa) electron acceptor. **Figure 3** shows a few examples of the DNA sequences studied.

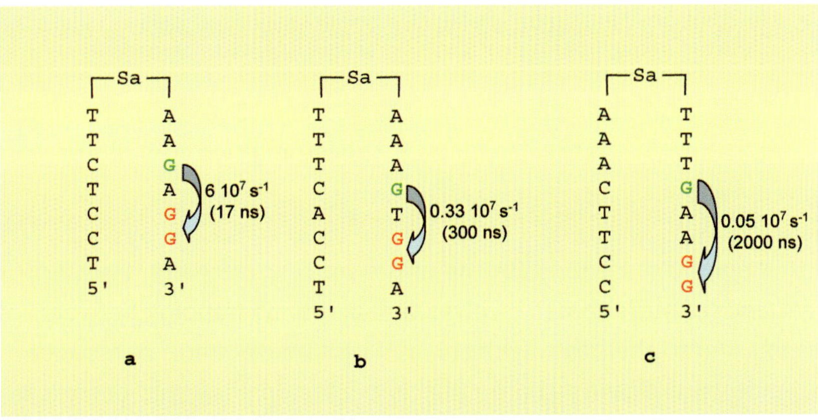

Figure 3 : *Examples of DNA sequences containing a stilbenedicarboxamide electron acceptor Sa. The rates of charge transfer between the primary G (green) and the distal GG doublet (red) strongly depend on the intervening base sequence as explained in the text. For all sequences, hole transitions between G and GG are shown by arrows. Numbers next to the arrows are the values of rate constants for the hole transfer between G and GG. Corresponding average times for this process are given in parentheses.*

In the experiments, the Sa molecular unit is photoexcited using a laser pulse. Upon photoexcitation Sa accepts an electron from DNA. This initially leads to the formation of the anion radical $Sa^{\bullet-}$ and a hole on the G site nearest to Sa (colored green in **Fig. 3**). The hole generated on this proximal G is able either to recombine with $Sa^{\bullet-}$ or to jump to a distal GG doublet (colored red in **Fig. 3**). The latter process leads to the formation of $(GG)^{\bullet+}$, but does not terminate the motion of positive charges, since in the case under consideration a hole can be back transferred to the primary G site. The decay kinetics of the transient of the measured transient optical absorption of the anion radical $Sa^{\bullet-}$ depends on the rate of recombination of $Sa^{\bullet-}$ with a hole on the primary G, and on the rates for reversible charge transfer between the primary G and the distal GG doublet. The absolute values of these rates could be obtained from a theoretical analysis of the measured decay kinetics of the transient optical absorption due to $Sa^{\bullet-}$ anions. The values obtained for the rate constants for hole transfer from the primary G to the distal GG are $6 \times 10^7\ s^{-1}$, $0.33 \times 10^7\ s^{-1}$ and $0.05 \times 10^7\ s^{-1}$ for sequences **a**, **b** and **c**, respectively. The inverse values of these rates correspond to the average charge transfer times, which are equal to 17 ns, 300 ns and 2000 ns for sequences **a**, **b** and **c**, respectively. These very different values show that the process of charge transfer is strongly sequence dependent. Comparison of the rate constants for hole transfer in sequences **a** and **b** demonstrates that hole transfer between the proximal G site and the distal GG doublet via an intervening adenine (A) is almost twenty times faster than via thymine (T). The rate constant for hole transfer decreases more than six fold upon introduction of an additional A:T base pair (sequences **b** and **c**).

In order to provide theoretical insight into the factors governing the rate of charge transfer, quantum chemical calculations have been carried out, which are described in detail in [8]. As discussed below, a qualitative explanation of the relative rates in the different DNA sequences can be obtained from a Hückel (tight-binding) type model. For a quantitative description of the charge transfer rates, it is necessary to include the effects of the charge-induced structural reorganization of DNA, and the aqueous environment used in the experiments.

In the Hückel type calculations the hole is described quantum mechanically by a wave function, which is a time-dependent linear combination of orbitals on the individual nucleobases. Charge transfer is affected by the structural conformation of the nucleobases in the DNA sequence. In particular, twisting of base pairs around the axis of the DNA helix strongly affects the rate of charge transfer. Therefore effects of twisting were included in the calculations. Initially the wave function is localized on the proximal G site. The time evolution of the wave function is then obtained by solving numerically the Schrödinger equation. In the calculations the charge is forced to decay at the distal GG doublet. The rate of charge transfer from the proximal G site to the distal GG doublet can be obtained from the probability for the hole to survive trapping at the GG doublet.

Figure 4 shows the calculated decay of population of charge on the initial G site, due to charge transfer from this site to the distal GG doublet, for the three different

DNA sequences in **Figure 3**. Hole transfer is seen to be fastest for DNA sequence **a** and becomes slower on going to sequences **b** and **c**. This trends in the kinetics of hole transfer is consistent with the results deduced from the experiments discussed above. The calculated energy of a hole localized on the proximal G site in sequence **a** is 0.3 eV lower than it would be if it were localized at the A nucleobase between the proximal G and distal GG doublet. This energetic difference is much larger than thermal energy at room temperature ($k_BT \sim 0.025$ eV). Hence, the hole migrates from the G site to the GG doublet by quantum mechanical tunneling through the barrier provided by the intervening A nucleobase. The height of the barrier due to the T nucleobase intervening the G site and the GG doublet in sequence b is equal to 0.8 eV. The much higher energetic barrier for tunneling *via* a T nucleobase as compared with A causes the charge transfer rate for sequence **b** to be much smaller than for sequence **a**. The reduction of the charge transfer rate on going from a barrier consisting of a single A nucleobase in sequence **a** to the barrier of two adjacent adenines in sequence **c** is a consequence of the increased length of the barrier in the latter case.

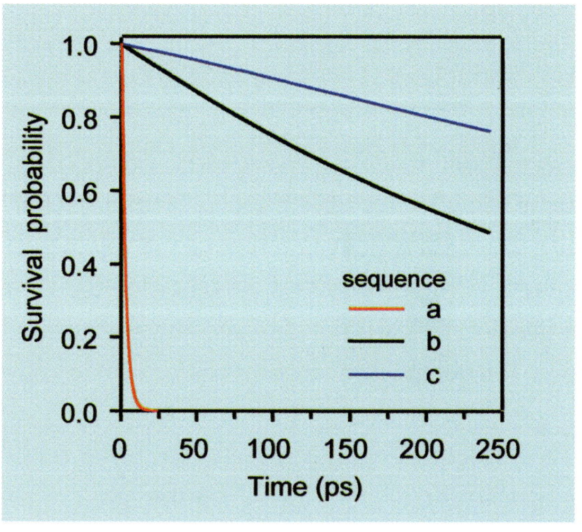

Figure 4 *: Probability for the hole to survive trapping by the GG doublet in the DNA sequences* **a**, **b**, *and* **c** *shown in* Figure 3.

The Hückel type calculations offer a qualitative explanation of the trend observed for the experimental rates of hole transfer between the proximal G and the distal GG doublet. However, as can be seen in **Figure 4**, the calculated timescales for charge transfer range from a few picoseconds for sequence **a** to several hundreds of picoseconds for sequence **c**. These times are about three orders of magnitude smaller than the average charge transfer times obtained from the experiments. The difference between the calculated and experimental timescales for charge transfer is due to charge-induced structural deformations, which are not included in the Hückel type calculations. An excess charge in DNA induces an internal structural reorganization of the nucleobases, which reflects the fact that the equilibrium structure of a cation differs from that of a neutral molecule. An additional effect is due to the fact that the experiments involve DNA in aqueous solution. The charge in DNA causes the surrounding dipolar water molecules to reorganize and to adopt an energetically favorable orientation around the charge. The effects of structural reorganization were taken into account by using a semiclassical superexchange model based on the Marcus theory for electron transfer. The absolute values of the experimental rates for charge transfer in the sequences in **Figure 3** could be reproduced by invoking a reorganization energy with a magnitude about 1 eV, which is close to literature data [21,22].

The information on the kinetics and mechanism of charge transfer obtained from the combination of the experimental and theoretical studies discussed above can be used to obtain insight into the mobility of holes in DNA. This information is important in order to understand the conducting properties of DNA. The results can be utilized to predict oxidative cleavage patterns in long sequences of DNA with several multiple guanine-containing sites. The results of experimental studies on the conductance of DNA are still highly controversial, and a large variety of possible electronic behavior has been suggested, ranging from DNA as an insulator to a superconductor (see the review in [19] and references therein). The experimental studies usually focus on current-voltage dependence measurements. The charge carrier mobility (*i.e.* the velocity of the charge per unit strength of an external electric field) cannot be deduced from these measurements since the charge carrier density is unknown. To the knowledge of the authors, the only study, from which a value of the mobility for hole hopping along stacks of A:T base pairs can be deduced is the work of Takada *et al.* [23], who determined the rate for A-hopping between adjacent nucleobase pairs to be $2 \times 10^{10}\ s^{-1}$. The mobility of holes undergoing hopping motion between A bases is then calculated to be $9 \times 10^{-4}\ cm^2\,V^{-1}s^{-1}$.

The mobility of a charge moving along a stack of either G:C or A:T base pairs with the Gs or As on the same strand of the double helix is of particular interest, since for these stacks charge motion can occur by hopping between identical and adjacent bases. In this case, the mobility will be much higher than for a DNA consisting of both G:C and A:T base pairs, since in the latter case charge transfer occurs by relatively slow tunneling steps between G nucleobases that are separated by A:T base pairs. Using the results from the quantum chemical calculations discussed above the mobility of a hole along a stack of either G:C or A:T base pairs can be calculated using Marcus type (polaronic) hopping theory for charge transfer. For a reorganization energy equal to 1 eV the mobility of holes is calculated to be $10^{-4}\ cm^2\,V^{-1}s^{-1}$ and $2 \times 10^{-5}\ cm^2\,V^{-1}s^{-1}$ for stacks of G:C and A:T base pairs, respectively. A slightly different value of the reorganization energy equal to 0.63 eV must be used to reproduce the mobility of 9 10^{-4} $cm^2 V^{-1} s^{-1}$ for holes on a stack of A:T base pairs, as deduced from the experiments by Takada *et al.* [23] More accurate calculations of the mobility require knowledge of the reorganization energy for the specific sequences studied in experiments on the conductance of DNA. Nevertheless, the estimated mobility values given above are more than six orders of magnitude smaller than the relative high mobility values of charges in conventional semi-conductors such as crystalline silicon. However, the estimated mobility values for stacks of G:C or A:T base pairs are comparable to that found for structurally disordered polymers in opto-electronic devices. Although the mobility of charges in DNA is strongly limited by the relatively large reorganization energy, DNA may find application in future electronic devices based on organic molecules.

Epilogue

Quantum mechanical calculations of the energy landscape, together with theoretical modeling of charge dynamics, offer the detailed mechanistic picture and quantitative kinetic information on charge transfer in DNA over distances as long as 50-300 Å. Biological implications of the obtained results pertain to the generation of DNA damage induced by radiation and by chemical reactions. In both cases, specific radical reactions with nucleobases occur, which finally lead to the formation of oxidized guanine followed by cation (hole) migration. For such situations, the present theoretical approach based on the multi-step hopping kinetic scheme with the individual hopping rates being inferred from quantum mechanical calculation is particularly relevant.

Due to the existence of the fundamental relation between conducting and charge transfer properties of the same nanostructured systems [14], the same kinetic-quantum approach helps to explain electrical conductance through DNA placed in the nanogap between two electrodes, and is therefore relevant to DNA nanoelectronics. Similar to other molecular wires, the current through the DNA molecule connecting two electrodes is determined by two processes. These include the injection of charge carriers onto the stack of base pairs and their transport along the stack. Our theoretical studies focused on the transport aspect of the problem provide information on the key characteristics of charge carriers in DNA, namely their drift mobility, and enable us to establish main factors affecting the value of this important quantity. We show that the drift mobility of holes and, hence, the DNA electrical conductance strongly depend on the reorganization of nucleobases and the water surroundings, on static and dynamic disorders in the stack of base pairs as well as on the base pair sequence. Certainly, the application of DNA in molecular electronic devices must take into consideration these and some other factors that promote or inhibit DNA-mediated charge transport.

The sequence-dependence of DNA conductance predicted by theory and verified experimentally is probably the most important outcome of our investigations, thus making them relevant for a number of electrochemical DNA-based technologies. The latter ones address primarily such fields as biosensors in which DNA is used as a conducting spacer, molecular recognition and sequencing methods that rely on interfacing of DNA to macroscopic electrodes. For all these applications, the kinetic-quantum approach serves as a useful and quite general theoretical background. More importantly, this approach allows us to demonstrate the high sensitivity of DNA conductance to mistakes in base pairing (the so-called mismatches). This lays the groundwork for new highly reliable, fast and accurate method for the detection of mismatches in a single DNA molecule that relies on the measurement of molecular electric conductance and does not require such time-consuming procedures as labeling and target amplification.

Acknowledgements

The Netherlands Organization for Scientific Research (NWO) is acknowledged for financial support. Yuri Berlin is grateful to the Chemistry Division of the ONR, the NASA URETI Program, to DoD/MURI and DURINT Programs for their support of the research. We also wish to thank many colleagues, particularly A.L. Burin, E.M. Conwell, and J. Jortner for useful discussions.

References

[1] See collection of articles published in special section "Double helix" of Nature, 2003, **421**, 396-453.

[2] Eley D.D., Spivey D.I., Semiconductivity of Organic Substances. Part. 9 – Nucleic Acid in the Dry State, Trans. Faraday Soc., 1962, **58**, 411-415.

[3] Grinstaff M.W., How do charges travel through DNA? – An update on a current debate, Angew. Chem. Int. Ed., 1999, **38**, 3629-3635.

[4] Wagenknecht H.-A., Reductive electron transfer and transport of excess electrons in DNA, Angew. Chem. Int. Ed., 2003, **42**, 2454-2460.

[5] Cai Z., Li X., Sevilla M.D., Excess electron transfer in DNA: Effect of base sequence and proton transfer, J. Phys. Chem. B, 2002, **106**, 2755-2762.

[6] Meggers E., Michel-Beyerle M.E., Giese B., Sequence dependent long range hole transport in DNA, J. Am. Chem. Soc., 1998, **120**, 12950-12955.

[7] Senthilkumar K., Grozema F.C., Guerra C.F., Bickelhaupt F.M., Siebbeles L.D.A., Mapping the sites for selective oxidation of guanines in DNA, J. Am. Chem. Soc., 2003, **125**, 13658-13659.

[8] Senthilkumar K., Grozema F.C., Guerra C.F., Bickelhaupt F.M., Lewis F.D., Berlin Y.A., Ratner M.A., Siebbeles L.D.A., Absolute rates of hole transfer in DNA, J. Am. Chem. Soc., 2005, **127**, 14894-14903.

[9] Voityuk A.A., Jortner J., Bixon M., Rösch N., Energetics of hole transfer in DNA, Chem. Phys. Lett., 2000, **324**, 430-434.

[10] Steenken S., Telo J.P., Novais H.M., Candeias L.P., One-electron-reduction potentials of pyrimidine bases, nucleosides, and nucleotides in aqueous solutions – consequences for DNA redox chemistry, J. Am. Chem. Soc., 1992, **114**, 4701-4709.

[11] Voityuk A.A., Michel-Beyerle M.E., Rösch N., Energetics of excess electron transfer in DNA, Chem. Phys. Lett., 2001, **342**, 231-238.

[12] Bixon M., Giese B., Wessely S., Langenbacher T., Michel-Beyerle M.E., Jortner J., Long-range charge hopping in DNA, Proc. Natl. Acad. Sci. USA, 1999, **96**, 11713-11716.

[13] Berlin Y.A., Burin A.L., Ratner M.A., On the long-range transfer in DNA, J. Phys. Chem. A, 2000, **104**, 443-445.

[14] Berlin Y.A., Ratner M.A., Intra-Molecular electron transfer and electric conductance via sequential hopping: Unified theoretical description., Radiat. Phys. Chem., 2005, **74**, 124-131.

[15] Berlin Y.A., Burin A.L., Ratner M.A., Charge hopping in DNA, J. Am. Chem. Soc., 2001, **123**, 260-268.

[16] Berlin Y.A., Burin A.L., Ratner M.A., Elementary steps for charge transport in DNA: Thermal activation vs. tunneling, Chem. Phys., 2002, **275**, 61-74.

[17] Berlin Y.A., Beratan D., Kurnikov I.V., Ratner M.A., Burin A.L, DNA electron transfer processes: Some theoretical notions, Top Curr. Chem., 2004, **237**, 1-36.

[18] Giese B., Long-distance electron transfer through DNA, Annu. Rev. Biochem., 2002, **71**, 51-70.

[19] Treadway C.R., Hill M.G., Barton J.K., Charge transport through a molecular π-stack: double helical DNA, Chem. Phys., 2002, **281**, 409-428.

[20] Lewis F.D., Letsinger R.L., Wasielewski M.R., Dynamics of photoinduced charge transfer and hole transport in synthetic DNA hairpins, Acc. Chem. Res., 2001, **34**, 159-170.

[21] Siriwong K., Voityuk A.A., Newton M.D., Rösch N., Estimate of the reorganization energy for charge transfer in DNA, J. Phys. Chem. B, 2003, **107**, 2595-2601.

[22] LeBard D.N., Lilichenko M., Matyushov D.V., Berlin Y.A., Ratner M.A., Solvent reorganization energy of charge transfer in DNA hairpins, J. Phys. Chem. B, 2003, **107**, 14509-14520.

[23] Takada T., Kawai K., Cai X., Sugimoto A., Fujitsuka M., Majima T., Charge separation in DNA *via* consecutive adenine hopping, J. Am. Chem. Soc., 2004, **126**, 1125-1129.

Chapter 15

Genome maintenance mechanisms in response to radiation-induced DNA damage

Evelyne SAGE and Bertrand CASTAING

Introduction

Maintenance of DNA integrity (the molecular support of genome) is essential to minimize heritable mutations and to promote healthy survival of organisms. Indeed, loss of genome integrity has long been implicated in genetic disorders, ageing and cancer. The genome is under constant threat arising from environmental agents (such as solar ultraviolet radiation, ionizing radiation and chemicals), from endogenous metabolic byproducts (such as reactive oxygen species) or from replication errors and arrests, that can alter its physical and chemical structure and corrupt its encoded message. A plethora of damages arises in DNA and their consequences are diverse and generally adverse. Many lesions block transcription, thus inactivating genes. They also interfere with DNA replication, triggering cell division and cell death. When left unrepaired or repaired inaccurately, DNA lesions result in irreversible mutations or genetic rearrangements, contributing to carcinogenesis as long-term effects in higher organisms. So, DNA repair systems appeared early in life (they exist in Archae bacteria) and have been maintained during evolution. In addition, stability of DNA is so important that cells evolved a whole strategy in response to DNA damage, as summarized in **Figure 1** [1]. Damage on DNA constitutes a signal which is transduced by a cascade of events (protein phosphorylations) to protein effectors that mediate cell cycle arrests in order to give time to the cells for repairing their DNA. Indeed, the activation of the thus-called DNA integrity checkpoints prevents cell cycle progression when DNA is damaged, and limits the fixation of mutations and their transmission to the offspring. Inability to properly face a genotoxic threat leads to genomic instability, and eventually to tumoral transformation. In multicellular organisms, a cell which is too heavily damaged induces its own death by apoptosis, thus ensuring a protection for the organism.

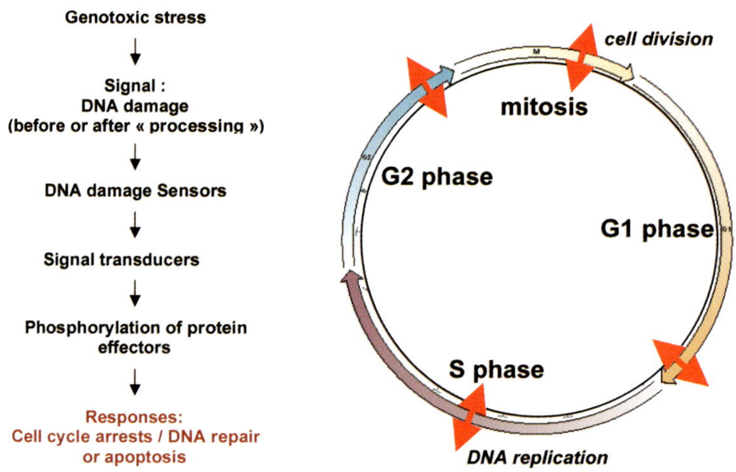

Figure 1 : *Cellular response to DNA damage. Sensor proteins are able to detect the presence of DNA damage and alert the cellular machinery so that the cell can face the stress and protects itself. The sensor then activates a transducer and through a series of proteins activation by phosphorylations, DNA integrity checkpoints (red arrow), in charge of the cell cycle control are activated and play their role which is to stop the cell cycle progression in order to allow repair to fully operate. Indeed, DNA replication should not occur on damaged DNA because of possible mistakes when copying modified bases, so the G1/S interphase checkpoint prevents the cell to enter into S-phase and to start replication. The G2 checkpoint verifies DNA for the presence of double strand breaks and incomplete replication, so that an altered DNA copy is not transmitted to the daughter cell.*

A wide diversity of lesions arises in DNA. For instance, over hundred oxidative modifications produced by reactive oxygen species (ROS) have been identified. Also, ionizing radiations produce over hundred of different DNA modifications (Chapters 12, 13, 14). Meanwhile, DNA lesions have been classified in two main groups, base modifications and strand breaks. Base modifications comprise addition of or substitution by bulky residues forming bulky adducts (such as cyclobutane pyrimidine dimers produced by ultraviolet radiation, or benzo(a)pyren adducts), as well as base alkylation, or base oxidation and reduction produced by ROS or ionizing radiation. Strand breaks comprise single and double strand breaks, typically produced by ionizing radiation and radiomimetics. Most of sugar modifications generate single strand breaks. Inter- and intra-strand crosslinks implicate covalent bounds between the crosslinking agent and DNA bases; they enter into the category of base damage. DNA-proteins crosslinks induced, for example, by ionizing radiation are yet poorly understood. Notably, one agent does produce neither a unique DNA lesion, nor a unique class of lesions. Typically, ionizing radiation induces predominantly base damage, *i.e.* oxidized and reduced bases and abasic sites, single and double strand breaks, as well as clustered lesions (**Fig. 2**). Clustered lesions, also called Multiply Damaged Sites, result from heterogeneous energy deposition by ionizing radiation in DNA. They consist of two or more closely spaced oxidized bases, abasic sites and single strand breaks distributed on both strands within 1-2 helical turns (Chapter 12) [2-3]. They are thought to largely contribute to the deleterious effect of ionizing radiation, since they may be difficult to repair [4-5].

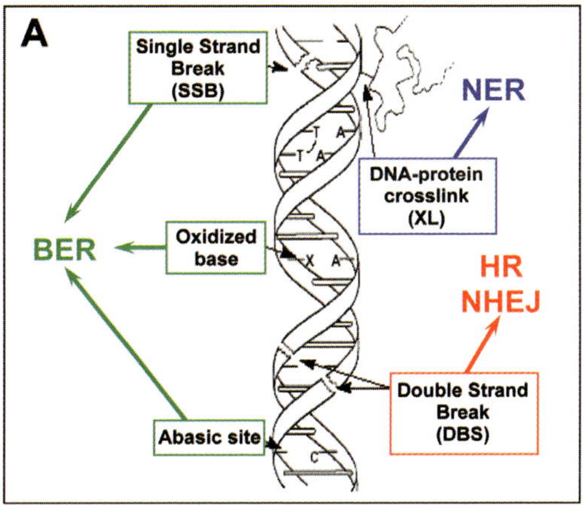

B

	radioinduced per Gy per cell	endogenous per Gy per cell
Single Strand Breaks	1000	10000-55000
Oxidized bases	2000	3200
Abasic sites	250	12600
Double Strand Breaks	40	8
DNA-protein XL	150	nd
Clustered lesions	150	nd

Figure 2 : ***DNA damage induced by ionizing radiation.*** A) *DNA damage and repair. All the constitutive elements of DNA (sugar-phosphate backbone and bases) are possibly modified by ionizing radiation. Single strand breaks (SSB), oxidized bases and abasic site are processed by base excision repair (BER), double strand breaks (DSB) by homologous recombination and non homologous end joining (HR and NHEJ) and DNA-protein crosslinks by nucleotide excision repair (NER).* B) *Quantitative measurement of radiation-induced and spontaneous DNA damage.*

DNA repair processes have to face quite a diversity of chemical and physical alterations of DNA. No single repair process can cope with all kinds of damage, but several DNA repair systems have evolved, adapted to the class of lesions to be eliminated. Excision repair takes care of base damage (**Fig. 2**); it is divided into several sub-pathways, including nucleotide excision repair (NER) and base excision repair (BER), depending on the type of base modification. Double strand breaks are mainly eliminated by two repair modes, homologous recombination (HR) and non-homologous end-joining (NHEJ). We will further focus on the description of pathways involved in the processing of radiation-induced DNA damage.

Excision repair pathways for removing base damage

Most base modifications in the cellular genome are cleared by base excision repair (BER) or nucleotide excision repair (NER) pathways [6-7]. The first step in these processes is the recognition of the damage by proteins which have to detect one base damage among 10^4 to 10^6 normal base pairs. Then, the excision step allows the removal of a short single stranded DNA fragment that contains the damage. Then DNA repair synthesis by a DNA polymerase and its accessory proteins which fills the gap is carried out, and then, ligation of the repaired strand which restores DNA continuity (**Fig. 3**). Those are error-free repair processes, meaning that an original and intact DNA, on the chemical, structural and functional point of view, is restored.

The type of base modification will determine which protein will come first and which system, BER or NER, will be used to cure the damaged DNA. NER is responsible for the removal of bulky adducts such as dimerized pyrimidine bases produced by UV radiation from sunlight, or benzo(a)pyren or psoralen adducts and others (**Tab. 1**). Bulky adducts are

Table 1. *Radiation-induced DNA damage recognition proteins in human cells.*

Repair protein	Function/Substrates*
Base excision repair	
- Uracil DNA glycosylase Ung	- glycosylase for U
- Uracil DNA glycosylase Smug1	- glycosylase for U, 5-hmU, 5-fU
- 8-oxoguanine glycosylase Ogg1	- glycosylase for 8-oxoG and FapyG opposite C
- Myh	- glycosylase for A opposite 8-oxoG
- Nth1	- glycosylase for thymine glycol, FapyG, 5-OHU, 5- OHC
- Neil1	- glycosylase for 5-OHU, 5-OHC, DHT, DHU, thymine glycol, 8-ox oG, FapyG, FapyA
- Neil2	- glycosylase for 5-OHU, 5-OHC
- Ape1/Hap1	- hydrolytic AP endonuclease for abasic sites
- PARP1	- Poly(ADP-ribose)polymerase recognizes single strand breaks
Nucleotide excision repair	
- XPC-HR23B	- damage induced by UV, benzo(a)pyren, cisplatin, psoralen, acetylaminofluoren
- XPE	- damage induced by UV, cisplatin, MNNG, abasic sites
Double strand break repair	
- Ku70/80	- double strand break ends

**U, uracil; 5-hmU, 5-hydroxymethyluracil; 5-fU, 5-formyluracil; 5-OHU, 5-hydroxyuracile; 5-OHC, 5-hydroxycytosine; DHT, dihydrothymine; DHU, dihydrouracil; 8-oxoG, 8-oxoguanine; FapyG, FapyA, formamidopyrimidine from guanine or adenine; AP or abasic site, apurinic and apyrimidinic site; MNNG, N-methy, N'-nitro, N-nitrosoguanidine.*

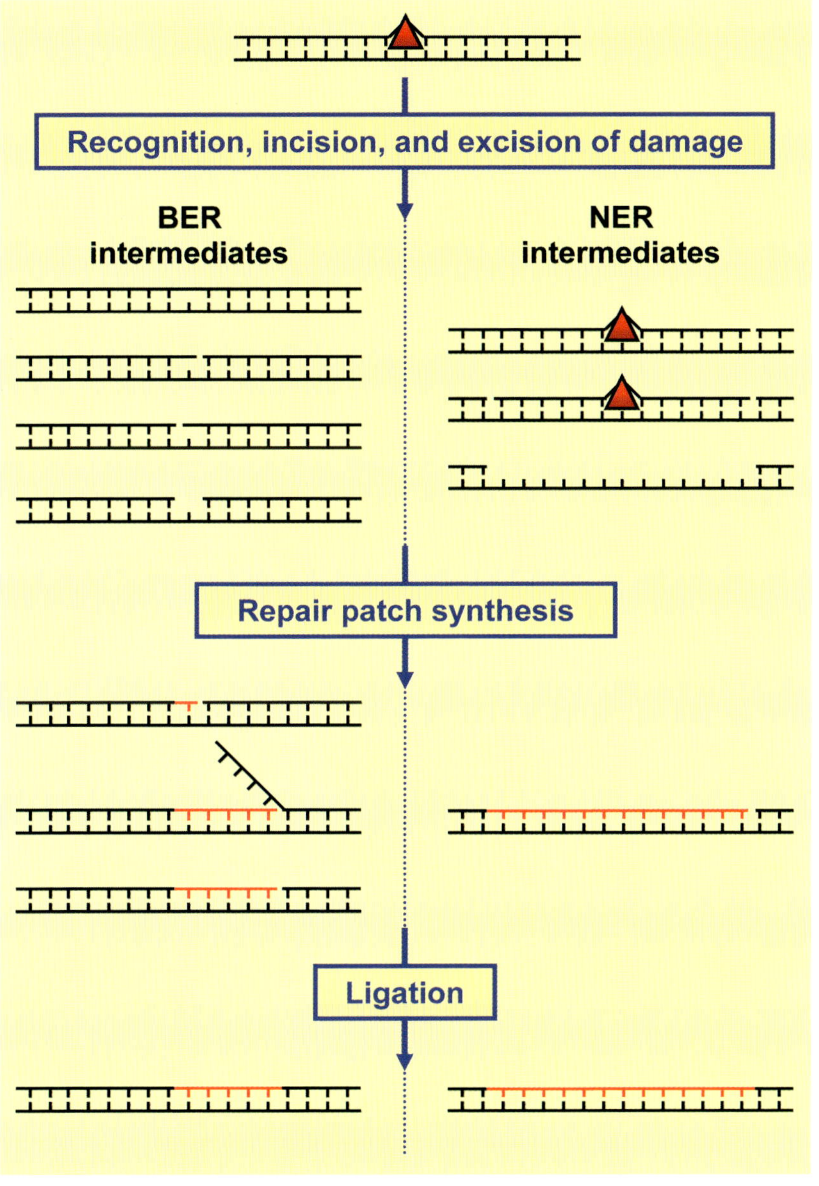

Figure 3 : ***Excision repair pathways for removing base damage.*** *The first step in base excision repair (BER) and nucleotide excision repair (NER) is the recognition of the damage (red triangle) by two protein complexes (NER) or a damage specific N-glycosylase (BER). The second step is the incision and excision of damage. In NER, the incision of the DNA phosphodiester backbone on both sides of the bulky adduct is followed by the removal of a short single stranded DNA fragment, 29-32 nucleotides long, that contains the damage. In BER, the N-glycosylase excises the damaged base, the DNA phosphodiester backbone is then incised and the ends are processed, leaving a gap of 1-7 nucleotides. The gap has to be filled by DNA polymerases in conjunction with accessory proteins: repair patch synthesis. The last step performed by a ligase, restores the DNA continuity. Different proteins are involved in the two last steps, repair replication and ligation, regarding BER and NER.*

recognized by two protein complexes, which sense the distortion of the double helix induced by the adduct, rather than the adduct itself. In contrast, relatively minor modifications to the bases of DNA are removed by BER. They are alkylated bases, oxidized and reduced bases, abasic sites, which do not drastically disturb the DNA double helix structure (**Tab. 1**). Interestingly, NER pathway may serve as back-up when BER is not available.

In regard with NER, at least 32 proteins, forming several complexes, are necessary and sufficient in a successful repair of a bulky lesion. In the case of BER, 5-6 proteins are necessary to reconstitute the entire process on a single base damage [7]. For both systems, the action of the proteins or complexes is coordinated and protein interactions allow passing from one step to the next one. Thus, free repair intermediates which could be entry sites for nucleases that would degrade DNA, are avoided.

Base excision repair

Ionizing radiation induces the formation of base damage, *i.e.* oxidized and reduced bases and abasic sites (Chapter 12), which are repaired by BER. Single strand breaks (SSB) which are produced by ionizing radiation but also as repair intermediates at the early steps of BER (**Fig. 3**), are processed by BER.

As seen from **Table 1** and in contrast to the DNA damage proteins involved in NER which recognize a broad variety of substrates, in BER the damage recognition proteins have a rather narrow substrate specificity. In consequence, a defined number of different proteins exist to deal with the numerous small modifications to bases. Those proteins are DNA glycosylases, which selectively bind with a high affinity to an altered base and hydrolyse the N-glycosylic bond to release the altered base from the sugar-phosphate backbone. The resultant abasic deoxyribose (apurinic, and apyrimidinic (AP) site) is processed further by an AP lyase or an AP endonuclease, which cleaves the phosphodiester backbone. The single base gap, single strand nick or flap generated as a repair intermediate, has to carry a 3'OH terminus in order to allow a DNA polymerase to fill the patch (1 up to 7 nucleotides long). The DNA polymerase adds nucleotides to this extremity using the undamaged strand as template and a ligase reseals the DNA molecule.

SSB induced by ionizing radiation typically possess "damaged" termini, in other words they lack the conventional 5'-phosphate and 3'-hydroxyl moieties that could easily be religated by a ligase. Instead, the 3' termini possess monophosphate or phosphoglycolate end groups, and 5' termini may carry a hydroxyl group. The unligatable (dirty) 3' termini are processed by a second activity of the AP endonuclease APE1. The polynucleotide kinase

PNK phosphorylates the 5′-OH ends. The specialized BER polymerase and the ligase can then proceed. Anyhow, an early step in the repair of SSB appears to be the rapid binding by poly(ADP-ribose) polymerase-1 (PARP1), a nick sensor that is activated at SSB. PARP1 synthesizes negatively charged polyADP-ribose that may serve as scaffold for the binding of the repair proteins involved later in the process [8].

BER has been poorly investigated in radiobiology, although it is a major repair process for ionizing radiation-induced DNA damage. It is also the main guardian against DNA damage due to cellular metabolism that results in ROS, methylation, deamination, hydroxylation.

Double strand breaks repair by homologous recombination and non-homologous end joining

For simplicity, we will describe double strand breaks (DSB) repair in mammalian cells [7]. When a DSB occurs, if the ends are ligatable, it will quickly be repaired at low cost by non-homologous end joining pathway (**Fig. 4**). The first step is the recognition of the DSB by the proteins Ku70/80 which bind at the ends and thus protect DNA from degradation by nucleases. Next, the cell has to maintain the two ends of the broken molecule in a physical proximity to avoid the loss of genetic information. This is realized by protein-protein interactions, DNA-PKcs is recruited and interacts with the complex Ku70/80. Then, ligation may occur by the complex XRCC4/Ligase4. Most of DSB created by ionizing radiation have unligatable (dirty) ends. The cell must then "clean up" the end to generate 3'-hydroxyl groups and 5'-phosphate groups. The repair process is more complicated, requires other proteins and it will take more time. So, after recognition of the DSB by Ku70/80, a signal is transduced to the cell-cycle apparatus to delay the progression through the cell cycle and up-regulate the transcription of the DNA repair genes (see Introduction). The complex rad50/Mre11/Nbs1 is recruited as well as Artemis protein which cleans and resects the ends. The core end joining proteins proceed to end joining. Unfortunately, this process may be associated with gain or loss of a few nucleotides and this is why non-homologous end joining pathway is considered as an error-prone repair process. Non-homologous end joining is the major DSB repair pathway used by mammalian cells, and the only one operating when cells are in G1 phase.

If a DSB is produced when DNA is replicating (S-phase) or in G2 phase when cells carry two gene copies on a chromosome (**Fig. 1**), it will be repaired by homologous recombination. This repair mechanism involves a strand exchange between the damaged DNA duplex and the same sequence from the intact homologous double stranded DNA molecule. Firstly, the complex rad50/Mre11/Nbs1 is recruited and resects the ends. The 3′ single stranded end

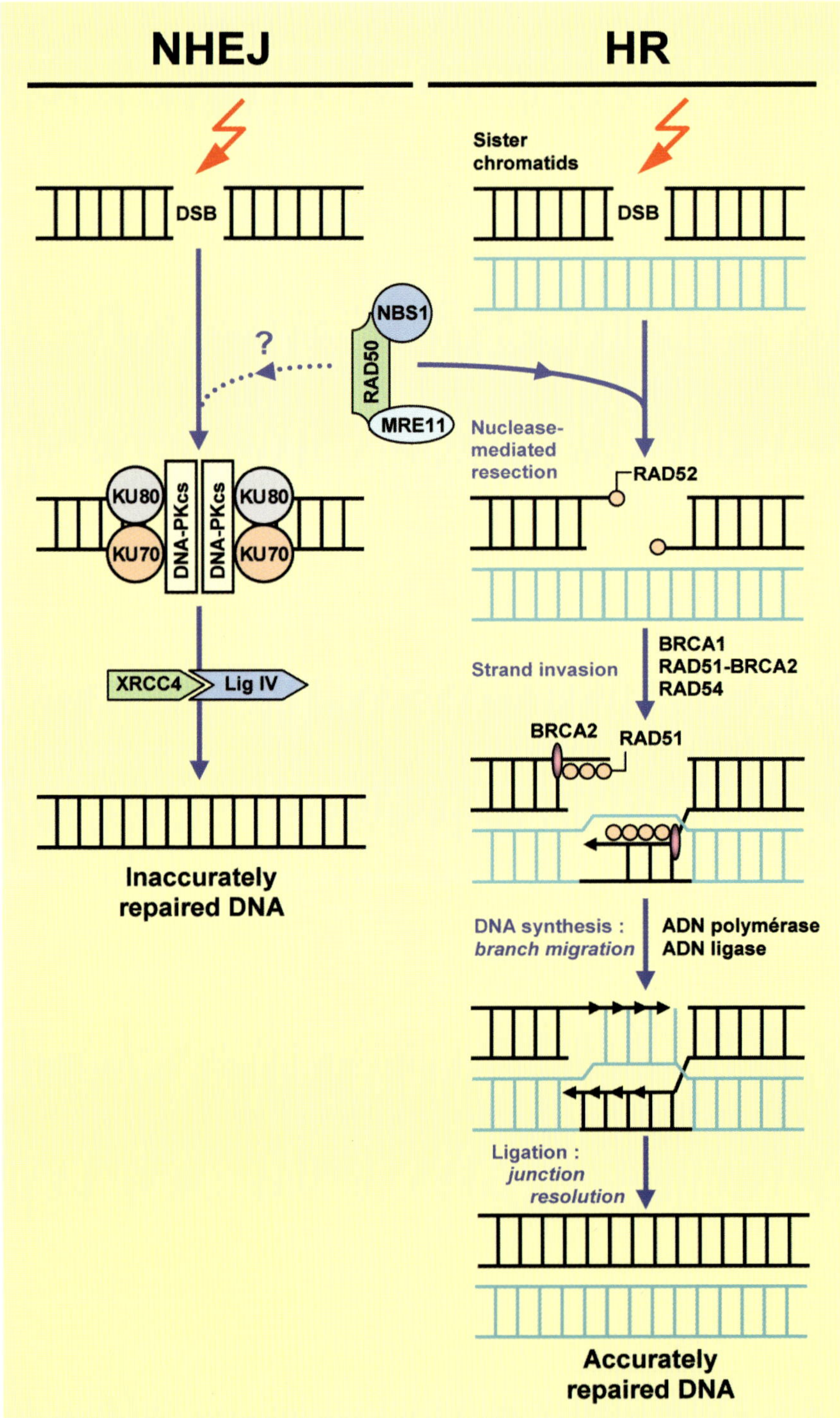

Figure 4 : *Double strand breaks repair by non-homologous end joining (NHEJ) and homologous recombination (HR).*

thus created are then coated with rad51 protein forming a filament with the help of other proteins. A correct positioning of the duplexes allow invasion of the homologous sequence by the coated single stranded end. The intact double stranded copy is used as template to properly heal the broken ends by DNA synthesis. The cross-junctions are resolved and ligation occurs to restore DNA integrity. The whole transaction is orchestrated by a number of known proteins (**Fig. 4**) leading to an accurately repaired DNA.

Mutagenesis

In the absence of repair, a transient DNA damage event may be transformed into irreversible, heritable mutational event. Left unrepaired, base damage may lead to mutation during DNA replication when the replicative DNA polymerase inserts the incorrect base opposite the lesion. For base damage which blocks the replication machinery, specialized DNA polymerases evolved to bypass the blocking lesion and performed the so-called translesion synthesis (TLS), thus overcome the replication blockage, allowing cell to pursue its cycle [7]. Unfortunately, TLS DNA polymerases are often error-prone and insert the wrong nucleotide opposite the lesion and the price to pay for a cell to survive is a mutation. In most cases, those mutations are base substitutions or gain or loss of 1-2 nucleotides. The best known case is the change of a GC pair into a TA pair at 8-oxoG, which is the mutational signature of 8-oxoG. Most of repair processes are error-free, while, as mentioned earlier, DSB repair by end joining is often error-prone and lead to loss of a few nucleotides. Unrepaired DSB lead to a more drastic loss of genetic information, so that ionizing radiation may induce 100 to 30 kilo base-pair deletions and loss of parts of chromosomes. Mutations lead to inactivation or absence of proteins. Depending on which gene is mutated and so which protein is inactivated, in which organ, the biological consequences may drastically differ: cancer upon inactivation of a tumor suppressor gene, genetic disease in the offspring.

Chemistry and structural biology of DNA glycosylases: the case of the formamidopyrimidine-DNA glycosylase (Fpg)

The prokaryote Fpg protein (the functional homologue of the eukaryotic Ogg1), (**Tab. 1**) catalyses the removal of oxidized purines such as imidazole-ring opened purines (Fapy) and the major oxidized product of purines, the 8-oxoguanine (8-oxoG) [9]. Both lesions can be generated in DNA by irradiation (UV and gamma) and are associated with replication arrests (Fapy residues) and G to T transversions (8-oxoG). Fpg cleaves the N-glycosidic bond between the damaged base and its associated sugar (DNA glycosylase) thus generating an AP site in DNA. The enzyme is also associated with an AP lyase activity which eliminates the resulting AP site by successive cleavages at 3′ and 5′ sides by a βδ-elimination reaction. Then, the one nucleotide gap is filled in by DNA polymerase and finally, the DNA strand is rejoined by DNA ligase (**Fig. 3**). Fpg achieves its catalytic process using its P1 N-terminal amino group to perform a nucleophilic attack at the C1′ of the damaged nucleoside. During this process, the enzyme forms an imino enzyme-DNA intermediate (**Fig. 5**). Fpg is probably one of the BER enzymes the more documented at the atomic level (for review, see [10]).

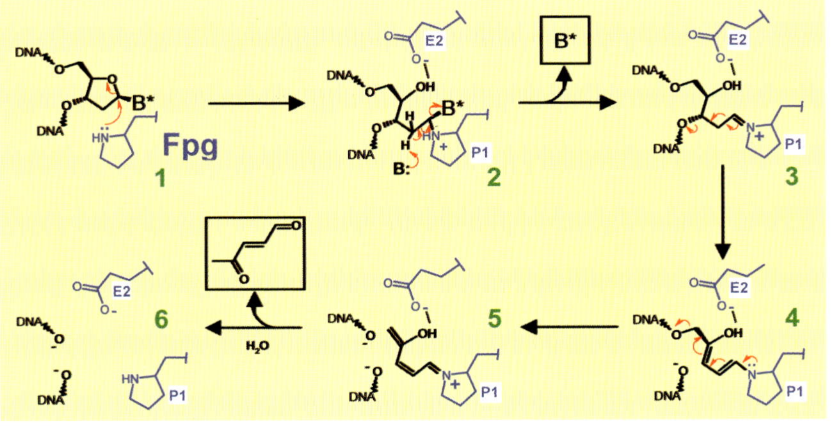

Figure 5 : *Reaction mechanism catalised by of the Fpg protein-DNA glycosylase and AP lyase process. The deprotonated N-terminal proline P1 of Fpg (in blue) is exquisitely positioned inside the active site for nucleophile attack (*1*) at the C1′ anomeric position of the damaged nucleoside (in black). The initial condensation step led to the formation of a transient N,N-disubstituted hemiaminal (*2*). This unstable intermediate looses the damaged base (B*) by the hydrolysis of the N-glycosidic bond and forms a protonated Schiff base intermediate (*3*) between P1 and the C1′ of the resulting abasic site (DNA glycosylase activity). From the Schiff base, β-elimination (cleavage at the 3′ side of the sugar) produced the covalent intermediate (*4*). After δ-elimination and the hydrolysis of the imino enzyme-DNA intermediate, the sugar is released as the unsaturated 4-oxo-2-pentenal (*5, 6*). (AP lyase activity). Repair is achieved by DNA polymerase and DNA ligase.*

Nucleoside chemistry to obtain stable complexes of Fpg bound to DNA

Two general approaches have been used so far to generate stable Fpg/DNA complexes enabling to study at the molecular level the interaction of this enzyme with its substrates, reaction intermediates and products. One of these relies on site-directed mutagenesis of the active site to abolish catalysis, independently from the Fpg specific DNA binding [11]. An alternative approach relies on the design and synthesis of uncleavable substrate analogues or inhibitors (**Fig. 6**). One class of inhibitors recently used consists of nucleobase analogues with a stabilized *N*-glycosydic bond, that are not processed by Fpg [12,13]. The second class of inhibitors more extensively used contains cyclic or non cyclic AP site analogues that mimic the AP site structure in the catalytic transition state, preventing the Fpg end products synthesis (**Fig. 6**) [14,15].

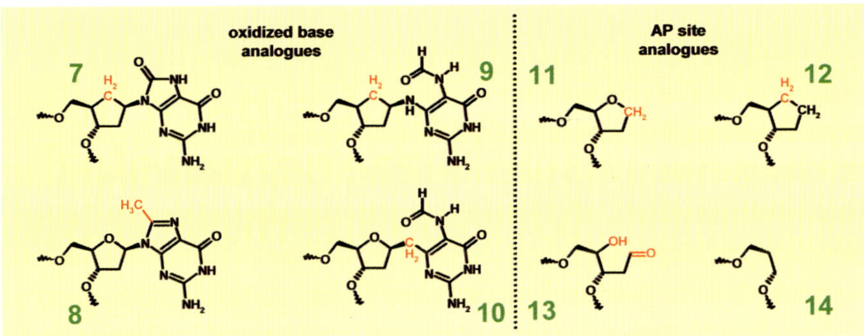

Figure 6 *: Substrate analogues of the Fpg protein. The synthetic derivatives of oxidized guanine (*7, 8, 9, 10*) and abasic site analogues (*11, 12, 13, 14*) are high affinity ligands for Fpg. Chemical "mutations" of natural damages are indicated by red atoms.*

DNA binding and nucleobase flipping by Fpg

The challenge of Fpg is to detect and excise one lesion among million of normal bases, all the more this lesion does not disturbed the local DNA structure. Recently, the crystal structure of Fpg interrogating undamaged DNA was solved using disulfide cross-linking technology [16]. The protein searches its substrate by sliding along DNA. During this process Fpg protein inserts a probe residue into the DNA helical stack and severely buckles the normal target base pair. Crystal structures of Fpg bound to analogue-containing DNA have revealed how Fpg, after recognizing the damage in DNA, exposed the C1′ anomeric centre of the damaged nucleobase to the nucleophilic attack of N-terminal proline of the enzyme (**Fig. 7**). Fpg solved this problem by flipping the target nucleobase out of the DNA helix and stabilizing it in an extrahelical conformation into the active site of the protein [17,18, 19]. This extrusion of the damage results firstly from a strong curvature of the DNA helix centred on the target site induced by the Fpg binding to the damaged strand, and secondly from the intercalation in the minor groove of three strictly conserved residues which fill in the hole resulting from the nucleobase extrusion (**Figs. 7A** and **7B**). The amino acids (Met75, Arg109 and Phe111) inserted into DNA prevent the double helix to collapse locally and allow specific contacts with the orphan pyrimidine on the opposite strand, maintaining it in an intrahelical conformation (**Fig. 7B**). In addition to elucidate the recognition of the damaged base by the enzyme, structural studies indicate that Fpg specifies a pyrimidine opposite the lesion. This structural observation is also supported by biochemical experiments indicating that Fpg negatively discriminates a purine opposite the lesion [20].

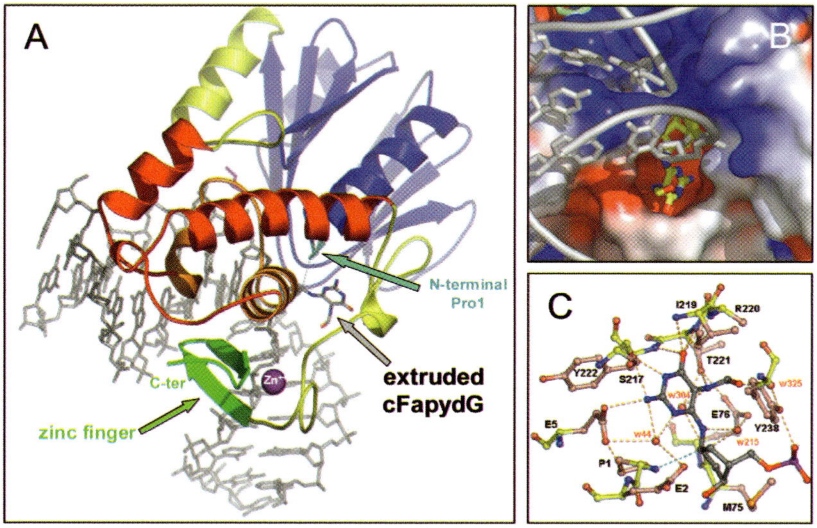

Figure 7 : *Crystal structure of Fpg bound to cFapydG-containing 14-mer DNA duplex* A) ***Overview of the complex showing the strong DNA torsion induced by Fpg binding.*** *DNA double helix is in grey and the peptide main chain fold of Fpg is represented by a ribbon multicolour. The side chain of the active site Pro1 is indicated in magenta and the zinc binding domain of the protein in green.* B) ***Enlargement of the structure at the DNA target site showing the intercalation of the protein by the minor groove.*** *The phophodiester backbone of DNA strands are represented by white cylinders. Fpg is represented by its solvent accessible surfaces (blue, red and white areas indicate basic, acidic and neutral regions, respectively). In the enzyme binding pocket, the damaged nucleoside (its C, N and O atoms are coloured in yellow, blue and red, respectively) is stabilized in an extrahelical conformation while the estranged cytosine on the strand opposite the lesion in an intrahelical conformation.* C) ***Network of hydrogen bonds made by the enzyme with the FapydG lesion inside the catalytic site.*** *The peptide main chain of Fpg is shown in yellow ball-and-sticks, the side chains of amino acids residues involved in the recognition of the damage in pink and carbon atoms of DNA in grey balls. N, O and P atoms are indicated by blue, red and purple balls, respectively. Covalent links and hydrogen bonds are indicated by sticks and red dashed lines, respectively. A red "w" letter indicates a structural water molecule.*

Fpg recognition mode of FapyG residue and mechanism of catalyzing N-glycosidic bond cleavage

The extruded FapydG residue in an *anti*-conformation is inserted into the binding pocket of the enzyme active site [19]. This pocket consists in aromatic and hydrophobic residues that can modulate the size and the shape of the pocket, and in polar residues that make specific hydrogen bonds with the damaged guanine (**Figs. 7B** and **7C**). Fpg specifies an oxidized purine, establishing hydrogen bonds with all the hydrogen donors and acceptors of the Watson-Crick face of the FapydG residue. In addition, Fpg recognizes, through water molecule-mediated interactions, all the structural determinants of the imidazole ring-opened moiety of FapydG (especially, the protonated N7 and the C8O groups of the damage which distinguish clearly the normal guanine from FapyG; **Fig. 7C**). All these interactions constrain the nucleobase in an optimal conformation inside the active site of the enzyme for catalysis. In this conformation, the C1′ anomeric centre of the extrahelical damaged nucleoside is exposed to the nucleophilic attack of the N-terminal proline which results in an imino enzyme-DNA complex (**Figs. 5** and **7C**). This intermediate can be easily trapped by its irreversible reduction [17].

Conclusion

During the past decade, our understanding of the molecular mechanism by which Fpg (and other DNA glycosylases) recognizes and excises base lesions has advanced greatly by combining organic chemistry, biochemistry and structural biology approaches. Despite our understanding of Fpg process, several aspects must be now best documented, especially substrate specificity and catalysis, and how this enzyme recognizes its substrate in DNA before the damaged nucleoside flipping.

Acknowledgements

We are very grateful to all people who participated to our respective laboratory work. Our work is supported by CNRS, *Institut Curie*, EDF, CNES, ARC, the *Région Centre* and the *Ligue nationale contre le cancer* (Indre).

References

[1] Iliakis G., Wang Y., Guan J., Wang H., DNA damage checkpoint control in cells exposed to ionizing radiation, Oncogene, 2003, **22**, 5834-5847.

[2] Ward J., DNA damage produced by ionizing radiation in mammalian cells: identities, mechanisms of formation and reparability, Prog. Nucleic Acid Res. Mol. Biol., 1988, **35**, 95-125.

[3] Nikjoo H., O'Neill P., Terrissol M., Goodhead D., Modelling of radiation-induced DNA damage: the early physical and chemical event, Int. J. Radiat. Biol., 1994, **66**, 453-457.

[4] Blaisdell J., Harrison L., Wallace S., Base excision repair processing of radiation-induced clustered DNA lesions, Radiat. Prot. Dosimetry, 2001, **97**, 25-31.

[5] Eot-Houllier G., Eon-Marchais S., Gasparutto D., Sage E., Processing of a complex multiply damaged DNA site by human cell extracts and purified repair proteins, Nucleic Acids Res., 2005, **33**, 260-271.

[6] Cline S., Hanawalt P., Who's first in the cellular response to DNA damage?, Nature reviews, 2003, **4**, 361-372.

[7] Hoeijmakers J., Genome maintenance mechanisms for preventing cancer, Nature, 2001, **411**, 366-374.

[8] Caldecott K., Mammalian DNA single-strand break repair: an X-ra(y)ted affair, BioEssays, 2001, **23**, 447-455.

[9] Zaika EI, Perlow R.A., Matz E., Broyde S., Gilboa R., Grollman A.P., Zarkov D.O., Substrate discrimination by formamidopyrimidine-DNA glycosylase, J. Biol. Chem., 2004, **279**, 4849-4861.

[10] Huffman J.L., Sundheim O., Tainer J.A., DNA base damage recognition and removal: new twists and grooves, Mutation Res., 2005, **577**, 55-76.

[11] Fromme JC, Verdine G.L., DNA lesion recognition by bacterial repair enzyme MutM, J. Biol. Chem., 2003, **278**, 51543-51548.

[12] Wiederholt C.J., Delaney M.O., Greenberg M.M., Interaction of DNA containing FapydA or its C-nucleoside analogues with base excision repair enzymes. Implications for mutagenesis and enzyme inhibition, Biochemistry, 2002, **41**, 15838-15844.

[13] Ober M., Linne U., Gierlich J., Carell T., The two main DANN lesions 8-oxo-7,8-dihydroguanine and 2,6-Diamino-5-formamido-4-hydroxypyrimidine exhibit strongly different pairing properties, Angew. Chem. Int. Ed., 2003, **42**, 4947-4951.

[14] Castaing B., Boiteux S., Zelwer C., DNA containing a chemically reduced apurinic site is a high affinity ligand for the *E. coli* formamidopyrimidine-DNA glycosylase, Nucleic Acids Res., 1992, **20**, 389-394.

[15] Castaing B., Fourrey J.L., Hervouet N., Thomas M., Boiteux S., Zelwer C., AP site structural determinants for Fpg specific recognition, Nucleic Acids Res., 1999, **27**, 608-615.

[16] Banerjee A., Santos W.L., Verdine G.L., Structure of DNA glycosylase searching for lesions, Science, 2006, **311**, 1153-1157.

[17] Serre L., Pereira de Jésus K., Boiteux S., Zelwer C., Castaing B., Crystal structure of the *Lactococcus lactis* formamidopyrimidine-DNA glycosylase bound to an abasic site analogue-containing DNA, EMBO J., 2002, **21**, 2854-2865.

[18] Fromme J.C., Verdine G.L., DNA lesion recognition by the bacterial repair enzyme MutM, J. Biol. Chem., 2003, **278**, 51543-51548.

[19] Coste F., Ober M., Carell T., Boiteux S., Zelwer C., Castaing B., Structural basis for the recognition of the FapydG lesion (2,6-diamino-4-hydroxy-5-formamidopyrimidine) by the Fpg DNA glycosylase, J. Biol. Chem., 2004, **279**, 44074-44083.

[20] Castaing B., Geiger A., Seliger H., Nehls P., Laval J., Zelwer C., Boiteux S., Repair and binding of a DNA fragment containing a single 8-oxoguanine by wild type and mutant Fpg protein, Nucleic Acids Res., 1993, **21**, 2899-2905.

Chapter 16

Pulse radiolysis studies of free radical processes in peptides and proteins

Chantal HOUÉE-LEVIN and Krzysztof BOBROWSKI

Introduction

The story of free radical reactions in biological systems began many years ago, when Harman postulated that they played a prominent role in ageing [1]. Since the sixties, the relevance to biological as well as industrial processes became more and more clear every year. It is now current to invoke free radicals in ordinary life, in cooking, in prevention of ageing processes, etc. and the scientific knowledge gave a basis to these assertions [2]. It is now beyond doubt that free radicals processes in proteins are involved in all steps of life, going from conception to death induction. They are believed to be part of the cellular defence against oxidative stress and at the same time responsible of severe damage like atherosclerosis [3] Protein free radicals are also enzyme active sites [4].

The studies by the methods of radiolysis provided a wealth of knowledge about the kinetic and thermodynamic controls of radical reactions, the importance of which is no more questioned. Indeed, it is known that the chemical events initiated by ionizing radiation, are the same as those that take place in normal and deleterious events of every day's life. In this review we focus on some of the major knowledge that was acquired by the use of pulse radiolysis and steady-state gamma radiolysis of aqueous solutions of amino acids, peptides and proteins (**Inset**). The potential role of pulse radiolysis (Chapter 2) for studying biomolecules has been acknowledged rather early. In most cases, pulse radiolysis method has been very valuable in identification of radicals, establishing their structures and exploring their reactivity (Chapter 1). We would like also to show how the reactions that were evidenced took their place in biological events, thus how a synergy between pure chemistry and biology was established.

Inset : Protein structure

A protein is a polymeric chain. (A) Monomers are amino acids. (B) In polymerization, residues are linked by the peptidic bond, which is an amide function.

A/ An amino acid. R_1 will be the residue 1 in scheme B.

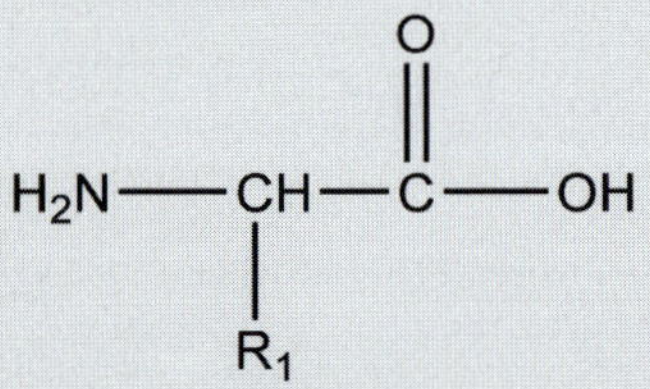

B/ The protein

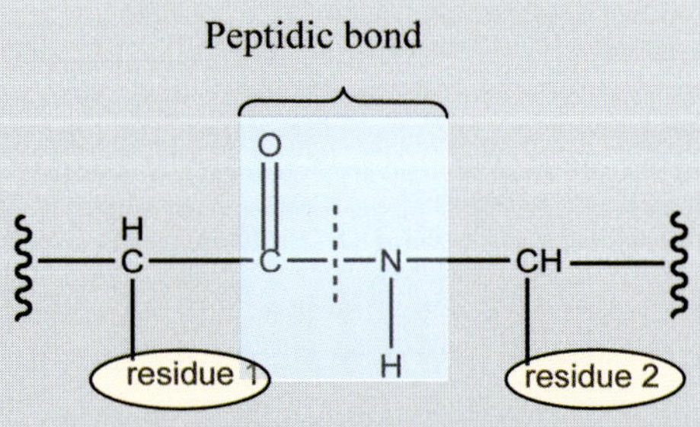

However this polymeric chain is folded. The folding has a tremendous importance in the action of the protein and in its recognition by biological partners and/or by the degradation pathways. The different levels of structure are presented in **Figure 1**. (For more details, see for instance [i]).

[i] Voet D., Voet J.G., Biochemistry 2nd edition, John Wiley New York.

The simplest amino-acid, glycine

One of the more recent results concerns $^{\bullet}OH$-induced oxidation of glycine, the simplest amino acid [5]. Two main radical products $^{+}H_2N^{\bullet}\text{-}CH_2\text{-}CO_2^{-}$ and aminyl radicals $HN^{\bullet}\text{-}CH_2\text{-}CO_2^{-}$ have been identified and their subsequent reaction pathways including decarboxylation with parallel formation of $^{\bullet}CH_2NH_2$ and β-fragmentation into the respective

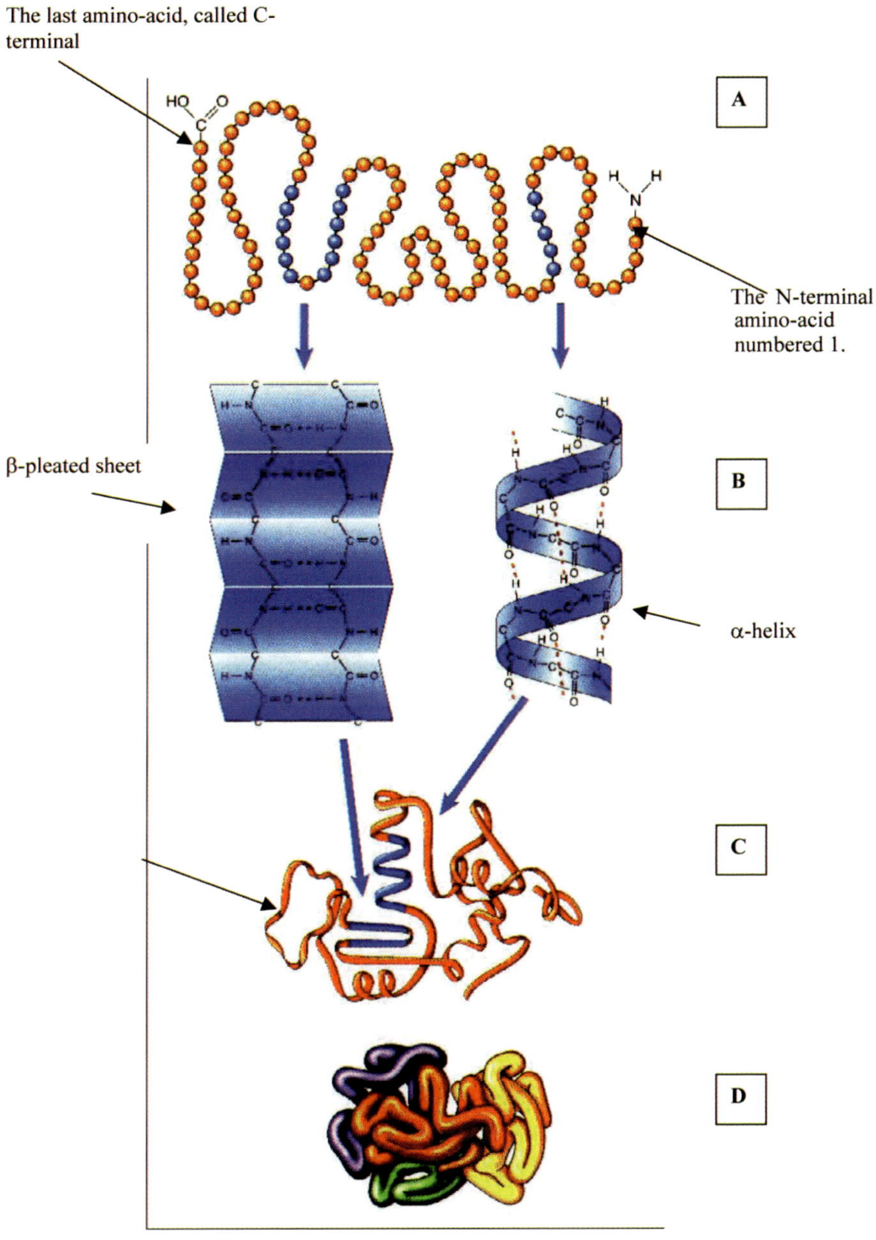

Figure 1 : *The several levels of structure of a protein.*

imine and carboxyl radical $CO_2^{\bullet-}$, respectively. The possible initiation of amino acid decarboxylation by C-centred radicals are considered to be of general significance and interest in chemical and biological systems.

From amino acids to proteins

It was, however, soon recognized that in aqueous solutions peptides and proteins (**Inset**) and free radicals derived from them behaved differently from amino acids and free radicals derived from them. These differences arise from the following:

- The folding (**Fig. 1**) renders some amino acids not available to free radicals. Indeed, since the reactions are fast, their control is kinetic and not thermodynamic. One can imagine that the amino acids that are on the surface are more reactive than those in the interior.

- Some amino acids are charged; thus the protein creates an electrostatic field around itself, which orients the reactivity of the charged radicals.

- The reduction potentials of amino acids are sensitive to the environment.

- Functional groups present in side chains of neighbouring amino acids can be involved in reaction mechanism and thus may affect reaction pathway of radicals formed.

Thus kinetics as well as thermodynamics aspects of the free radical reactions are modified. Some examples are detailed in what follows.

Thiyl radicals

Thiyl radicals are important reactants in several enzymes and form *in vivo* during conditions of oxidative stress [6]. They have been considered for a long time as rather unreactive species. However, recently several reactions of thiyl radicals with biomolecules have been described (catalysis of *cis-trans* isomerization of unsaturated fatty acids, addition to the pyrimidine bases C5-C6 double bonds, and hydrogen abstraction from polyunsaturated fatty acid, thymine and peptide C_α-H and side chain C-H bonds) [7]. More recently, the intramolecular addition of peptide cysteine thiyl radicals ($CysS^•$) to phenylalanine (Phe) yielding alkylothio-substituted cyclohexadienyl radicals was demonstrated in the peptides Phe-Cys and Phe-Gly-Cys-Gly (**Fig. 2**) [8].

This addition reaction might be of great biological significance since the intramolecular addition of $CysS^•$ to aromatic ring of Phe can compete with addition of O_2 to $CysS^•$, assuming biologically relevant tissue concentration of O_2. It presents a possible free radical pathway to thioether-containing peptide and protein cross-links.

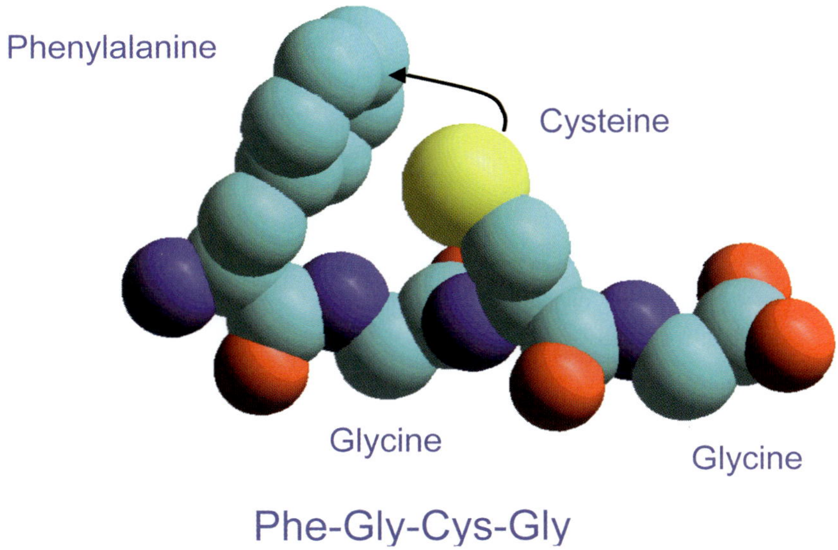

Figure 2 : *In the peptide Phe-Gly-Cys-Gly (formed by Phenylalanine, Glycine, Cysteine and Glycine amino acids linked by peptidic bonds), the sulphur atom is very close to the phenyl ring. An intramolecular addition of the CysS$^{\bullet}$ radical on the aromatic ring of Phe (marked with an arrow) leads easily to the alkylthio-substituted cyclohexadienyl radicals. Atom colors: Cyan: carbon; blue: nitrogen; red: oxygen; yellow: sulphur. H atoms are not shown.*

Disulfide radicals

Disulfide radical anions might play an important role in oxidative stress. In cellular media, they can be formed by oxidation of protein thiol functions (by $^{\bullet}OH$ radicals, for instance, Chapter 1), followed by dimerization:

$$\text{Protein/SH} + {}^{\bullet}OH \rightarrow \text{Protein/S}^{\bullet} + H_2O \quad (1)$$
$$\text{Protein/S}^{\bullet} + \text{Protein/SH} \rightarrow \text{Protein /S—S}^{\bullet -}\text{/Protein} + H^{+} \quad (2)$$

These disulfide radical anions are believed to be strong reductants. Thus they might counterbalance the action of oxidizing free radicals.

To study their properties, the easiest way is to form them by one-electron reduction of disulfide bonds. Among radicals from water radiolysis, hydrated electron is the most powerful reductant. It reacts with almost all amino acids and especially with the disulfide groups. Using less powerful reductants such as $COO^{\bullet -}$ radicals, some selectivity in the attack appears. An example is displayed in **Figure 3**.

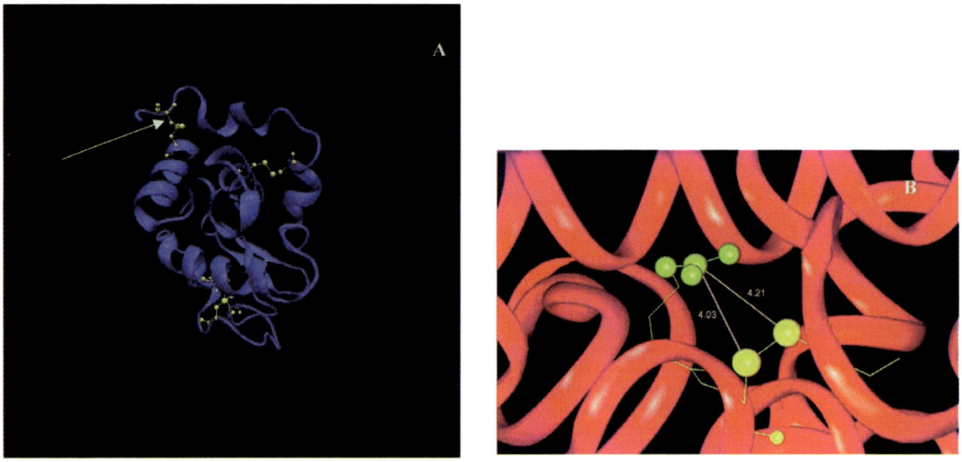

Figure 3 : ***Hen egg white lysozyme has 4 disulfide bridges.*** *However, only one of them is easily reduced* A) *marked with an arrow. The detail of its structure is shown in* B) *The disulfide radical anion is stabilized by interaction with the charged end of an Arginine residue (in green). In white: the distances (in Å) between the central carbon atom and the sulphur atoms. In red: the polypeptidic chain.*

In hen egg white lysozyme, out of the four disulfide bridges, one is much more easily reduced than the others (**Fig. 3 A** and **B**) [9]. A study by quantum chemistry could justify this selectivity by the stabilization of the resulting disulfide free radical by the positively charged end of arginine. The distances between sulphur atoms and the central carbon of arginine confirm the strong interaction between these groups (**Fig. 3**).

Pulse radiolysis studies demonstrated that all disulfide radicals do not have the same chemical properties. An example is given by the study of protonation equilibrium (**Scheme 1**).

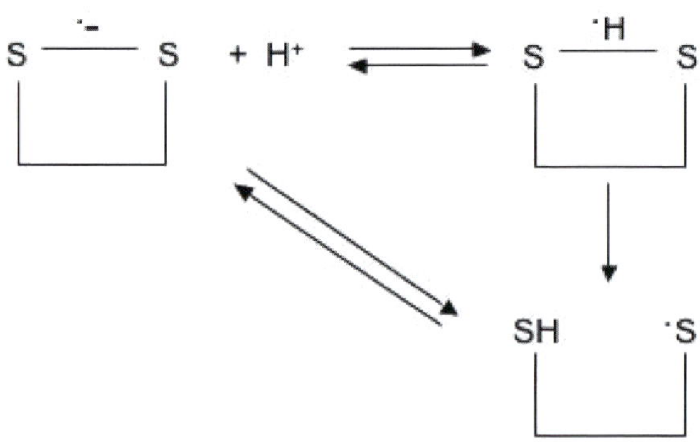

Scheme 1: *Protonation-deprotonation equilibrium of disulfide radical anion.*

The resulting protonated disulfide radical undergoes cleavage of the S-S bond. Since the thiyl radical is a strong oxidant, the protonation equilibrium appears like a switch between oxidant/reductant radical. The pK_a values of some of these radicals, measured by pulse radiolysis, are between 5 and 6 (**Fig. 4**). However in thioredoxin, it is below 3 [10].

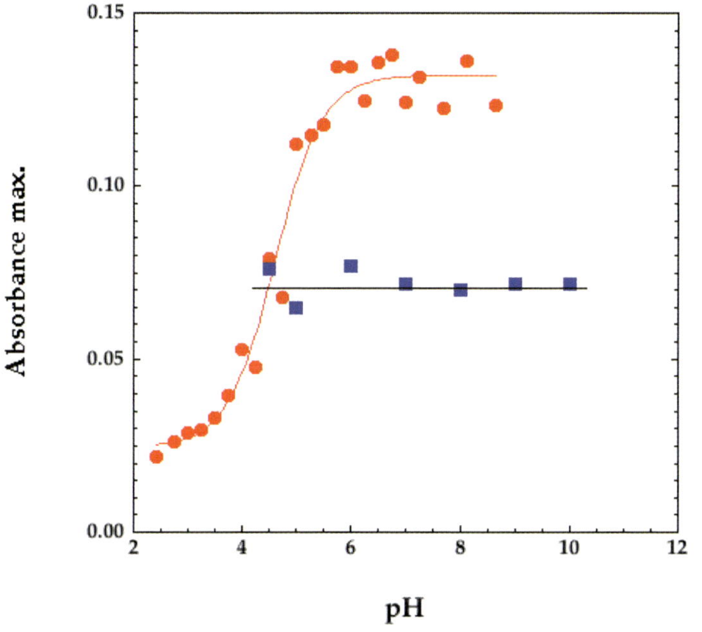

Figure 4 : *Titration curves for hen egg white lysozyme and thioredoxin disulfide radical. Red : lysozyme; blue : thioredoxin.*

It means that sulphur free radical from this protein are mostly reductant, which enlightens the role of thioredoxin in restoring sites which were oxidized and thus as protection against oxidative stress.

One-electron oxidation of methionine

Methionine (Met) is one of the sensitive sulphur-containing amino-acids toward one-electron oxidation. However its ease of oxidation is also modulated by the structure. The one-electron oxidation of Met in peptides yields sulfide methionine radical cations ($MetS^{\bullet+}$) which convert into intermediates that obtain catalytic support from neighbouring groups containing electron rich heteroatoms (S, N, O) and thus stabilize electron deficient sulphur centres in S∴S, S∴N, and S∴O-three-electron bonded complexes (**Fig. 5**) [11].

Interaction with particular peptide or protein domains would likely involve N- and C-terminal nucleophilic functionalities (NH_3^+, COO^-), and nucleophilic functionalities in the side chains of amino acid residues (Asp, Glu, Lys, Val, Thr). However, very often heteroatoms in peptide bonds are the only nucleophiles present in the vicinity of the $MetS^{\bullet+}$. It was

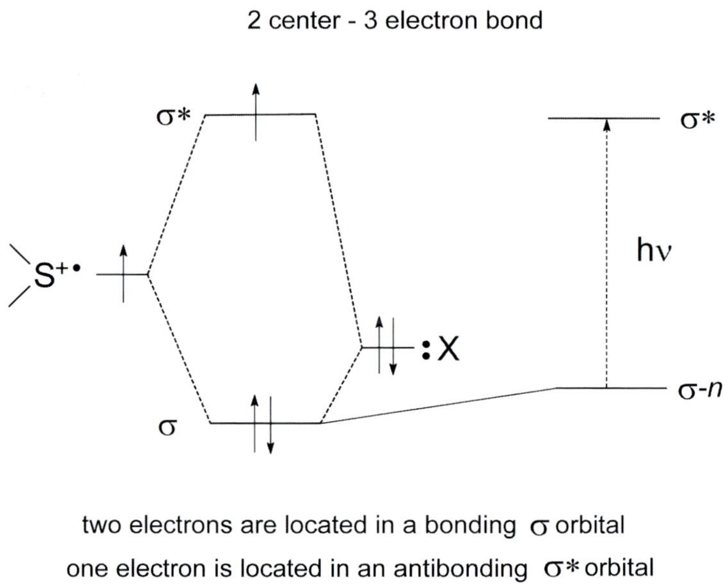

Figure 5 : *Electronic energetic diagram of the molecular orbitals of three–electron bonds in sulphur radicals and optical transition. X: S, N, or O atom.*

recently shown that such interactions play an important role in oligopeptides of the form N-Ac-Gly-Met-Gly and N-Ac-Gly-Gly-Gly-Met-Gly-Gly-Gly [12]. Pulse radiolysis studies with UV/Vis spectrophotometrical and conductometric detection showed for the first time that $MetS^{\bullet+}$in peptides can be stabilized through bond formation with either the oxygen or the nitrogen atoms of adjacent peptide bonds (**Fig. 6**).

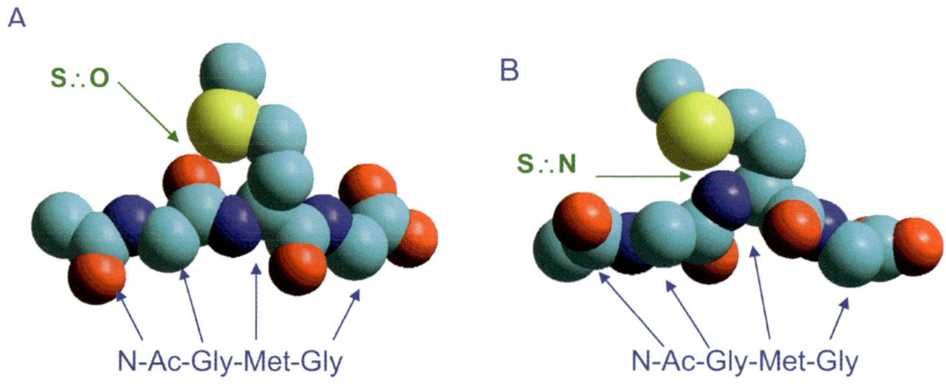

Figure 6 : *Structure of N-Ac-Gly-Met-Gly peptide showing interactions of the sulfur atom (S) in Met with the oxygen atom (O) (A) and the nitrogen atom (N) (B) (marked with arrows) located in the N-terminal adjacent peptide bond that lead to the S∴O and S∴N-bonded radicals, respectively.* *Atom colors: Cyan: carbon; blue: nitrogen; red: oxygen; yellow: sulphur. H atoms are not shown.*

Moreover, formation of radical transients with S∴O bonds is kinetically preferred, but on longer time scale they convert into transients with S∴N bonds in a pH dependent manner. Ultimately transients with S∴N bonds transform intramolecularly into C-centred radicals located on the $^{\alpha}$C moiety of the peptide backbone. Another type of C-centred radicals located in the side chain of Met-residue, α-(alkylthio)alkyl radicals, are formed *via* deprotonation of $MetS^{\bullet+}$. C-centred radicals are precursors for peroxyl radicals ($ROO^{\bullet}$) that might be involved in chain reactions of peptide and/or protein oxidation. Stabilization of $MetS^{\bullet+}$through formation of S∴O- and S∴N-bonded radicals might potentially accelerate oxidation and autooxidation processes of Met in peptides and proteins. Considering that methionine sulfoxide, which is the final product coming from all radicals centred on sulphur, is restored by the enzyme methionine sulfoxide reductase into MetS, stabilization of $MetS^{\bullet+}$appears as a protection against an eventual peroxidation chain that would develop from a carbon centred radical.

The amyloid β peptide (βAP), a 39- to 43-amino acid-long peptide is the major constituent of the neuritic plaques and neurofibrillary tangles in brain. Their progressive formation is characteristic for the development of the Alzheimer disease. The integrity of Met35 is very important for the constitution of the senile plaque. In the structure of βAP, Met35-$S^{\bullet+}$ formation can be facilitated by a pre-existing close sulphur-oxygen (S-O) interaction between the Met35 sulphur and the carbonyl oxygen of the peptide bond C-terminal to isoleucine (Ile31) that might lower one-electron reduction potential of $MetS^{\bullet+}$/Met couple. The first experimental evidence that Met is more easily oxidized than in other peptides and proteins comes from one-electron oxidation of βAP1-40 using azide radicals ($N_3^{\bullet}$) produced by pulse radiolysis [13]. One-electron oxidation of the natural peptide was compared to that of the non-natural peptide of reverse sequence (βAP40-1). Circular dichroism showed that whereas Aβ1-40 is highly structured, Aβ40-1 has no regular structure. It appeared that the oxidation patterns of these two compounds are different: in βAP (1-40) Met35 is mostly the target of free radicals, whereas in βAP40-1, Tyr10 gets dimerized. This observation underlines the role of structure in driving the free radical reactions and seems to be indeed relevant to explain specificity of the Aβ1-40 in the development of Alzheimer disease.

The pathogenesis of another well-known neurodegenerative disease (Jacob Creutzfeld disease) seems to be strongly linked to the presence of prion proteins in the brain. These macromolecules contain multiple Met residues, some of them in close vicinity. Such structure should favour stabilization of $MetS^{\bullet+}$as intramolecular $(S∴S)^{\bullet+}$ complexes. Since weak intramolecular non-bonded S•••O and S•••N interactions have been recently suggested in proteins [14], stabilization of $MetS^{\bullet+}$ through formation of S∴N- and/or S∴O-complexes might potentially accelerate oxidation processes in proteins. The first experimental evidence for the stabilization of $MetS^{\cdot+}$as intramolecular complex with N-atom in adjacent amide group was obtained during one-electron oxidation of calmodulin ($CaM\text{-}Ca_4$, wild type), studied on the microsecond time domain by pulse radiolysis [15]. Calmodulin is a regulatory "calcium sensor" protein that contains nine Met residues. Stabilization of $MetS^{\cdot+}$ by peptide bonds might be a general phenomenon in proteins.

There are different reaction pathways of decay of the S∴N-complex depending on the structure of the peptide. In model peptides containing N-terminal glutamic acid residue (γ-Glu-Met, γ-Glu-Gly-Met-Gly), an intramolecular electron transfer from the carboxylate group to the electron-deficient center at the nitrogen within S∴N-bond followed by homolytic bond-breakage of the carbon-carboxylate bond leads to formation of α-amino radicals and CO_2 [16]. On the other hand, in model peptides Thr-$(Gly)_n$-Met, n = 0-4, the intermediary S∴N-bonded radicals do not decarboxylate but undergo homolytic cleavage of the C_α-C_β bond of Thr, yielding the highly toxic acetaldehyde [17]. Once again, the fate and the consequences of peptide oxidation depend strongly on the stabilization of $MetS^{\cdot+}$.

Intramolecular long range electron transfer in peptides and proteins

Intramolecular long-range electron transfer (LRET) plays a major role in many biological processes, including fundamental energy storage processes such as photosynthesis and respiration, redox-mediated enzyme catalysis, and in various pathologic processes such as radiation damage, oxygen toxicity and cellular aging. Most of these processes occur on ground-state potential energy surfaces, making pulse radiolysis an effective and truly unique tool for these studies. The typical LRET pulse radiolysis experiment begins with the rapid selective oxidation or reduction of one-redox site on a macromolecule (formation of the donor-acceptor complex) followed by the intramolecular LRET. The advantage of pulse radiolysis is that using either an oxidizing or a reducing radical can generate the donor-acceptor complex. These radicals have redox potentials covering the approximate range from −1.1 to +1.8 V (*vs.* SCE) [18]. The entire subject of LRET is much too wide to be covered in detail. Numerous comprehensive reviews are available which present and discuss this topic in a more detailed manner (see for instance [19] and references therein). Therefore, a few selected important achievements emerging in this field will be highlighted.

Most LRET processes in biological systems are nonadiabatic. In quantum-mechanical electron-transfer theory, the rate constant for nonadiabatic ET from a donor to acceptor can be expressed as the product of the square of an electronic coupling matrix element (H_{DA}) and a nuclear Franck-Condon factor (FC): $k_{ET} = (2\pi/h)[H_{DA}]^2(FC)$. The $[H_{DA}]$ is a measure of the coupling or the interaction between the orbitals of the donor (D) and the acceptor (A) and is influenced by the distance and structure of the medium separating the electron donor and acceptor. The role of the distance, standard free energy change, and reorganization energy was examined. Influence of secondary structural features of the peptide bridge has been probed by applying flexible oligoglycine bridges, conformationally more rigid oligoproline bridges and helical bridges (for review see [20] and references therein) Pulse radiolysis studies of simple model synthetic peptides have demonstrated intramolecular ET involving radicals located on the side chains of aromatic, histidine and sulfur containing (methionine, cysteine) amino acid residues.

Elaboration of LRET mechanism by resolving the parameters that determine specific rates of LRET has stimulated pulse radiolysis studies in proteins. Examples include generation of metastable electron donor and acceptor complexes in (1) native and mutant proteins, (2) proteins with the directed single-site specific mutations, (3) native and mutant multi-site redox proteins, (4) proteins with the site specific modification with transition metal complexes covalently attached to a specific surface amino acid residues.

Pulse radiolysis investigations related to oxidative stress

The methods of gamma and pulse radiolysis provided the basis for the understanding of oxidative stress. Indeed, one cannot imagine how the biological importance of short-lived transients could have been demonstrated otherwise and how their chemical properties could have been studied.

Control of superoxide radical anion steady state

A striking example is control of the steady state level of superoxide radical anion ($O_2^{\bullet-}$). Although superoxide anions are not very reactive, they play a prominent role in oxidative stress by triggering formation of peroxynitrite. The control of their steady state level is thus vital for cells. In most living organisms the well-known metalloenzyme superoxide dismutase (SOD), present in almost all aerobic cells catalyzes the disproportionation of $O_2^{\bullet-}$ into H_2O_2 and O_2:

$$O_2^{\bullet-} + O_2^{\bullet-} + 2H^+ \rightarrow H_2O_2 + O_2 \qquad (3)$$

SODs are differentiated mainly by the redox-active metal in the active site and by the localization: copper-zinc (in the cytosol), manganese (in mitochondria) and iron. The iron and manganese SODs are structurally similar. However, some bacteria can live without the Cu-Zn SOD, whereas no organisms can live without Mn-SOD. It stresses the vital importance of these enzymes close to the place where $O_2^{\bullet-}$ are produced. The mechanisms by which Mn-SOD reacts with $O_2^{\bullet-}$ are of interest as knowledge of the kinetic parameters, and the reaction pathways may allow the synthesis of model compounds with specific chemical features. The kinetics of the dismutation of $O_2^{\bullet-}$ by *Escherichia coli* MnSOD were measured by pulse radiolysis, and shown to fit a mechanism involving the rapid reduction of Mn^{3+}SOD by $O_2^{\bullet-}$ followed by both the direct reoxidation of Mn^{2+}SOD by $O_2^{\bullet-}$ and the formation of a $Mn^{II}SOD(O_2^{\bullet-})$ complex [21].

The second system was discovered in prokaryotic cells a few years ago [22]. It is the recently characterized non-heme iron superoxide reductase (SOR) that catalyzes the reduction of $O_2^{\bullet-}$ into H_2O_2 by an intracellular reductant:

$$O_2^{\bullet-} + 1e^- + 2H^+ \rightarrow H_2O_2 \tag{4}$$

For studies about the mechanisms of action of this enzyme, pulse radiolysis also played a prominent role. Current research on improving knowledge about these enzymes, creating more performant ones by site directed mutagenesis or testing the ability of destruction of superoxide by metal complexes also uses pulse radiolysis as a method of screening.

Haem peroxidases are globular proteins with an iron-porphyrin complex as a prosthetic group. These enzymes are widespread among prokaryotes and eukaryotes. They catalyze the oxidation of substrates by organic peroxides or hydrogen peroxide. During the past decades, considerable scientific effort has been put into elucidation of the mechanisms of reactions catalyzed by these enzymes. Pulse radiolysis technique has made an important contribution by providing information on the redox states of the enzymes and their interconversion, as well as on the properties of the free radical intermediates involved [23].

Pulse radiolysis investigations of the reaction of superoxide radical anions ($O_2^{\bullet-}$) with radicals derived from various amino acids

The following results were all obtained using pulse radiolysis.

Methionine. The reaction of superoxide radical anions ($O_2^{\bullet-}$) with sulfide radical cation-nucleophile complexes might represent an efficient sulfoxide-forming process in peptides and proteins containing methionine under conditions where significant amounts of sulfide radical cation complexes and superoxide are formed simultaneously. The rate constant for the reaction of $O_2^{\bullet-}$ with the $(S\therefore N)^+$ complex was found to be *ca.* 3-fold slower as compared to that of the reaction with the $(S\therefore S)^+$ complex. This drop in reactivity may, in part, reflect the lower probability of $O_2^{\bullet-}$ to encounter S-atom in the $(S\therefore N)^+$ complex as compared to the symmetrical $(S \therefore S)^{\bullet+}$ complex. It is important to note that the reactions of $O_2^{\bullet-}$ with the sulfide radical cation complexes proceed 2.5 to 8-fold faster than the reaction of $O_2^{\bullet-}$ with superoxide reductases SOD [24]. From a biological point view, it means that sulfide radical cation-$O_2^{\bullet-}$ reactions might represent a potential source for sulfoxide formation when the system is exposed to high concentrations of reactive oxygen species (ROS).

Tryptophan. The reaction of $O_2^{\bullet-}$, with the semi-oxidized tryptophan neutral radical ($Trp^{\bullet}$) generated from tryptophan (Trp) by pulse radiolysis has been studied in a variety of functionalized Trp derivatives and in lysozyme [25]. These studies demonstrate that the reactivity of $O_2^{\bullet-}$ with the $Trp^{\bullet}$ radical is significantly dependent on the charge in the vicinity of the Trp residue from which it originates, by the electrostatic field generated by neighbouring groups and/or other charged cell constituents (*e.g.* DNA and lipids). The kinetics of $O_2^{\bullet-}$ reaction with the semi-oxidized tryptophan neutral radicals ($Trp^{\bullet}Lyz$) has been investigated at various pHs and conformational states. It was found that at pH lower than 6.2, the apparent bimolecular rate constant is about $2 \times 10^8\ M^{-1}s^{-1}$ but drops to $8 \times 10^7\ M^{-1}s^{-1}$ or less above pH 6.3 and in CTAC micelles. Interestingly, at all pHs the rate constants of the reaction of $O_2^{\bullet-}$ with $Trp^{\bullet}$ radicals are more than an order of magnitude greater than rate constants characterizing the intermolecular recombination of $Trp^{\bullet}$ radicals. The well-established LRET from Tyr residues to Trp radicals-leading to the repair of the $Trp^{\bullet}Lyz$ radicals is inhibited by the $Trp^{\bullet}Lyz + O_2^{\bullet-}$ reaction. In conclusion, the reaction of $O_2^{\bullet-}$ with reactive radicals derived from amino acids in proteins may have important consequences for the turnover of proteins involved in metabolic reactions.

Conclusion

The studies summarized here brought important knowledge at several levels. In fundamental science, the evidence of phenomena such as Long range Electron Transfer, LRET, has stimulated experiments and discussions all over the world. It was discovered that LRET concerns all polymers. In DNA, it may explain the localization of base lesions induced by oxidative stress. As for biology, it helped to understand the chemical basis of initiation and development of the inflammation processes that take place in all diseases. Moreover this fundamental research had consequences on everyday life: a search for evidences of the importance of antioxidants in food in relation to quality and duration of life was undertaken. The results were so positive that the content of antioxidants is now a criterion of quality of food. The redox reactions of the sulphur amino acids have triggered a research on new anti-inflammatory drugs, and some of them (such as N-acetyl cysteine) are currently used in many pathologies. There is no doubt that in the close future other comprehension and applications of the radiolysis results will appear.

References

[1] Harman D., Aging: a theory based on free radical and radiation chemistry, J. Gerontol., 1956, 298-300.

[2] Houée-Levin C., Sicard-Roselli C., Bergès J., Chimie et Biochimie radicalaires, Belin coll. Ellipses, 2005.

[3] Aruoma O. I., Halliwell B., Molecular biology of free radicals in human diseases, OICA international, London, 1998.

[4] Stubbe J. A., Van der Donck W. A., Protein radicals in enzyme catalysis, Chem. Rev., 1998, **98**, 705-762.

[5] Bonifačić M., Armstrong D. A., Carmichael I., Asmus K.-D., β-Fragmentation and other reactions involving aminyl radicals from amino acids, J. Phys. Chem. B., 2000, **104**, 643-649.

[6] Jourd'heuil D., Jourd'heuil F. L., Feelisch M., Oxidation and nitrosation of thiols at low micromolar exposure to nitric oxide. Evidence for a free radical mechanism, J. Biol. Chem., 2003, **278**, 15720-15726.

[7] Ferreri C., Kratzsch S., Landi L., Brede O., Thiyl radicals in biosystems: effects on lipid structures and metabolisms, Cell. Mol. Life Sci., 2005, **62**, 834-847.

[8] Nauser T., Casi G., Koppenol W. H., Schöneich C., Intramolecular addition of cysteine thiyl radicals to phenylalanine in peptides : formation of cyclohexadienyl type radicals, Chem. Commun., 2005, 3400-3402.

[9] Bergès J., Kassab E., Conte D., Adjadj E., Houée-Levin C., Ab initio calculations on arginine-disulfide complexes modelling the one-electron reduction of lysozyme. Comparison to an experimental reinvestigation, J. Phys. Chem., 1997, **101**, 7809-7817.

[10] El Hanine, Lmoumène C., Conte D., Jacquot J.-P., Houée-Levin C., Redox properties of protein disulfide bond in oxidized thioredoxin and lysozyme. A pulse radiolysis study, Biochemistry, 2000, **39**, 9295-9301.

[11] Bobrowski K, Holcman J., Formation and stability of intramolecular three-electron S∴N, S∴S, and S∴O bonds in one-electron-oxidized simple methionine peptides. Pulse radiolysis study, J. Phys. Chem., 1990, **93**, 6381-6387.

[12] Schöneich C., Pogocki D., Hug G. L., Bobrowski K., Free radical reactions of methionine in peptides : mechanisms relevant to β-amyloid oxidation and Alzheimer's disease, J. Am. Chem. Soc., 2003, **125**, 13700-13713.

[13] Kadlčik V., Sicard-Roselli C., Kodicek M., Houée-Levin C., One-electron oxidation of beta-amyloid peptide: sequence modulation of reactivity, Free Rad. Biol. Med., 2004, **37**, 881-891.

[14] Iwaoka M, Takemoto S., Tomoda S., Statistical and theoretical investigations of directionality of nonbonded S•••O interactions. Implications for molecular design and protein engineering, J. Am. Chem. Soc., 2003, **124**, 10613-10620.

[15] Nauser T., Jacoby M., Koppenol W.H., Squier T.C., Schöneich C., Calmodulin methionine residues are targets for one-electron oxidation by hydroxyl radicals : formation of S∴N three-electron bonded radical complexes, Chem. Commun., 2005, 587-589.

[16] Bobrowski K, Schöneich C., Decarboxylation mechanism of the N-terminal glutamyl moiety in γglutamic acid and methionine containing peptides, Radiat. Phys. Chem., 1996, **47**, 507-510.

[17] Pogocki D., Ghezzo- Schöneich E., Schöneich C., Conformational flexibility controls proton transfer between methionine hydroxy sulfuranyl radical and the N-terminal amino group in Thr-$(X)_n$-Met peptides, J. Phys. Chem. B, 2001, **105**, 1250-1259.

[18] Wardman P., Reduction potentials of one-electron couples involving free radicals in aqueous solution, J. Phys. Chem. Ref. Data, 1989, **18**, 1637-1755.

[19] Pecht I., Farver O., Pulse radiolysis: A tool for investigating long-range-electron transfer in proteins, *in* "Photochemistry and radiation chemistry. Complementary methods for the study of electron transfer", Wishart J. F., Nocera D. G. (eds.), Adv. Chem. Ser., 1998, **254**, ACS, Washington, D.C., 65-69.

[20] Houée-Levin C., Sicard-Roselli C., Radiation chemistry of proteins, *in* "Radiation chemistry: present status and future prospects", Jonah C., Rao B.M. (eds.), Elsevier, 2001, 53-84.

[21] Hsu J.L., Hsieh Y., Tu C.K., O'Connor D., Nick H.S., Silverman D.N., Catalytic properties of human manganese superoxide dismutase, J. Biol. Chem., 1996, **271**, 17687-17691.

[22] Jenney F.E. Jr., Verhagen M.F.J. M., Cui X., Adams, M.W.W., Anaerobic microbes: oxygen detoxification without superoxide dismutase**,** Science, 1999, **286**, 306-309.

[23] Candeias L.P., Gebicka L,. Investigation of mechanisms of peroxidase-catalyzed reactions by radiation-chemical techniques, J. Radioanal. Nucl. Chem., 1998, **232**, 29-34.

[24] Bonifacic M., Hug G.L., Schoeneich C., Kinetics of the reaction between sulfide raducal cation complexes [SS]+ and [SN]+ and superoxide or carbon dioxide radical anion, J. Phys. Chem. A, 2000, **104**, 1240-1245.

[25] Santus R., Patterson L.K., Hug G.L., Bazin M., Mazière J.C., Morlière P., Interactions of superoxide anion with enzyme radicals : kinetics of reaction with lysozyme tryptophan radicals and corresponding effects on tyrosine electron transfer, Free Rad. Res., 2000, **33**, 383-391.

Chapter 17

Radiation-induced damage of membrane lipids and lipoproteins

Monique GARDES-ALBERT

Introduction

If radiation-induced damages to DNA have long been considered as the only critical events for the cell machinery, it is now admitted that cell membranes are also significant biological targets of ionizing radiation [1]. Lipids are the major constituents of biomembranes and of lipoproteins. Their amphiphilic structure, namely hydrophilic polar heads and hydrophobic hydrocarbon tails, allowing the build up of dense networks of intermolecular bonds, favours the packing of membrane lipids into bilayers. In addition, proteins are inserted into the lipidic matrix where they are associated with lipid domains. The cell membrane is not a simple barrier defining the boundaries of the cell, since it modulates signals from the extracellular medium into the cell and it also controls intercellular communications [2].

Ionizing radiations such as γ-rays, are able to directly ionize the lipid/protein network of biomembranes, but also to ionize the water molecules surrounding all the cell constituents. Molecules of H_2O which are present in both intra- and extracellular compartments, undergo radiation-induced decomposition into free radical species. The proportion of directly ionized biological targets versus indirectly-mediated damages, via free radicals from water radiolysis, remain unclear and controversial, often depending on the studied systems. In this chapter, it will be focused on the indirect effects coming from the reactions of lipids and lipoproteins with oxygenated free radicals produced by radiation-induced decomposition of water. These effects are related to numerous works devoted to free radical-induced damages on aqueous lipidic model systems (micelles, liposomes, lipoproteins) submitted to ionizing radiation. Moreover, the mechanism of formation of free radicals from water radiolysis being well known, together with the radiolytic yields of production of each radical species (Chapter 1), it is relatively easy to obtain from radiation dose effects, kinetic results on the chemical reactions of these free radicals with biological targets (such as lipids) dissolved in water [3]. For these reasons, water radiolysis can be seen as a powerful tool for studying radical mechanisms on defined biological systems.

Lipid peroxidation has been found as the main damage of membrane lipids and lipoproteins, initiated by oxygenated free radicals. A key feature of this phenomenon is its propagating capacity into the lipidic network by a chain reaction, due to the tight packing of the lipid molecules, leading to lipid hydroperoxides as major initial products. In addition, radiation-induced oxidative fragmentation of lipids has been described [4], together with changes in the physical state (fluidity/rigidity, permeability, ...) and as a function of biomembranes [1,2]. In the present review, recent knowledge about the chemical nature and mechanism of free radical-induced lipidic damages will be examined, from simple models of lipid aggregates submitted to γ-rays. The consequences of the presence of new peroxidized lipidic products and short-chain fragments into the cell membranes will be discussed. Oxidative stress phenomena can be indeed responsible for the *in vivo* peroxidation of lipoproteins leading to atherosclerosis [5].

Composition of membrane lipids

Fatty acids, glycerophospholipids and sphingolipids, are the three major classes of lipids entering in the composition of biological membranes. Fatty acids are carboxylic acids (RCOOH) with long-chain hydrocarbon side groups, usually ranged between 14 and 20 carbon atoms. In **Table 1** are listed some common biological fatty acids. In higher plants and animals, the predominant fatty acid residues are those of the C16 and C18 species [6] named palmitic, oleic, linoleic and stearic acids (**Tab. 1**). Over half of the fatty acid residues of plant and animal lipids are often polyunsaturated (containing two or more double bonds). They are symbolized by PUFAs (Poly Unsaturated Fatty Acids). For example, 18:2 means that the hydrocarbon chain contains 18 carbon atoms and 2 carbon-carbon double bonds (see linoleic acid in **Tab. 1** and **Fig. 1A**). Almost all fatty acid double bonds have the *cis*-configuration as it can be seen for linoleic acid as an example, in the **Figure 1A**.

Table 1. *Names and formulas of fatty acids.*

Name[1]	Symbol[2]	Structure
Unsaturated fatty acids		
Palmitoleic acid (9-Hexadecenoic acid)	16:1	$CH_3(CH_2)_5CH{=}CH(CH_2)_7COOH$
Oleic acid (9-Octadecenoic acid)	18:1	$CH_3(CH_2)_7CH{=}CH(CH_2)_7COOH$
Linoleic acid (9,12-Octadecadienoic acid)	18:2	$CH_3(CH_2)_4(CH{=}CHCH_2)_2(CH_2)_6COOH$
γ-Linolenic acid (6,9,12-Octadecatrienoic acid)	18:3	$CH_3(CH_2)_4(CH{=}CHCH_2)_3(CH_2)_3COOH$
Arachidonic acid (5,8,11,14-Eicosatetraenoic acid)	20:4	$CH_3(CH_2)_4(CH{=}CHCH_2)_4(CH_2)_2COOH$
Saturated fatty acids		
Palmitic acid (Hexadecanoic acid)	16:0	$CH_3(CH_2)_{14}COOH$
Stearic acid (Octadecanoic acid)	18:0	$CH_3(CH_2)_{16}COOH$

1*) Common name (systematic name in brackets).* 2*) Number of carbon atoms : number of carbon-carbon double bonds. The systematic name of fatty acids derives from the total number of carbon atoms of the chain (for example, "hexadeca" for 16 atoms, "octadeca" for 18 atoms,…), the other numbers designing the carbon atoms bearing the carbon-carbon double bonds. For example, linoleic acid is the 9,12-Octadecadienoic acid which has 2 double bonds beared on the 9th and 12th carbon atoms, the 1st being attributed to the carboxylic acid function (**Fig. 1A**).*

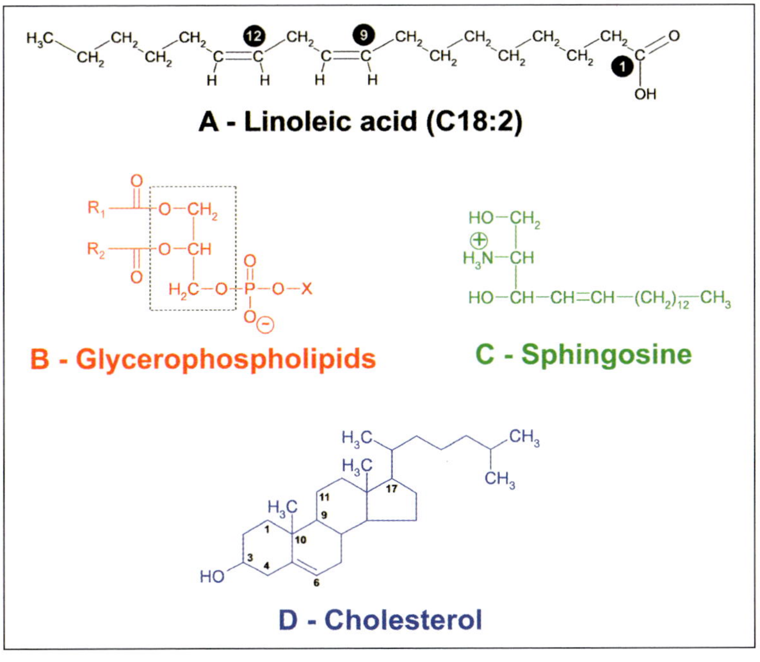

Figure 1 *: Structures of some membrane lipids.* A*) Linoleic acid as an example of PUFA.* B*) General structure of glycerophospholipids, R1, R2 being long chain fatty acid residues and X a polar group; the glycerol backbone is inside the dashed line.* C*) Formula of sphingosine, the simplest member of the sphingolipids family.* D*) Cholesterol which can be esterified to long chain fatty acid on its C3 OH group.*

Glycerophospholipids are, with sphingolipids, the major lipid components of biological membranes. Glycerophospholipids (general formula in **Figure 1B**) derive from glycerol (CH_2OH-CHOH-CH_2OH) by esterification to two long-chain fatty acid residues (R1 and R2 in **Figure 1B**), the third alcohol function being transformed into a phosphoryl-X group. X is a polar group such as an alcohol or an amine. Sphingolipids are derivatives of the amino alcohol sphingosine (**Fig. 1C**). In this latter class of lipids, there are several different families such as ceramides, sphingomyelins and sphingoglycolipids with complex head groups of up to four sugar residues.

Cholesterol (**Fig. 1D**), which is the metabolic precursor of steroid hormones, is a major component of plasma membranes. Its fused ring system brings a greater rigidity than other membrane lipids. Cholesterol can be esterified on the C3 OH-group to long-chain fatty acids to form cholesteryl esters which are major components of lipoproteins.

Lipid aggregates and model systems

The common feature of lipids is their amphiphilic structure, characterized by a hydrophobic part (the non-polar aliphatic acid residue(s)) and a hydrophilic part (for example, the polar carboxylic acid function in fatty acids or the phosphoryl head in glycerophospholipids, ...). When lipids are dissolved in water, they spontaneously aggregate, the non-polar tails being associated by weak Van der Waals interactions. Single tailed lipids as PUFAs tend to form micelles (**Fig. 2A**) which are spheroidal aggregates where the polar carboxylate heads are in contact with water. Micelles appear in aqueous solution when the fatty acid concentration surpasses the critical micelle concentration (symbolized by cmc), namely the upper limit concentration above which monomers aggregate. The cmc depends on the nature of the fatty acid and on the solution conditions (pH, temperature, ...). For example, at room temperature, the value of linoleate cmc is close to 2×10^{-3} mol l^{-1} in aqueous solution at pH = 10.5 [7]. This means that at concentration below 2×10^{-3} mol l^{-1}, linoleate anions cannot form micelles but only very small aggregates (oligomers) and monomers, whereas above 2×10^{-3} mol l^{-1} they are mainly associated into micelles. The basic pH (10.5) of the aqueous medium is necessary in order to deprotonate the carboxylic acid function of PUFAs giving carboxylate groups COO^-, namely polar charged heads (surrounded by polar water molecules). PUFAs micelles in aqueous solution are the simplest lipidic models allowing a mechanistic approach of lipid peroxidation under ionizing radiations (see section "Quantitative determination of hydroperoxides").

Glycerophospholipids and sphingolipids are biological lipids which have two large hydrophobic tails, one of them being often a PUFA chain. They exhibit a very low cmc,

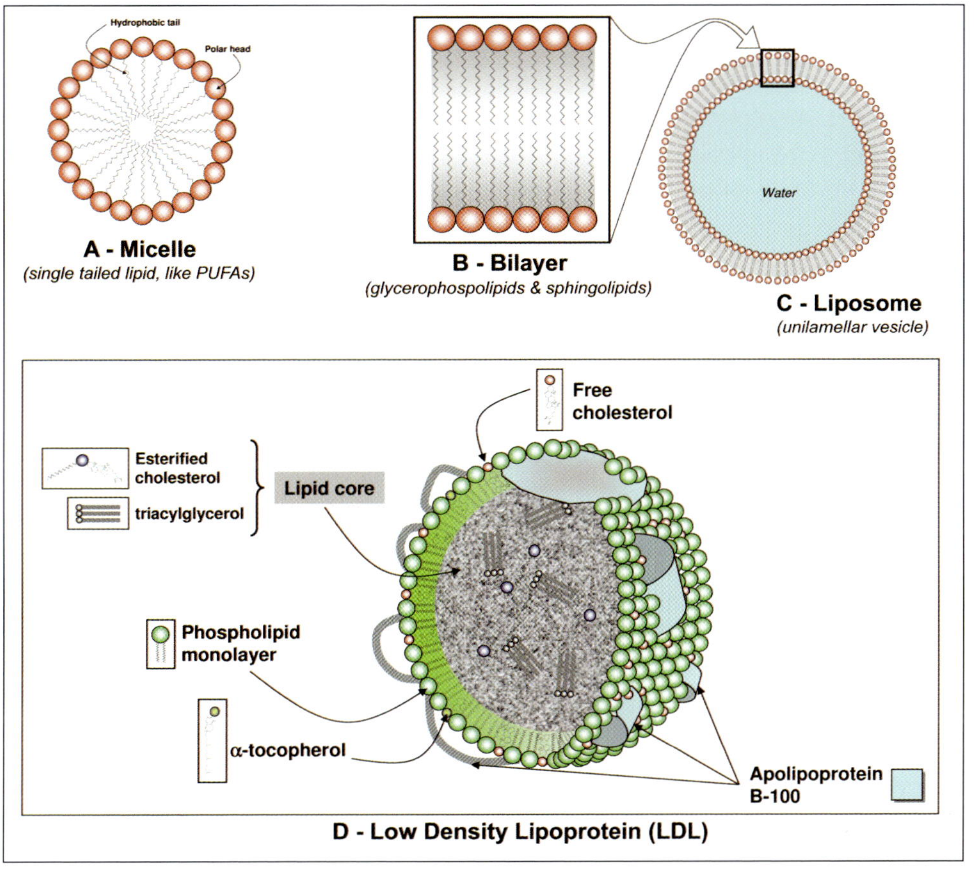

Figure 2: *Lipid aggregates.* A) *Micelle, spheroïdal aggregate of single tailed fatty acid, the polar heads being in contact outside with water molecules.* B) *Bilayer, two dimensional fluid, composed of glycerophospholipids or sphingolipids, each molecule possessing two hydrophobic tails.* C) *Liposome, spheroïdal vesicle filled with water, bounded by a single bilayer.* D) *LDL, lipids/protein/antioxidants aggregate, carrier of cholesterol in blood plasma.*

generally lower than 10^{-6} mol l^{-1}, and tend to form bilayers (**Fig. 2B**). Lipid bilayers are two dimensional fluids in which lipid molecules can diffuse lateraly but not across the bilayer [6]. Liposomes are vesicles filled with water, constituted of several bilayers (multilamellar vesicles, MLV), which can rearrange to form unilamellar vesicles bounded by a single bilayer (**Fig. 2C**). According to the size distribution, one considers the small unilamellar vesicles (SUV) with a diameter of about 50 nm and the large unilamellar vesicles (LUV) whose diameter is superior to 100 nm. The liposome size distribution depends largely on the method of production of the vesicles, the sonication (agitation by ultrasonic vibrations) giving mainly SUVs, whereas the extrusion (submission to high pressures) generates LUVs. SUVs being highly curved, the unsaturated fatty acyl chains have a better exposure to water and consequently SUVs are highly oxidizable *via* radiation-induced water free radicals. By contrast, LUVs (high diameters) characterized by a tight packing of phospholipid molecules are less penetrable by water molecules and consequently less oxidizable than SUVs [8]. Liposomes can be considered as rather good model systems since they reflect the arrangement of lipid bilayers of cell membranes.

However, to get still closer the living membranes, it is necessary to add proteins into the lipid matrix. Lipoproteins are suitable candidates to study lipid peroxidation and protein oxidation under ionizing radiation, allowing to specify lipid/protein interactions. Low Density Lipoproteins (LDLs) (**Fig. 2D**) isolated from human plasma, are composed of a single big protein (apolipoprotein B-100) surrounded by lipids (choline phospholipids, cholesteryl esters and triacylglycerols). In addition, antioxidants (mainly α-tocopherol and β-caroten) are located in the lipidic part. LDLs are now considered as very good membrane models allowing to determine the reciprocal influence of the oxidation of both lipid and protein moieties at the molecular level, and to specify the role of antioxidants under oxidative stress conditions, specially those of water radiolysis [5, 9]. It can be noticed that other membrane models such as erythrocyte membranes have been investigated under radiation exposure, providing interesting results on the post-irradiation effects to the lipid and protein moieties, together with structural changes in the membrane arrangement (see for example [1]).

Water radiolysis

When liposome or micelle suspensions in diluted aqueous medium, are irradiated by γ-rays (from a radioactive source of ^{137}Cs or ^{60}Co), they are attacked by the free radicals generated by water radiolysis. Indeed, there is no direct ionisation of the lipids by the γ-rays if their concentration remains lower than 10^{-2} mol l^{-1}, whereas the molecules of water (in high concentration of 55 mol l^{-1} in aqueous solutions) are the only molecular targets to be ionized (Chapter 1).

The radiolysis of water leads to the formation of the three radical species $^{\bullet}OH$ (hydroxyl radical), $H^{\bullet}$ (hydrogen atom) and e_{aq}^{-} (hydrated electron), within the nanosecond time scale [3]. In aerated medium (that is in the presence of dioxygen, O_2 concentration dissolved in water being 2×10^{-4} mol l^{-1}), the free radicals $H^{\bullet}$ and e_{aq}^{-} are replaced by $HO_2^{\bullet}$ and $O_2^{\bullet -}$ radical species, respectively, which are related by an acid/base equilibrium (pK_a $(HO_2^{\bullet}/O_2^{\bullet -}) = 4.8$). The radiolytic yields (G-values expressed in moles per Joule) of each radical species are well known: 2.8×10^{-7} mol J^{-1} and 3.4×10^{-7} mol J^{-1}, respectively for $^{\bullet}OH$ and $O_2^{\bullet -}$ free radicals at pH = 7 (Chapter 1) [3].

Hydroxyl ($^{\bullet}OH$) and superoxide ($O_2^{\bullet -}$) radical species are thus the protagonists of the oxidative stress initiated by water radiolysis to diluted biological targets [10]. But they exhibit very different properties against lipids. Hydroxyl free radical is the strongest oxidant species because its one-electron reduction potential is very high (E° $^\bullet OH/H_2O = 2.34\ V_{NHE}$ at pH = 7, versus the potential of normal hydrogen electrode (NHE) as reference, under standard conditions) and its second order rate constants are diffusion controlled, namely close to the upper limit of 10^{10}-10^{11} $l\ mol^{-1}\ s^{-1}$ (Chapter 1). In contrast, superoxide free radical is not an efficient initiator of lipid oxidation. Indeed, even if its one-electron reduction potential is high enough ($E^\circ\ O_2^{\bullet -}/H_2O_2 = 0.93\ V_{NHE}$ at pH = 7), its second order rate constants against lipids are very low, generally lower than 10^2 $l\ mol^{-1}\ s^{-1}$ [11]. In other words, in aqueous medium, superoxide free radicals do not attack lipids. However, it has been shown that their protonated form, $HO_2^\bullet$, can react more rapidely with PUFAs (10^3 $l\ mol^{-1}\ s^{-1}$ [12]) than does $O_2^{\bullet -}$. Note that at pH = 7, $HO_2^\bullet$ radicals represent less than 1% of the total amount of $O_2^{\bullet -}$ radicals (pK_a ($HO_2^\bullet/O_2^{\bullet -}$) = 4.8) in the irradiated medium. Hence, it can be assumed that at pH = 7, the predominant initiating species are hydroxyl radicals.

Radical-chain mechanism of lipid peroxidation

Hydroxyl radicals are able to initiate the one-electron oxidation of unsaturated acyl chains symbolized here by LH (as linoleic acid (18:2), for example), by abstracting a single H-atom from the aliphatic chain leading to a carbon-centred radical $L^\bullet$ (alkyl radical) (reaction (1)).

$$LH + {}^\bullet OH \rightarrow L^\bullet + H_2O \qquad (1)$$

The removed hydrogen atom is preferentially located on a methylene group (CH_2) between two adjacent C=C double bonds (bis-allylic position) (see **Fig. 3** for linoleic acid). Indeed, such C-H bonds are weaker than others along the saturated carbon chain and they represent critical target sites [13,14].

The fate of the carbon-centred radical $L^\bullet$ (which is a resonant pentadienyl radical in the case of linoleic acid, **Fig. 3**) is to lead, in a very rapid reaction with dioxygen, to the formation of a peroxyl radical, $LO_2^\bullet$ (reaction (2)).

$$L^\bullet + O_2 \rightarrow LO_2^\bullet \qquad (2)$$

In the case of linoleic acid, two isomeric peroxyl radicals are formed (**Fig. 3**). Such peroxyl radical species are expected to abstract H-atoms from other fatty acid molecules (reaction (3)) giving lipid hydroperoxides LO_2H (9 and 13-hydroperoxyoctadecadienoic acid in the case of linoleic acid (**Fig. 3**)), which are the initial non-radical products formed during lipid peroxidation.

$$LO_2^\bullet + LH \rightarrow LO_2H + L^\bullet \qquad (3)$$

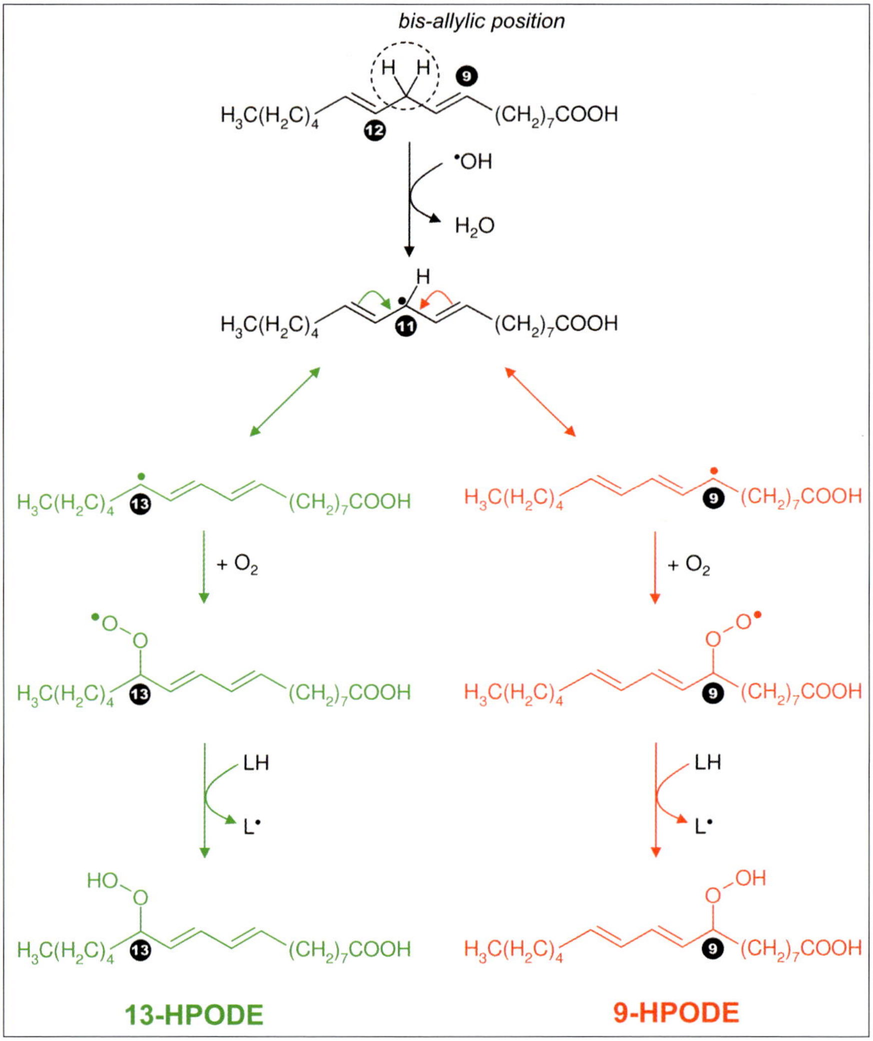

Figure 3: *Mechanism of peroxidation of linoleic a cid (18:2) initiated by $^{\bullet}OH$ free radicals from water radiolysis. LH symbolizes another molecule of linoleic acid in micelles. 9- and 13-HPODE are the main hydroperoxides formed, possessing a conjugated dienic structure.*

The resulting alkyl free radical $L^{\bullet}$ can in turn react with dioxygen (reaction (2)), propagating the radical-chain mechanism (reactions (2) and (3)) inside the lipid aggregates. These radical-chain reactions amplify the peroxidation process since one single $^{\bullet}OH$ radical can lead to several alkyl radicals ($L^{\bullet}$), peroxyl radicals ($LO_2^{\bullet}$) and hydroperoxide molecules (LO_2H). To stop the radical-chain mechanism, two radical species have to react together (for example reactions (4) and (5), reaction (4) being the most probable in an oxygenated medium) leading to non-radical products.

$$LO_2^{\bullet} + LO_2^{\bullet} \rightarrow \text{non-radical products} \quad (4)$$
$$LO_2^{\bullet} + L^{\bullet} \rightarrow \text{non-radical products} \quad (5)$$

From this free radical-chain mechanism it appears that hydroperoxides (LO_2H) are the major early reaction products of lipid peroxidation. Their structure are very diversified since it depends on the number of allylic hydrogen atom(s) to be abstracted along the acyl chain of the PUFA (see next section). However, the lipid degradation process does not stop after the formation of hydroperoxides. Indeed, hydroperoxides are not very stable products (apart from irradiation) and they undergo spontaneous decomposition (oxidative cleavage) that generates numerous breakdown products (see section "Oxidative fragmentation of lipids").

Quantitative determination of hydroperoxides, initial markers of lipid peroxidation

In micelles

Hydroperoxides (ROOH) are the next non radical products to be formed in aerated medium, after $^{\bullet}$OH-attack on the unsaturated carbon chain. Hence their quantification as a function of radiation doses is particularly of interest. They can be specifically titrated by reverse-phase HPLC (High Pressure Liquid Chromatography) using detection by chemiluminescence [15,16]. As it can be shown on a simple lipidic model such as linolenate (18:3) micelles, increasing radiation doses lead to increasing hydroperoxide concentrations (**Fig. 4**). This ROOH formation is linolenate concentration-dependent, *i.e.* at a given radiation dose, hydroperoxide concentration is higher in the presence of 25 mM than of 7.5 mM linolenate (**Fig. 4**). The highest part of hydroperoxides comes generally from the H-abstraction on the bis-allylic positions of PUFAs (see section above). Thus conjugated dienes are simultaneously formed on the unsaturated carbon chain. Hence conjugated dienes (CD) are also initial markers of lipid peroxidation. They can be determined by spectrophotometric absorption at 234 nm (maximum absorption wavelength), with a molar extinction coefficient of 28000 l mol^{-1} cm^{-1} [17]. As it can be seen in **Figure 4**, the concentration of conjugated dienes formed in linolenate micelles are increasing with the radiation dose, and they are, like hydroperoxides, linolenate concentration-dependent. However, for a given linolenate concentration, the CD concentration is lower than that of hydroperoxides whatever the radiation dose, meaning that some hydroperoxides do not possess a dienic structure.

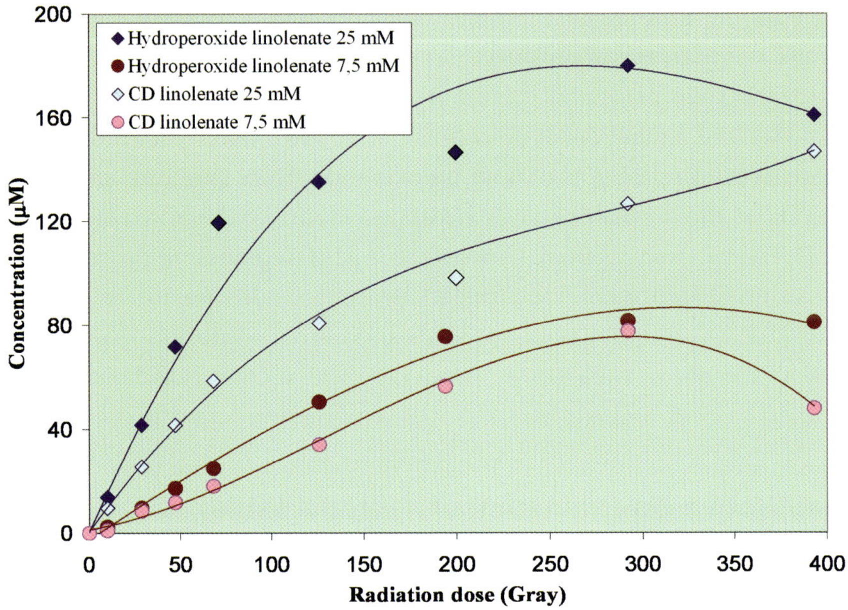

Figure 4: *Formation of hydroperoxides and conjugated dienes (CD) as a function of the radiation dose (dose rate = 9.8 Gy.min^{-1}) in linolenate (18:3) aqueous micellar solutions at pH = 10.5. Influence of the concentration of linolenate (25 mM and 7.5 mM) on the peroxidation process [16].*

The radiolytic yields of formation of hydroperoxides (G(hydro)) are determined from the slope of the initial tangent of the curves (concentration versus radiation dose, as in **Fig. 4**). An important lipid concentration dependency is observed in micellar medium where, above the cmc, the yields of oxidized products formation increase steeply with PUFA concentration. This means that a chain-mechanism is involved in the formation of hydroperoxides and conjugated dienes. As an example, in the case of a 25 mM linolenate concentration (**Fig. 4**), the G-values are 20.0×10^{-7} mol J^{-1} and 9.3×10^{-7} mol J^{-1} for hydroperoxides and conjugated dienes, respectively. These yields are considerably larger than $G_{\cdot OH} = 2.8 \times 10^{-7}$ mol J^{-1}, since the hydroxyl radicals are initiators of the oxidation chain mechanism.

It has been demonstrated, not only with linolenate (18:3) aqueous solutions but also with linoleate (18:2), and more recently with arachidonate (20:4) solutions [18], that the hydroperoxides were composed of two different kinds of molecules according to the concentration of PUFA in the irradiated medium. Indeed, for a given PUFA, above the critical micellar concentration, namely when micelles are formed, one type of hydroperoxide predominates, whereas when monomers (or very small aggregates) are dispersed in solution (below the cmc), other types of hydroperoxide are formed. This phenomenon is illustrated in **Figure 5** for arachidonate. We can see that monohydroperoxides are produced in micelles whereas in monomers, either a cyclic hydroperoxide or an aliphatic dihydroperoxide is obtained [18].

Such differences seem to arise from the mechanism of their formation which privilegiates a self-rearrangement in monomers, instead of an intermolecular reaction in aggregates. The ratio of "micellar" hydroperoxide concentration over "monomer" hydroperoxide concentration appears to be a signature of the balance micelles/monomers in aqueous medium [18].

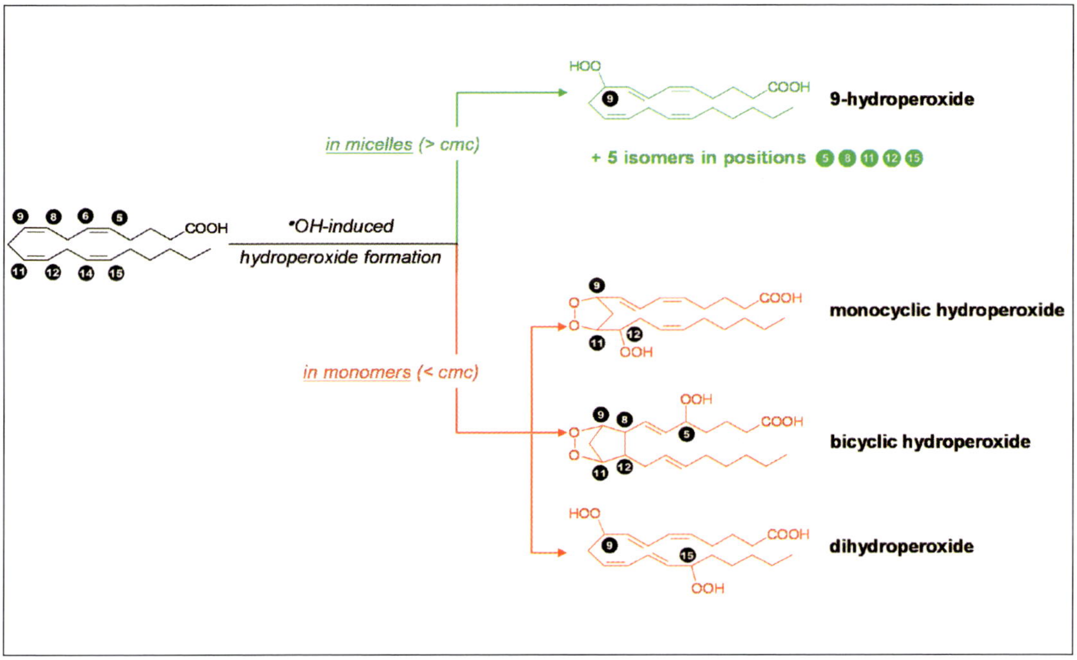

Figure 5: *Proposed structures of hydroperoxides formed in arachidonate (20:4) aqueous solutions (pH = 10.5) submitted to γ-rays. Monohydroperoxides (6 isomers in positions 5, 8, 9, 11, 12, 15) are formed in micellar medium (high arachidonate concentration) whereas monocyclic and bicyclic hydroperoxides, together with dihydroperoxides are produced in dispersed monomers (low arachidonate concentration)* [18].

In liposomes

In the case of liposomes of PLPC (1-palmitoyl-2-linoleoyl-sn-glycero-3-phosphocholine) submitted to free hydroxyl radicals attack (generated by γ-radiolysis of water), phosphatidylcholine hydroperoxides are the initial markers of lipid peroxidation. Since the molecule of PLPC contains only one bis-allylic position located on the linoleyl acyl chain, the •OH-induced hydroperoxides are characterized by the OOH groups at positions 9 and 13 on the linoleyl chain, as in the case of linoleate micelles (**Fig. 3**). Moreover, these hydroperoxides exhibit a conjugated dienic structure. A G-value of total hydroperoxide formation has been determined to be equal to 6×10^{-7} mol J^{-1}, in γ-irradiated small unilamellar vesicles (SUVs) of PLPC 250 µmol l^{-1} [19]. Such a G-value which is 2 fold the G-value of •OH-formation, also indicates a chain-mechanism of hydroperoxide production. In addition, other products such as hydroxides, epoxides and fragments have been identified using HPLC/MS analysis. In LUVs (high diameters), because of the tight packing of PLPC molecules, $^{\bullet}OH$ radicals have to react first with the polar head before reaching the linoleyl chains, and for this reason kinetic results are very different between LUVs and SUVs peroxidation, LUVs being less oxidizable than SUVs [8].

Oxidative fragmentation of lipids

Lipid hydroperoxides are not very stable products and their decomposition can lead to the formation of a lot of breakdown products such as short-chain aldehydes (**Fig. 6**), epoxides, ketoaldehydes, and alkanes (pentane, hexane, for example). These hydroperoxide decomposition may result from the action of traces of reducing metal cations (Fe^{2+}, Cu^{+}) which catalyses the cleavage of hydroperoxides according to "Fenton like reactions" (reaction (6)), in which LO_2H acts like H_2O_2. The alkoxyl radicals $LO^{\bullet}$ so formed are able to abstract H-atom leading to hydroxylated products (LOH) or to generate epoxy-allyl radical by a self-addition on a double bond or still to undergo a scission giving aldehydic products [13]:

$$LO_2H + Fe^{2+} \rightarrow LO^{\bullet} + {}^{-}OH + Fe^{3+} \qquad (6)$$

Consequently, in addition to hydroperoxides, a lot of secondary oxidized lipidic compounds, mainly short-chain aldehydes, may appear and represent late markers of lipid peroxidation. As an example, malondialdehyde (MDA, **Fig. 6**) is known to be the most abundant lipid peroxidation aldehyde whose determination by 2-thiobarbituric acid (TBA) is one of the most common assays in lipid peroxidation studies [20]. However, it can be noticed that the TBA assay method [21] is not specific of MDA titration since it also can detect a variety of peroxides and secondary degradation products of lipid peroxidation called

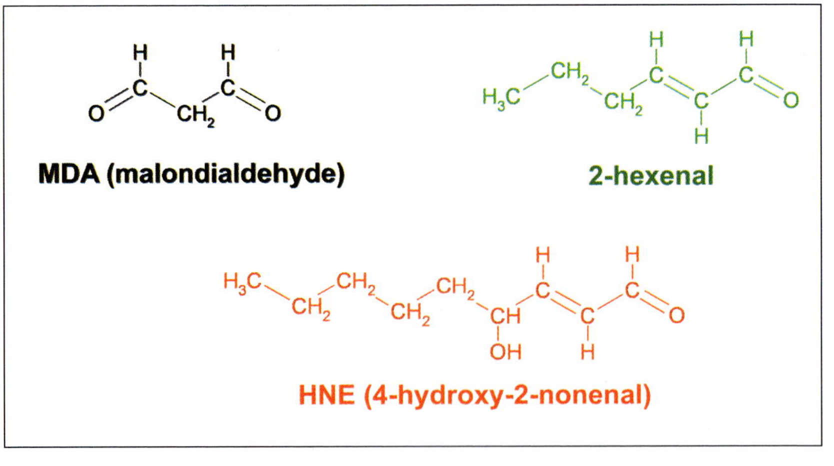

Figure 6: *Short-chain aldehydic products resulting from the oxidative fragmentation of lipid hydroperoxides.*

2-thiobarbituric acid-reactive substances (TBARS). Hence, the determination of TBARS must be used as a global index of lipid peroxidation. Aldehydic molecules are involved in cytotoxic processes because they can easily react with DNA, protein, and phospholipids leading to covalent adducts. In the case of the 4-hydroxy-2-nonenal (HNE, **Fig. 6**) which is produced during peroxidation of PUFAs such as linoleic and arachidonic acids, HNE is known to react with amino acids (cystein, histidine, lysine) and with cellular target proteins, inactivating enzymes [20].

However, hydroperoxides decomposition is not the only process undergoing short-chain lipid fragments. Indeed, γ-radiolysis of multilamellar liposomes of glycerophospholipids such as cardiolipin, phosphatidylglycerol and phosphatidylinositol, has been characterized by the formation of phosphatidic acid resulting from the cleavage of the phosphatidylglycerol moiety, as the main product [4]. Hence the polar heads of glycerophospholipids appear to be targeted by $^{\bullet}OH$-attacks. The determination of phosphatidic acid has been performed using matrix-assisted laser desorption/ionisation time-of-flight (MALDI-TOF) mass spectrometry [4]. The formation of carbon-centred free radicals, in β-position with respect to the phosphoester bond and containing a hydroxyl group, seems to be a key step of this type of fragmentation process. A similar mechanism has been proposed to explain the formation of lysophosphatidylcholine in γ-irradiated liposomes of PLPC (LUVs) [8]. Radiation-induced free radical fragmentation of lipids in their polar moiety seems to play an important role as signaling molecules in living systems [4].

Peroxidation of low density lipoproteins

It has well been established that PUFAs contained in phospholipids and cholesteryl esters (see composition of LDLs in section "Lipid aggregates and model systems", and in **Fig. 2D**) are targeted sites of radiation-induced oxygen free radicals attack in LDL particles. Mechanisms of lipid peroxidation of LDLs, initiated by oxygen free radicals such as $^{\bullet}OH$, $^{\bullet}OH/O_2^{\bullet -}$, $O_2^{\bullet -}/HO_2^{\bullet}$ have been proposed, based on the determination of the radiolytic yields of consumption of α-tocopherol and β-carotene (endogenous antioxidants of LDL), and of formation of conjugated dienes and TBARS (for a review see [5]). However, as we have already seen, TBARS are not specific markers of lipid peroxidation (section above). Recently, the determination of defined hydroperoxides such as hydroperoxides of phosphatidylcholines (linoleate, arachidonate and docosohexaenoate) and of cholesteryl esters (linoleate and arachidonate) have been developed using HPLC coupled with a chemiluminescence detection [9]. These hydroperoxide formations are kinetically characterized by three steps, namely (1) at low radiation doses, the inhibition step resulting from the presence of α-tocopherol and β-carotene, able to protect LDL by scavenging radiolytic-induced oxygenated free radicals, (2) at higher radiation doses, the propagation step resulting from the lipid peroxyl radicals formation (LOO$^{\bullet}$) by a chain-mechanism, leading to hydroperoxides of phosphatidylcholines and of cholesteryl esters, and (3) the chain-termination and decomposition phase. The kinetic behaviour of conjugated dienes is similar to those of hydroperoxides, the propagation step beginning at the end of the inhibition step, when there is no more antioxidants. As an example of quantification of lipid markers for LDLs 3 g l^{-1}, the radiolytic yield of CD formation (3.4×10^{-7} mol J^{-1}) is found slightly higher than $G_{^{\bullet}OH}$ (2.8×10^{-7} mol J^{-1}) due to the presence of a short chain reaction, and the G-value of hydroperoxide formation is 10 times lower than the yield of CD production, meaning that conjugated dienes other than hydroperoxides (such as alcohols, epoxides and endoperoxides) are taken into account [9]. Besides, it has been shown that the phosphatidylcholines located on the outer surface of LDLs were more accessible than the cholesteryl ester molecules (in the lipidic core), to the free radicals attack.

In addition, the protein moiety contained in LDLs (apoB-100) is also targeted by $^{\bullet}$OH radicals, since western blot immunoassays allow to detect carbonyl groups (C=O) into protein side chains [9]. These assays involve the derivatization to 2,4-dinitrophenylhydrazone of the carbonyl groups by reaction with 2,4-dinitrophenylhydrazine (2,4-DNPH). It has been shown that carbonylated apoB-100 appeared at low radiation doses, during the inhibition phase of CD and hydroperoxides, meaning that the protein was not protected by the endogeneous antioxidants which are localized in the lipid part of LDLs and that the polypeptidic chain was attacked by $^{\bullet}$OH free radicals. This apoB-100 carbonylation was radiation dose-dependent, its radiolytic yield being relatively low (1.8×10^{-8} mol J^{-1}) because $^{\bullet}$OH free radicals simultaneously react on all other molecular components of LDLs (lipids, antioxidants, ...). Moreover, the fragmentation of the carbonylated apoB-100 may occur, but this process which only begins during the lipid peroxidation propagation phase, seems to be related to the interactions between the oxidized parts of lipids and protein. Hence, whereas radiolytic-induced apoB-100 carbonylation appears to be independent on lipid peroxidation, in contrast oxidative fragmentation seems to be dependent on it. All these results obtained from radiolysis conditions represent an accurate and rigourous approach of the oxidative stress phenomena which can be responsible of the *in vivo* peroxidation of LDLs leading to atherosclerosis [5].

Conclusion

Effects of ionizing radiation on lipid molecules have been understood by studying model systems which are simpler than the real biological membranes, such as PUFA micelles and liposomes. The formation of lipid oxidative modifications of PUFAs appears as a dynamic process initiated by hydroxyl free radicals generated by water radiolysis, amplified by a propagating-chain mechanism involving alkyl and peroxyl free radicals, and leading not only to hydroperoxides but also to a lot of other lipidic oxidized end-products. Kinetic data, such as radiolytic yields of production of hydroperoxides and conjugated dienes which are early products of lipid peroxidation, have allowed us to establish the reaction schemes of their formation. Knowing that hydroperoxides and other lipid oxygenated products are more hydrophilic than the initial hydrocarbon chains of PUFAs, their appearance can lead to morphological changes in the bilayers, together with disturbances in membrane functions. Moreover, hydroxyl free radicals may initiate the oxydative processes not only within the hydrophobic moiety of lipids but also in the hydrophilic polar heads leading to fragments such as phosphatidic acid and lysophosphatidylcholine residues. Such fragments seem to regulate a variety of cellular functions and to stimulate various signalling pathways. Hence, membrane constituents which are molecular targets of ionizing radiations have to be considered as important factors the damages of which are susceptible to largely control the viability of the cells.

Acknowledgments

I am especially indebted to Fabrice Collin for his talented designs of lipid aggregates and molecular mechanisms and for his judicious comments. I would like to acknowledge Prof. Dominique Bonnefont-Rousselot for her precious collaboration all along the story of oxidized lipoproteins (EA 3617, *Faculté de Pharmacie, Université Paris 5*).

References

[1] Benderitter M., Vincent-Genod L., Pouget J.P., Voisin P., The cell membrane as a biosensor of oxidative stress induced by radiation exposure: a multiparameter investigation, Radiat. Res., 2003, **159**, 471-483.

[2] Albanese J., Dainiak N., Modulation of intercellular communication mediated at the cell surface and on extracellular, plasma membrane–derived vesicles by ionizing radiation, Experimental Hematology, 2003, **31**, 455-464.

[3] Spinks J.W.T., Woods R.J., Water and inorganic aqueous systems, in "Introduction to Radiation Chemistry", 3rd ed., Wiley, New York, 1990, p. 243-313.

[4] Shadyro O.I., Yurkova I.L., Kisel M., Brede O., Arnhold J., Formation of phosphatidic acid, ceramide, and diglyceride on radiolysis of lipids: identification by MALDI-TOF mass spectrometry, Free Radical Biology and Medicine, 2004, **36**, 1612-1624.

[5] Bonnefont-Rousselot D., Gamma radiolysis as a tool to study lipoprotein oxidation mechanisms, Biochimie, 2004, **86**, 903-911.

[6] Voet D., Voet J.G., Lipids and membranes (Chapter 11), *in* "Biochemistry", 2nd ed., Wiley & Sons, INC., New York, 1995, p. 277-329.

[7] Hauville C., Rémita S., Thérond P., Rouscilles A., Couturier M., Jore D., Gardès-Albert M., Determination of the yield of radiation-induced peroxidation of sodium linoleate in aqueous monomeric and micellar solutions, Radiat. Res., 1998, **150**, 600-608.

[8] Vitrac H., Liposomes de phosphatidylcholine comme modèles de la peroxydation membranaire. Action des radicaux hydroxyles et mécanismes réactionnels, Thesis, Université René Descartes (Paris 5), 2004.

[9] Jedidi I., Thérond P., Zarev S., Cosson C., Massot C., Jore D., Gardès-Albert M., Legrand A., Bonnefont-Rousselot D., Paradoxical protective effect of aminoguanidine toward low-density lipoprotein oxidation: inhibition of apolipoprotein B fragmentation without preventing its carbonylation. Mechanism of action of aminoguanidine, Biochemistry, 2003, **42**, 11356-11365.

[10] Gardès-Albert M., Jore D., Aspects physicochimiques des radicaux libres centrés sur l'oxygène, in "Radicaux libres et stress oxydant", Delattre J., Beaudeux J-L., Bonnefont-Rousselot D., (eds), Lavoisier, 2005, 1-23.

[11] Bielski B.H.J., Cabelli D.E., Arudi R.L., Ross A.B., Reactivity of HO_2/O_2^- radicals in aqueous solution, J. Phys. Chem. Ref. Data, 1985, **14**, 1041-1051.

[12] Bielski B.H.J, Arudi R.L., Sutherland M.W., A study of the reactivity of HO_2/O_2^- with unsaturated fatty acids, J. Biol. Chem., 1983, **258**, 4759-4761.

[13] Bors W., Erben-Russ M., Michel C., Saran M., Radical Mechanisms in fatty acid and lipid peroxidation, *in* "Free Radicals, Lipoproteins, and Membrane Lipids", Crastes de Paulet A (eds), Plenum Press New York, 1990, p. 1-16.

[14] Wagner B.A., Buettner G.R., Burns C.P., Free-radical mediated lipid peroxidation in cells: oxidizability is a function of cell lipid bis-allylic hydrogen content, Biochemistry, 1994, **33**, 4449-4453.

[15] Therond P., Couturier M., Demelier J.F., Lemonnier F., Simultaneous determination of the main molecular species of soybean phosphatidylcholine or phosphatidyletanolamine and their corresponding hydroperoxides obtained by lipoxygenase treatment, Lipids, 1993, **28**, 245-249.

[16] Hindo J., Hauville C., Rémita S., Thérond P., Couturier M., Jore D., Gardès-Albert M., Evidence of the formation of different hydroperoxides in irradiated gamma-linolenate solutions: effect of micelle formation, Radiat. Res., 2000, **153**, 201-207.

[17] Pryor W.A., Castle L., Chemical methods for the detection of lipid hydroperoxides, Meth. Enzymol., 1984, **105**, 293-299.

[18] Vitrac H., Hauville C., Collin F., Couturier M., Thérond P., Delaforge M., Rémita S., Jore D., Gardès-Albert M., Hydroperoxide characterisation as a signature of the micelle/monomer balance in radiation-induced peroxidation of arachidonate, Free Radical Research, 2005, **39**, 519-528.

[19] Vitrac H., Courrègelongue M., Couturier M., Collin F., Thérond P., Rémita S., Peretti P., Jore D., Gardès-Albert M., Radiation-induced peroxidation of small unilamellar vesicles of phosphatidylcholine generated by sonication, Can. J. Physiol. Pharmacol., 2004, **82**, 153-160.

[20] Uchida K., 4-Hydroxy-2-nonenal: a product and mediator of oxidative stress, Progress in Lipid Research, 2003, **42**, 318-343.

[21] Patton S., Kurtz G.W., 2-Thiobarbituric acid as a reagent for detecting milk fat oxidation, J. Dairy Sci., 1951, **34**, 669-674.

Chapter 18

Predicting radiation damage distribution in biomolecules

Marie DAVIDKOVA and Melanie SPOTHEIM-MAURIZOT

Introduction

Because of the high proportion of water in living matter, the effect of low LET ionising radiation on biological systems occurs, as discussed for DNA in Chapter 12, at a large extent *via* the reactive species generated by water radiolysis. Among those, the hydroxyl radical $^{\cdot}OH$ (Chapter 1) attacks all biomolecules mainly by abstracting a H-atom from the biomolecule (from the sugar moiety of DNA or from the peptide chain of a protein) or by addition to the double bonds of aromatic moieties (DNA bases or aromatic moieties of protein side chains). Consequently, DNA strand breaks can occur on one strand (called single strand breaks, SSB) or on the two strands in spatially close (a few base pairs away) locations (called double strand breaks, DSB). All strand breaks that can be revealed at neutral pH by classical electrophoresis methods (agarose gel electrophoresis, sequencing gel electrophoresis) belong to the category of frank strand breaks (FSB), mainly resulting from the damage of deoxyriboses. Several base alterations can lead to strand breaks after a subsequent treatment with alkali and thus can also be observed by the classical electrophoresis methods. They are called alkali revealed breaks (ARB). In the case of proteins, the peptide chain breaks can also be revealed by classical electrophoresis methods (denaturing sodium dodecylsulfate-polyacrylamide gel electrophoresis), whereas the modification of amino-acids side chain can be observed by spectroscopic methods and mass spectrometry.

The lesions are non-homogeneously distributed along the biomolecules. The probability to get a lesion at a given nucleotide or amino-acid is determined by the chemical reactivity of the attack site toward the $^{\bullet}OH$ radical and by the accessibility of the reactive sites to this radical. The reactivity is different for the different entities of the biomolecules. For instance, a cysteine is more reactive than a tyrosine which is less reactive than an alanine. Moreover, the accessibility strongly depends on the tridimensional structure of the macromolecule. A tyrosine located in the interior of a folded protein is less reactive than one located on the surface of the protein (see Chapter 16). In the case of the classical double stranded DNA, the tridimensional structure of the molecules is modulated by the sequence of nucleotides. For instance, the probability that an adenosine localised in an AATT sequence reacts with an $^{\bullet}OH$ radical is lower than that of one located in a GCAT sequence, because the accessibility of the deoxyribose of that adenosine is lower in the narrow minor groove of the AATT sequence. The accessibility of the $^{\bullet}OH$ radicals to the reactive sites can also be modulated by the presence of covalently or non-covalently bound molecules. These molecules can physically protect their binding sites (masking effect) or can modify the structure of the "host" molecule (conformational effect). For instance, an electrostatically bound positively charged polyamine, such as spermine, that can compact DNA, locally and globally radioprotects the negatively charged DNA. The only commercial radioprotector of normal tissues in the case of the radiotherapy of human tumors, Ethyol® (Amifostine), once metabolised, gives rise to a positively charged ligand that binds and protects DNA by several effects: scavenging of free radicals at the binding site, chemical repair of radiation-induced DNA radicals and DNA structure modification.

But the most important DNA ligands that naturally radioprotect DNA are the proteins. Chromatin, the constituent of chromosomes in eukaryotes, is in fact a complex of DNA and of several structuring proteins (histones). The regulation of gene expression in prokaryotes involves the tight interaction of proteins (repressors) with specific DNA sequences. The repair of endogenous DNA damages is realised by a set of repair proteins that, in a first step, bind to the region of DNA bearing the lesion. In the DNA-protein complex protection is mutual: the protein radioprotects DNA and DNA radioprotects the protein.

Below here we will focus on an original computational method that accounts for the experimentally observed radioprotection of the partners in DNA-protein complexes. Validated by comparison between experiment and calculation, it can be used to predict the damage extent and distribution in any biomolecule or complex of biomolecules whose tridimensional structure is known.

RADACK, an original model of radiolytic attack to biomolecules

To account for the extent and location of lesions in the case of free DNA, several models of radiation damage were proposed. The main difference between these models concerns the representation of the DNA molecule. In the early models DNA was represented as a cylinder [1] or as a structured quasi-cylindrical volume in which bases and sugar-phosphate domains were delimited [2]. At the beginning of 90s, the first models taking into account the atomic structure of DNA emerged [3-5], shortly followed by higher order DNA structures (nucleosome, chromatin) [6-8]. All these models take into account the radiation track structure (spatial distributions of energy deposition events), which is obtained by using Monte Carlo method. Although these models take into account the DNA structure, they investigate only the canonical B-form of DNA.

The RADACK (contraction of **RAD**iation-induced att**ACK**) model, that we have developed [9,10], accounts for the experimentally determined probabilities of radiolytic damages caused by the OH$^•$ radical attack in all forms of DNA (B [11], Z [12], triplex [13], quadruplex [14]), in DNA-protein complexes [15] and has the potential to predict radiolytic attack probabilities in other molecules or assemblies. Direct ionisation effects are not taken into account. The determination of relative probabilities of reaction of the target with the $^•$OH radicals takes into account two factors: 1) the accessibility of the reactive sites of the target since it uses the exact tridimensional structure of the macromolecule or assembly as determined by NMR, crystallography or as built up by molecular modelling, and 2) the chemical reactivity of the residues (nucleotides or amino-acids).

The atoms of the target are represented by spheres with van der Waals radii, or, if the atom is reactive toward the $^•$OH radical, by a sphere with a larger radius (proportional to reactivity parameters) called a Smoluchowski's sphere. $^•$OH radicals, represented by spheres of 1.2 Å radius, are generated randomly in a volume of interest (a water volume surrounding the studied structure). The radical diffusion in random directions is simulated using Monte Carlo method. The radical is displaced by steps of 1 Å. When the radical encounters a non-reactive atom of the target, it restarts its movement. When it encounters a reactive atom, a reaction is counted. When the radical does not encounter any atom, it escapes the volume of interest and no reaction is counted (**Fig. 1**). A large number of radicals is generated for a good precision of the simulation procedure. The counting can be done separately for each reactive site of the molecule and one can thus obtain the relative probabilities of attack at each reactive site. In order to pass from such a reaction probability to the probability to get an observable lesion, one applies experimentally obtained coefficients representing the efficiencies of such transformation. They include all types of processes (*e.g.* charge transfer to preferential sites). Here comes into play also the chemical fate of the species resulting from the reaction between the target and the ·OH radical. Their evolution toward the observable lesions depends on the nature of these species (*e.g.* base radicals) and on their environment (radical transfer to a spatially proximal site can occur). Therefore the coefficients of transformation can be different for the one and same base radical in two different structures (guanine radicals in a B- or in a Z-form DNA do not have the same coefficient).

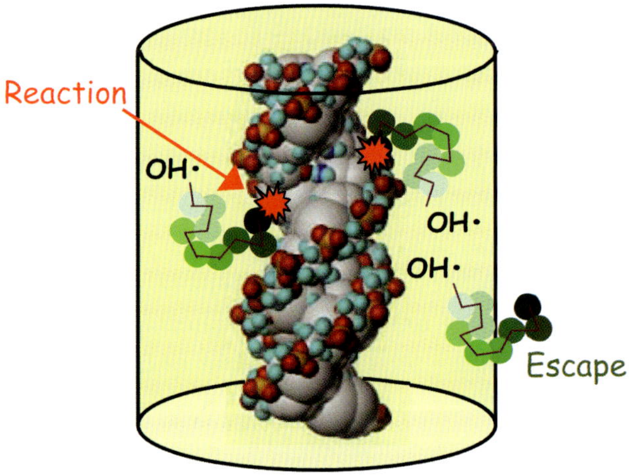

Figure 1 : *Principle of RADACK. DNA as considered by RADACK : DNA reactive atoms are represented by their Smoluchowski's sphere whereas the non-reactive ones, by van der Waals spheres. Colours : H atoms in cyan, C atoms in grey, N atoms in blue, P atoms in orange, O atoms in red. The ·OH radicals (in green) move by steps of 1 Å, react or escape from the volume of interest surrounding the biomolecule. They are presented with increasing intensity of green colour in the direction of diffusion.*

Sequence and structure-dependent damage distribution in free DNA

Sequencing gel electrophoresis of irradiated DNA has shown that the probability of FSB or ARB production is not identical at all nucleotides of the right handed helical double stranded B-form of DNA [11,15]. The variations are even stronger in the left handed helical double stranded Z-form [12]. In the triple stranded DNA [13] or along the DNA quadruplex [14] , variations of FSB and ARB production are also observed. We have applied RADACK to all these different forms of DNA, the exact structures of which are available in the database (Protein Data Bank PDB) or were obtained by molecular modelling. In all cases we have obtained a very satisfactory agreement between the relative probabilities of FSB and ARB determined by sequencing gel electrophoresis and the values calculated with RADACK. The modulation of FSB distribution is mainly due to the variation of the accessibility of H atoms of deoxyriboses, whereas the distribution of ARB is due both to variations of accessibility of the bases and to the higher reactivity of guanine as compared to adenine, tymine and cytosine. Moreover, the efficiency of the transformation of a DNA radical into a lesion depends on the chemical nature of the radical and also on the environment in which the radical is situated.

RADACK may also account for the experimentally observed reduced probability of getting a damage at the site of binding of a minor groove ligand that has a preference for a specific sequence (a natural polyamine [16] or a therapeutic radioprotector, Ethyol® [17]) as it does for the experimentally observed effect of an intercalator (ethidium bromide) [18].

Protection of DNA by a specifically bound protein (radiolytic footprinting)

Sequencing gel electrophoresis patterns of DNA fragments irradiated in presence of bound proteins reveal regions in which no FSB are occurring. They are called protein "radiolytic footprints" on DNA [15]. Such protected regions were observed when irradiating the operator-lactose repressor complexes (see **Inset 1**).

Inset 1 : Lactose repressor

The *E. coli* lactose operon (**Fig. 2**) is one of the most studied and best understood systems of gene expression and regulation. The *lac* repressor is a tetrameric protein (4 x 360 amino acids), a dimer of dimers, that binds tightly the *lac* operator, a DNA segment of about 25 base pairs, with a quasi-palindromic sequence. This binding prevents the expression of the structural genes of the *lac* operon, coding for enzymes involved in lactose metabolism. Thus, the repressor negatively regulates lactose metabolism. The repressor can also bind non-operator DNA, but with a strongly reduced (10^5 times) affinity. This protein is organised in domains: the tetrameric core (formed by the C-terminal parts of the four protomers), and the four headpieces (N-terminal peptides of about 60 amino-acids) connected by hinge regions. Its interaction with each operator sequence occurs by the binding of two headpieces (**Fig. 2**). Due to the mutual interaction in the complex, the conformation of the two partners (DNA and protein) changes with respect to their free state. DNA bends and the unstructured hinge region of the repressor monomers becomes helical.

The structure (complete or partial) of the free repressor and of the DNA-repressor complexes was solved by NMR or X-ray crystallography. They are available in the structural database PDB (Internet address http://www.rcsb.org/pdb/home/home.do) as for instance the entries 1LQC, 1LCC, 1L1M, 1CJG, 1LBG, 1EFA, 1LBI, 1JWL.

RADACK calculations were performed in which structural parameters of the target extracted from PDB entries were used. For the complex between the lactose repressor headpiece and a DNA fragment bearing the specific sequence for the repressor one observes the very good fit between the probabilities of experimental FSB production and the probability of attack to deoxyriboses (the sites involved in FSB production) (**Fig. 2**).

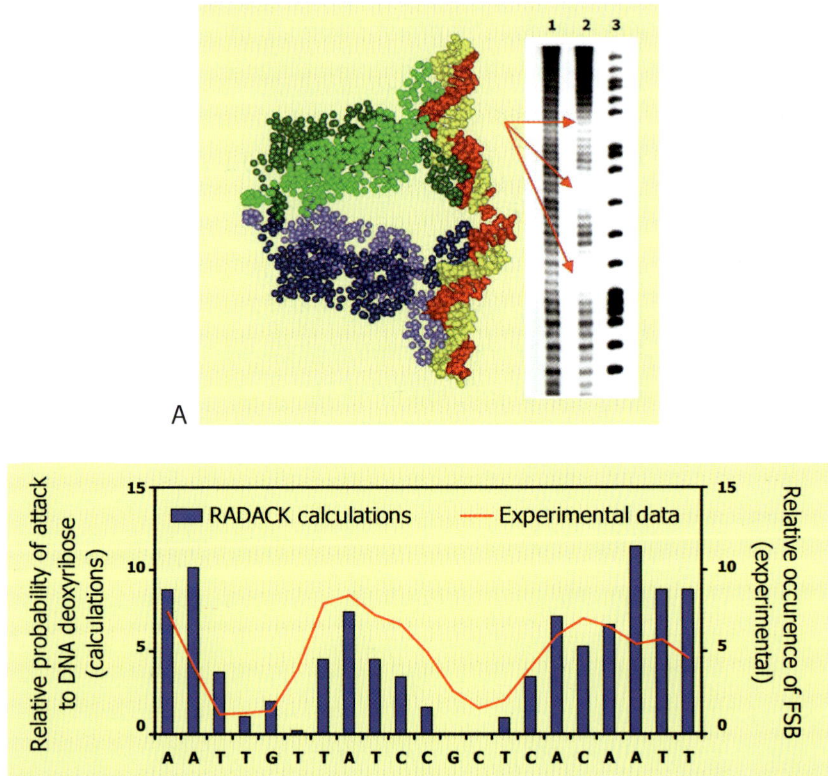

Figure 2 : *Footprints of lactose repressor on DNA in the irradiated DNA-repressor complex.* A) *Sequencing gel electrophoresis of the DNA fragment irradiated alone (lane 1) or in the complex with the repressor (lane 2) and the structure of the complex extracted from 1LBG entry of PDB Databank). In lane 3, the Maxam-Gilbert sequencing of purines (position of A and G). The four monomers of the tetrameric repressor are in blue and green and the DNA segments are in red and yellow. The arrows point toward the regions were no strand breaks are observed which are the footprints of the protein along DNA.* B) *Comparison between the relative probabilities of FSB production deduced from the experiment and the relative probabilities of attack to deoxyribose (leading to FSB) calculated with RADACK.*

Calculations were performed also for two other "objects" : the same fragment but in a linear form and for the same fragment bent as in the complex ("stripped" fragment). From the superposition of the three curves (**Fig. 3**) one can conclude that the "footprint" of the protein on DNA is almost entirely due to the masking by the protein of its binding site on DNA [19,20].

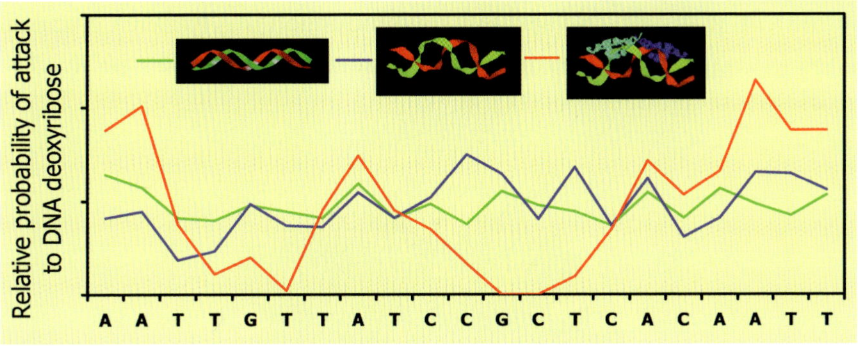

Figure 3 : *Explanation of the repressor footprint on DNA. Pattern of relative probability of attack to deoxyribose (leading to FSB) along the irradiated linear free DNA (green), along the fragment irradiated in complex with the protein (structure 1JWL from PDB) (red) and along the fragment bent as in the complex but "stripped" of the protein (blue).*

This result validates the experimental method of **radiolytic footprinting** as a very interesting tool for revealing the sites of "contact" between the protein and DNA when they are not known.

The same procedure was applied to another biologically very important DNA-protein complex, the nucleosome core particle (**Inset 2**). It is composed by a DNA fragment wrapped around a globular core constituted of several small proteins called histones (**Fig. 4A**).

Inset 2 : Nucleosome

The nucleosome core particle NCP (**Fig. 4**) is the first organisation level of chromatin, constituent of chromosomes. It is the "bead" in the "beads on a string" model of chromatine. A nucleosome core particle is formed by a 146 base pairs DNA fragment wrapped 1.65 times around the histone octamer (formed by two copies of H2A, H2B, H3 and H4 histones). NCP are connected by regions of naked DNA called linkers. The nucleotide sequences of the 146 bp DNA fragments involved in nucleosomes are polymorphic. The regions of contact between DNA and the core of histones are distributed along DNA with a periodicity of around 10.4 bp, which is the number of base pairs per helical turn of DNA in the nucleosome.
The structure (complete or partial) of the nucleosome core particle NCP was solved by X-ray crystallography. They are available in the structural database PDB (Internet address http://www.rcsb.org/pdb/home/home.do) as for instance the entries 1ID3, 1F66, 1AOI, 1KX3.

The sequencing gel electrophoresis of the DNA fragment irradiated in the nucleosome core particle NCP (DNA bound to the histones) show a regular alternation of regions (of around 10.4 base pairs, which is also the number of base pairs on one helical turn of β form-DNA) with high and with low FSB production [15]. RADACK calculations using structural parameters from a PDB entry revealed exactly the same periodicity of variation in the pattern of reaction probabilities (**Fig. 4**).

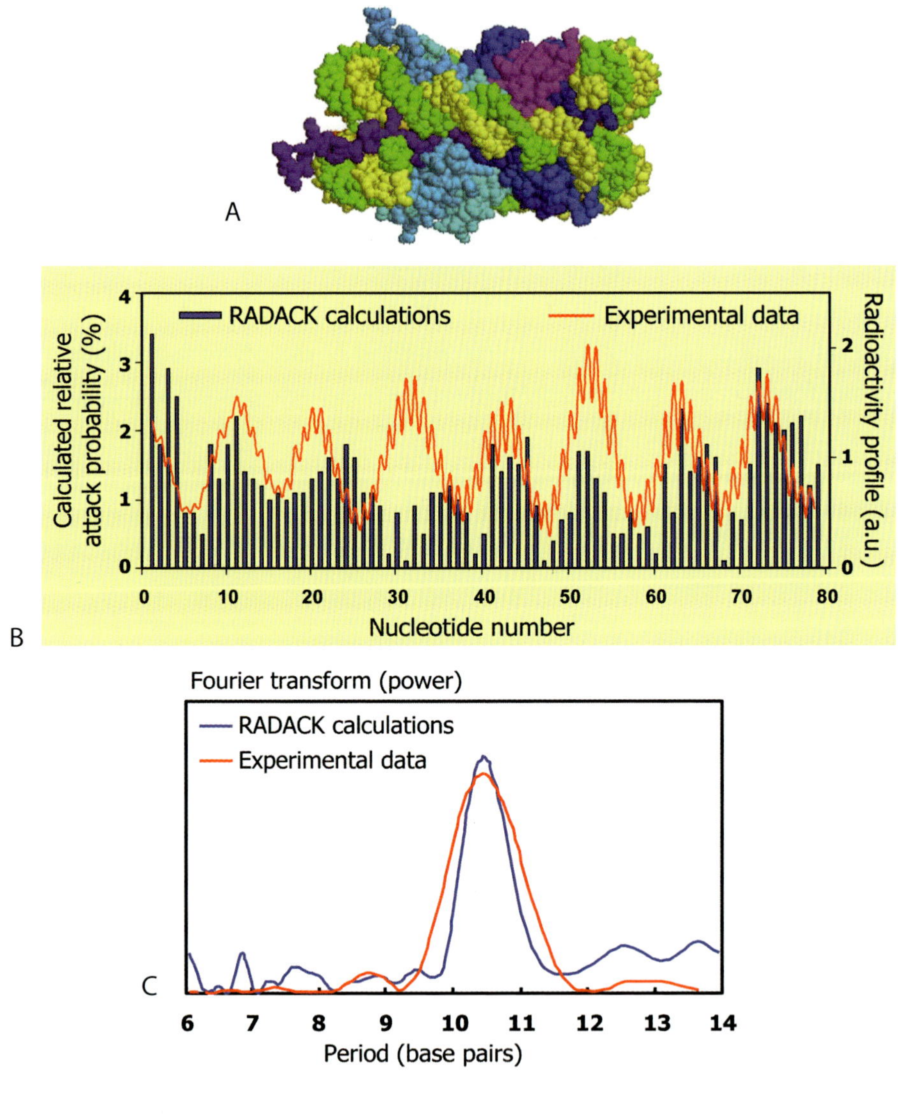

Figure 4 : ***Footprint of histones on the DNA of irradiated core nucleosome.*** A) *Nucleosome core particle structure extracted from 1AOI entry of PDB. The 146 base pairs DNA fragment is in green and yellow, and the histone octamer is in blue and magenta tones.* B) *Experimental relative probabilities of FSB deduced from the radioactive profile (electrophoresisi) and relative probability of attack to deoxyriboses leading to FSB (calculated with RADACK).* C) *The Fourier transform (power) of the probabilities showing that probability variation has a period of 10.4 base pairs both for the experiment and for the calculated values.*

Thus the good fit between calculations and experiments validates RADACK once more [21]. Here again calculations were performed for the straight DNA fragment and for the fragment bent as in the nucleosome. They show that the histones protect DNA from the attack by $OH^{\bullet}$ radicals at the contact points between DNA and the core of histones (located every 10.4 base pairs) by the masking effect. DNA bending does not bring an important contribution to the effect.

Protection of the protein by the specifically bound DNA

When the complexes between the lactose operator and operator DNA [22] or that between the repair protein and abasic site-bearing DNA [23] are irradiated, the complexes are destroyed mainly due to the damage to the protein. When irradiated alone the lactose repressor or the repair protein Fpg lose their ability to bind DNA at a dose that is much lower than that necessary for destroying the complexes. The explanation lies in the reverse radioprotection provided to the protein by the DNA. RADACK calculations using the structural parameters for a free and a complexed repressor headpiece (**Fig. 5**) reveals the most probable sites of protein lesions.

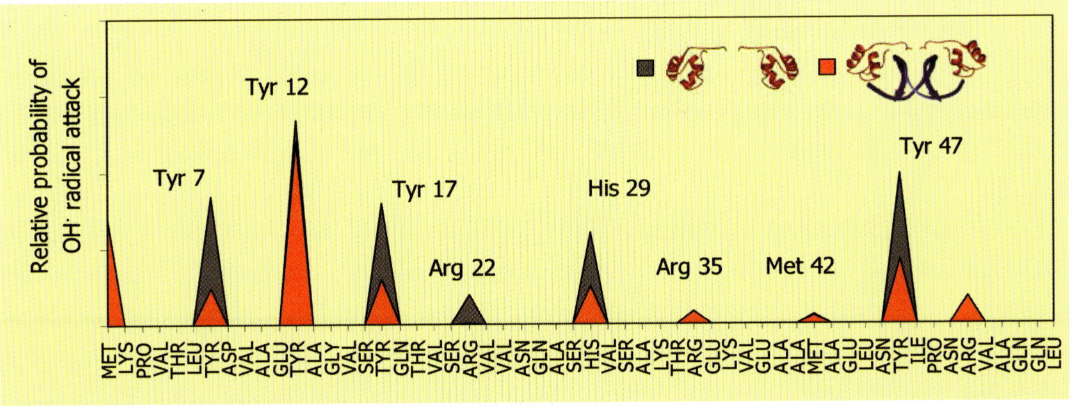

Figure 5 : ***RADACK applied to a DNA-binding protein.*** *Calculated relative probabilities of reaction of $OH^{\bullet}$ radicals with the amino-acids of lactose repressor headpiece along the free irradiated headpiece (structure extracted from 1LQC entry from PDB) (black) and along the headpiece irradiated in complex with DNA (structure extracted from 1CJG entry from PDB) (red).*

One can observe that for the same dose of radiation the probability at some reactive sites (the closest to DNA) is lower when the protein is irradiated in presence of DNA [24]. The same effect is obtained in the case of free and complexed Fpg [23]. Thus applying RADACK to free and complexed proteins allows the protection of the protein by the bound DNA to be revealed and quantified. It also shows that this protection is exerted at the contact points between the protein and DNA. To validate the applicability of RADACK to proteins, a good fit with experiments is necessary. Mass spectrometry experiments that will allow identification of the damaged and the protected amino-acids are in progress in our laboratory.

Conclusion

The potential of computational methods to account for radiolysis phenomena in biomolecule is proven by the good agreement between the calculated and experimental values of relative probabilities of damage to each entity (nucleotide, amino-acid) of free or assembled DNA and proteins irradiated in aerated solution. When the structure of the target of radiation is known, a model like RADACK can reveal the sites and the extent of damage. Thus, *in silico* methods can be used to predict the behaviour of a biomolecule or assembly of molecules exposed to ionising radiation with the condition that their tridimensional structure is known. The described good fit between the calculated and experimental results when taking into account only the reactions of $^{\bullet}OH$ radicals with DNA supports the idea that these radicals are the main damaging agents in the case of low LET radiation exposure of DNA or DNA complexes in aerated solutions.

Acknowledgements

All people having contributed to the research of our two teams on radiation damage to biomolecules (M. Charlier, former leader of the French team, V. Michalik, former leader of the Czech team, S. Eon, J. Franchet-Beuzit, N. Gillard, S. Goffinont, V. Isabelle, S. Ruiz, C. Saint-Marc, C. Savoye, L. Tartier) as well as our coworkers (B. Castaing, F. Culard, D. Sy, C. Prevost, S. Hugot, C. Flouzat, C. Durand, J. Gras, S. Giliberto) are warmly thanked. The research of our teams got the financial support of CNRS, INSERM, Czech Academy of Sciences, *Électricité de France, Ligue Nationale et Départementale contre le Cancer, Association pour la Recherche contre le Cancer*, Grants Agency of the Czech Academy of Sciences, bilateral French-Czech programs (CNRS-CRAS, Barrande-EGIDE) and the European action COST-P9.

References

[1] Charlton D.E., Goodhead D.T., Wilson W.E., Paretzke H.G., The Deposition of Energy in Small Cylindrical Targets by High LET Radiations, Radiat. Prot. Dosim., 1985, **13**, 123-125.

[2] Charlton D.E., Nikjoo H., Humm J.L., Calculation of initial yields of single- and double-strand breaks in cell nuclei from electrons, protons and alpha particles, Int. J. Radiat. Biol., 1989, **56**, 1-19.

[3] Holley W.R., Chatterjee A., Magee J.L., Production of DNA strand breaks by direct effects of heavy charged particles, Radiat. Res., 1990, **121**, 161-168.

[4] Pomplun E., A new DNA target model for track structure calculations and its first application to I-125 Auger electrons, Int. J. Radiat. Biol., 1991, **59**, 625-642.

[5] Zaider M., Bardash M., Fung A., Molecular damage induced directly and indirectly by ionizing radiation in DNA, Int. J. Radiat. Biol., 1994, **66**, 459-465.

[6] Tomita H., Kai M., Kusama T., Aoki Y., Ito A., Monte Carlo simulation of DNA strand breaks induced by monoenergetic electrons using higher-order structure models of DNA, Int. J. Radiat. Biol., 1994, **66**, 669-682.

[7] Andreev S.G., Khvostunov I.K., Spitkovskii D.M., Talyzina T.A., Biophysical modeling of radiation damage in DNA and chromatin induced by radiation of varying quality, Radiat. Biol. Radioecol., 1997, **37**, 533-538.

[8] Friedland W., Jacob P., Paretzke H.G., Stork T., Monte Carlo simulation of the production of short DNA fragments by low-linear energy transfer radiation using higher-order DNA models, Radiat. Res., 1998, **150**, 170-182.

[9] Michalik V., Spotheim-Maurizot M., Charlier M., Calculation of hydroxyl radical attack of different forms of DNA, J. Biomol. Struc. Dyn, 1995, **13**, 1-11.

[10] Begusova M., Spotheim-Maurizot M., Sy D., Michalik M., Charlier M., RADACK, a stochastic simulation of hydroxyl radical attack to DNA, J. Biomol. Struct. Dyn., 2001, **19**, 141-158.

[11] Sy D., Savoye C., Begusova M., Michalik V., Charlier M., Spotheim-Maurizot M., Sequence-dependent variations of DNA structure modulate radiation-induced strand breakage, Int. J. Radiat. Biol. , 1997, **72**, 147-155.

[12] Tartier L., Michalik V., Spotheim-Maurizot M., Rahmouni A.R., Sabattier R., Charlier M., Radiolytic signature of Z-DNA, Nucleic Acids Res., 1994, **22**, 5565-5570.

[13] Barone F., Begusova M., La Nave E., Matzeu M., Mazzei F., Sy D., Radiation damage to triplex DNA induced by gamma-rays: a footprinting study and Monte Carlo simulation, Int. J. Radiat. Biol., 2000, **76**, 731-740.

[14] Begusova M., Tartier L., Sy D., Michalik V., Spotheim-Maurizot M., Charlier M., Monte Carlo simulation of radiolytic attack to 5'-d[T_4G_4]$_4$ sequence in a unimolecular quadruplex, Int. J. Radiat. Biol., 1999, **75**, 913-917.

[15] Franchet-Beuzit J., Spotheim-Maurizot M., Sabattier R., Blazy-Baudras B., Charlier M., Radiolytic footprinting. β-rays, γ photons and fast neutrons probe DNA-protein interactions, Biochemistry, 1993, **32**, 2104-2110.

[16] Sy D., Hugot S., Savoye C., Ruiz S., Charlier M., Spotheim-Maurizot M., Radioprotection of DNA by spermine : a molecular modelling approach, Int. J. Rad. Biol., 1999, **75**, 953-961.

[17] Sy D., Durand C., Hugot S., Savoye C., Swenberg C., Charlier M, Spotheim-Maurizot M., Sequence dependence of DNA radioprotection by the thiols WR-1065 and WR-151326, Theor. Chem. Acc., 1999, **101**, 114-120.

[18] Begusova M., Spotheim-Maurizot M., Michalik V., Charlier M., Effect of ethidium bromide intercalation on DNA radiosensitivity, Int. J. Radiat. Biol., 2000, **76**, 1-9.

[19] Begusova M., Eon S., Sy D., Culard F., Charlier M., Spotheim-Maurizot M., Radiosensitivity of DNA in a specific protein-DNA complex: the *lac* repressor-*lac* operator complex, Int. J. Rad. Biol., 2001, **77**, 645-654.

[20] Begusova M., Giliberto S., Gras J., Sy D., Charlier M., Spotheim-Maurizot M., DNA radiolysis in DNA-protein complexes: a stochastic simulation of attack by hydroxyl radicals, Int. J. Rad. Biol., 2003, **79**, 385-391.

[21] Begusova M., Sy D., Charlier M., Spotheim-Maurizot M., Radiolysis of nucleosome core DNA. A modelling approach, Int. J. Radiat. Biol., 2000, **76**, 1063-1073.

[22] Eon S., Culard F., Sy D., Charlier M., Spotheim-Maurizot M., Radiation disrupts protein-DNA complexes through damage to the protein. The *lac* repressor-operator system, Radiat. Res. , 2001, **156**, 110-117.

[23] Gillard N., Begusova M., Castaing B., Spotheim-Maurizot M., Radiation affects binding of Fpg repair protein to an abasic site containing DNA, Radiat. Res., 2004, **162**, 566-571.

[24] Begusova M., Gillard D., Sy D., Castaing B., Charlier M., Spotheim-Maurizot M., Radiolysis of DNA-protein complexes, Radiat. Phys. Chem., 2005, **72**, 265-270.

Chapter 19

Chemical protection against ionizing radiation

Caroline PROUILLAC, Christine AMOURETTE and Ghassoub RIMA

Introduction

Since the discovery of radioactivity, applications in various domains have been developed. Among them, the study of the effects of radioactivity on living matter gave birth to a new science, radiobiology. In 1950s, when the memories of Hiroshima and Nagasaki were still in all minds, a new subject of research appeared concerning the development of chemical radioprotectors. Indeed, the protective action of some organic compounds against damages inflicted on biologic organisms by gamma radiation was discovered in 1942 by Dale [1]. He showed that addition of several substances, such as cysteine $NH_2CH(COOH)CH_2SH$ or thiourea $S{=}C(NH_2)_2$, to aqueous solutions of enzymes limits their inactivation by X-rays. Patt, in 1949, showed for the first time the radioprotective activity of cysteine *in vivo*. Administered into bone narrow or intravenously at 175 to 575 mg/kg, cysteine allowed 75 to 89% survival in rats exposed at a dose of 8 Gy [2]. Two years later, Bacq showed that cysteamine, the decarboxylated derivative of cysteine, has greater radioprotective activity than cysteine [3].

After these works, the intensive research in chemical radioprotection started by the synthesis of new compounds which will be efficient with lower toxicity. Several chemical products showed a fair radioprotective activity, especially sulphur compounds such as aminothiols, disulfides, phosphorothioates, thiazolidines, aminoethylisothiourea and dithiolanes. However, whereas these compounds have an interesting radioprotective activity, they also have much toxicity which does not allow clinical use, except in the case of WR 2721 [N-(3-aminopropyl)]-2-aminoethylphosphorothioate, ($H_2NCH_2CH_2CH_2NHCH_2CH_2SPO_3H_2$)]. Therefore WR 2721 was the subject of numerous studies [4-5]. It is marketed for a clinical use (Ethyol®) in radio- and chimio-therapy. It is currently considered as the most efficient radioprotector although it causes some adverse effects.

Therefore, several organosilicon, organogermanium and organophosphorus derivatives were recently developed in our laboratory. They showed interesting radioprotective activity compared to the starting organic compounds, owing to the presence of organometallic and organophosphorus groups which modify the electronic structure and the chemical and biological behaviour of these compounds.

Why a chemical protection against radiation?

Since the use of radioisotopes in nuclear power stations, in anti-cancer radiotherapy or in nuclear weapons, noxious effects of ionizing radiation on human cells are better known. According to the amount and distribution of exposure, ionizing radiation can locally eliminate tumors, but can also damage normal tissues. The biological effects of such radiation result from chemical processes as ionization or excitation of the biological macromolecules, such as DNA, either in a direct way or in an indirect way via water molecule radiolysis (Chapter 12). In both cases, many radical species appear then, which have various consequences on cellular scales such as mutations of DNA, cellular death, or cancer.

At present, only one way may constitute a real protection against radiation, that is the physical protection. This method uses screens in order to efficiently attenuate the radiation. The various types of radiation (alpha, beta, gamma- and X-rays) having different penetration capacity in the matter, then various screens are thus necessary (**Fig. 1**).

This method remains the most effective, and adequate shieldings are systematically used around research, industrial, and medical radiation sources or in space (concrete, water pools, lead, stainless steel, …) (Chapters 2, 6, 11). However, the method is difficult to implement to avoid any risk under spread hostile irradiation conditions, for example in the event of a nuclear accident. This is the reason why the objective of different laboratories was for a few years to set up a personal chemical protection against radiation. The aim of this method is to protect the system in a way, as effective as possible, by administering a molecule before exposure, which could limit the biological consequences of the ionizing radiation. Radioprotective agents act either by scavenging free radicals produced by irradiation or by restoring damaged biomolecules. In the event of a nuclear accident, this simple and practical method could thus be applied to the greatest number of people. Various types of radioprotectors have been developed in our laboratory. We present here only some of them.

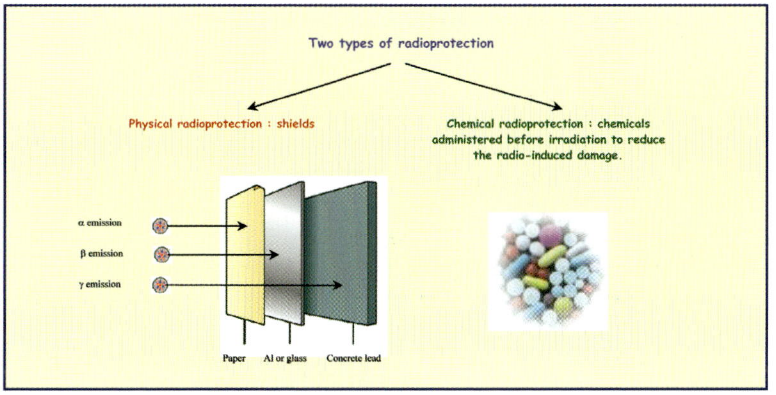

Figure 1: *Penetration of the different types of ionizing radiation.*

What are the criteria for an efficient radioprotector?

It is essential to establish certain criteria to evaluate the synthesized compounds, in particular their radioprotective effectiveness and their toxicity. For each of the molecules we synthesize, first we evaluate its acute (48 hours) and medium-term (2-30 days) toxicity in male Swiss CD1 mouse. The molecule is injected intraperitonally at different concentrations to determine the $LD_{50tox/48h}$, defined as the concentration which kills 50% of the animals at 48 hours. Then, radioprotective effectiveness is determined in the same animal model. In a preliminary study, the product is administered at the maximum tolerated dose (defined as ½ $LD_{50tox/48h}$) 15 or 90 minutes before radiation exposure at doses $LD_{100irr/30d}$ and $LD_{100irr/30d}$ + 2 Gy ($LD_{100irr/30d}$ is defined as the irradiation dose which kills 100% of the animals 30 days after exposure). Survival is observed for 30 days. In a second experiment we determine the Dose Reduction Factor (DRF) for the most radioprotective compounds. DRF is the ratio between the $LD_{50irr/30d}$ of treated mice and that of non treated mice.

$$DRF = \frac{LD_{50irr/30d} \text{ protected animals}}{LD_{50irr/30d} \text{ non protected animals}}$$

It must be higher than 1.3 for clinical use and ideally higher than 2 for management of a nuclear accident. Of course, an ideal radioprotector will have to demonstrate other characteristics: compatibility with other drugs, effectiveness for various types of radiation, low cost, possible synthesis in large quantity, stability… The aim of this work was to study the incorporation of potentially radioprotective organic compounds in organometallic structures such as the metallathiazolidines, metalladithioacetals and phosphorothioates in order to decrease their toxicity and to increase their radioprotective activity. It was then to vectorise the basic organic substances to their target in biological organism.

Organometallic radioprotectors

Metallathiazolidines

Many chemical structures have been studied in the field of chemical protection against radiation. Among them, cysteamine and methylcysteamine represent simple organic compounds having an interesting radioprotective activity due to their radical scavenging and chemical repair ability [3,6]. Nevertheless, these substances present a relative toxicity ($LD_{50tox/48h}$ ranging between 150 and 250 mg/kg). Thiazolidine structure is one of the most studied models [6,7], especially because of the prolonged or delayed action it confers to the radioprotector effect which is however not hardly better than that of cysteamine. The general structure of these modified cysteamine and methylcysteamine molecules is presented in **Figure 2**.

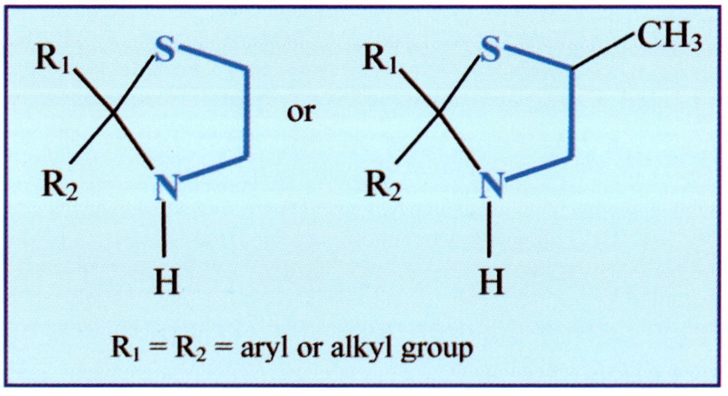

Figure 2 : *Structures of thiazolidines derived from cysteamine and methylcysteamine.*

In addition, previous works by Voronkov [8,9] have highlighted that the incorporation of silicon into certain biologically active organic molecules was of great interest by an exaltation of the required properties. Thus, for a few years, we have studied thiazolidines with radioprotective activity, of which we modified the structure, by replacing carbon in position 2 of the cycle by a metal of the group 14 (silicon or germanium). We drew aside tin and lead because of their intrinsic toxicity. In this spirit, we thus synthesized and studied sila- and germathiazolidines derived from cysteamine or methylcysteamine, and germaselenazolidines derived from selenocysteamine or methylselenocysteamine, with the aim of comparing them with their purely organic equivalents. The general structure of these compounds is presented in **Figure 3**.

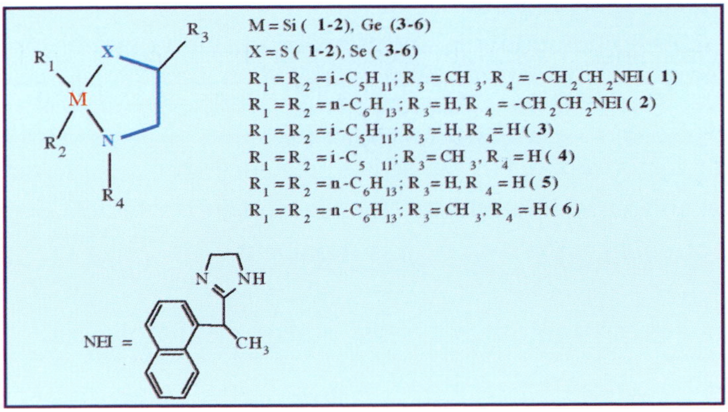

Figure 3 : *Structure of metallathiazolidines [NEI = 2-(1-naphthylethyl)-2-imidazoline].*

The metallathiazolidines are likely, by hydrolysis or cleavage, to release cysteamine, methylcysteamine or naphthylethylimidazoline. In **Table 1**, results of toxicological and radiopharmacological studies are represented.

Table 1 : *Toxicity and radioprotective activity of metallathiazolidines, germaselenazolidines and their parental organic derivatives [NEI= 2-(1-naphthylethyl)-2-imidazoline].*

Metallathiazolidines and Germaselenazolidines									
	M	**X**	**R_1**	**R_2**	**R_3**	**R_4**	**$LD_{50tox/48h}$ (mg/kg)[a]**	**Irr. dose (Gy)**	**Survival at 30 days (%) [t, min][b]**
1	Si	S	i-C_5H_{11}	i-C_5H_{11}	CH_3	$-CH_2CH_2NEI$	150	8.1	40 [15] 30 [90]
2	Si	S	n-C_6H_{13}	n-C_6H_{13}	H	$-CH_2CH_2NEI$	300	8.1	30 [15] 70 [90]
3	Ge	Se	i-C_5H_{11}	i-C_5H_{11}	H	H	255	8.1	40 [15] 60 [90]
4	Ge	Se	i-C_5H_{11}	i-C_5H_{11}	CH_3	H	236	8.1	30 [15] 50 [90]
5	Ge	Se	n-C_6H_{13}	n-C_6H_{13}	H	H	224	8.1	30 [15] 50 [90]
6	Ge	Se	n-C_6H_{13}	n-C_6H_{13}	CH_3	H	200	8.1	30 [15] 40 [90]

Parental organic derivatives				
	Compounds	**$LD_{50tox/48h}$ (mg/kg)[a]**	**Irr. dose (Gy)**	**Survival at 30 days (%) [t, min][b]**
7	$NEICH_2CH_2NHCH_2CH_2SH$	300	8.1	0 [15]
8	$NEICH_2CH_2NHCH_2CH(CH_3)SH$	212	8.1	0 [15]
9	$H_2NCH_2CH_2SeH$, HCl	17	8.1	0 [15]
10	$H_2NCH_2CH(CH_3)SeH$, HCl	10	8.1	0 [15]

a) *the administered dose is ½ $LD_{50tox/48h}$.* b) *(t) = time between administration of the compound and irradiation.*

From this study, if we compare homologous sila- and germathiazolidines, we observe that silathiazolidines are generally more protective against radiation (higher survival for the same delay), and have toxicities close to those of the germathiazolidines. These results are in agreement with the concept of vectorization, but the active substance seems not to act in the same way in the case of sila- and germathiazolidines. The sila- and germathiazolidines could thus have different modes of action in the biological environment.

Certain characteristic features can be attributed to sila- and germathiazolidines:

- silathiazolidine structure allows administration of the basic purely organic product, by exalting its radioprotective properties, and by decreasing its molar toxicity; *i.e.* the $LD_{50tox/48h}$ of silathiazolidines allows injection of more organic compound than for isolated basic derivative. The parental organic compound is thus potentiated.

- in the case of the silathiazolidines, it appears that the brittleness of the Si-N bond probably supports the vectorization of the organic radioprotector. In a previous work [10], it has been established that the purely organic thiazolidines are generally less active in protection against radiation than their silicon or germanium isologs. Moreover, previous research has highlighted the mechanism of action of the organic thiazolidines, a specific isotopic marking having allowed a pharmacokinetic follow-up of the products after their administration: thiazolidines are hydrolyzed in ketone and cysteamine which is partly responsible of the protection against radiation. This observation could, according to the concept of biodisponibility, explain the classification of the activities of the various thiazolidines, by the ease of cleavage of the chemical bonds: Si-L > Ge-L > C-L.

Because selenium plays a fundamental role as a biological cofactor of glutathion peroxidase which protects cellular membranes, nucleic acids and proteins against the degradation by free radicals, we have also synthesized and studied germaselenazolidines and germadiselenoacetals derived from selenocysteamine or methylselenocysteamine,

Metalladithioacetals

It is now well-known that the linear compounds having the sequence of atoms as S-C-C-S have an interesting radioprotective activity [10]. We thus synthesized metalladithioacetals for the same reasons as those which led us to consider the study of the metallathiazolidines (modification of a structure with interesting biological properties, to increase the activity and to decrease toxicity, incorporation of a metal of the group 14 (Si or Ge), to exalt the required effects). We were interested in the synthesis and the study of sila- and germadithioacetals derived from the cysteamine or the *N*-substituted methylcysteamine. We thus synthesized metalladithioacetals derived from the naphthylethylimidazoline (NEI) [11-12] and p-aminobenzophenone. The structure of these compounds is presented in **Figure 4** :

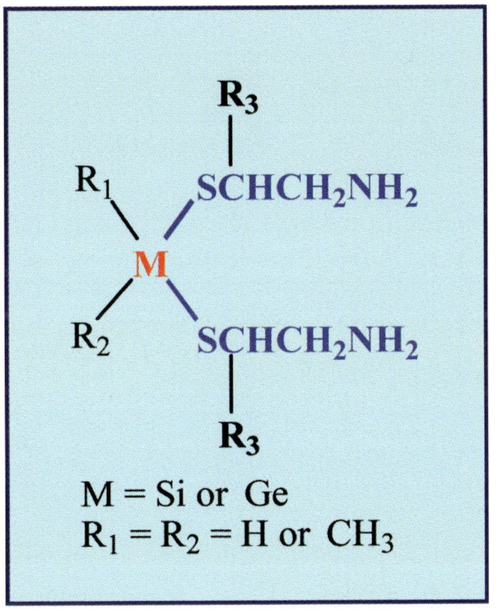

Figure 4 : *Structure of metalladithioacetals.*

The same interest which led us to study metallathiazolidines can be considered for the metalladithioacetals :

1) *in vivo* release of one or more substances known for their radioprotective properties, a delayed or prolonged release of active substances,

2) a fundamental modification of the electronic structure of these new derivatives by the incorporation of a metal which can be at the origin of new specific chemical or biological properties to these organometallic compounds.

A specific interest for the dithioacetals must be underlined ; these molecules make it possible to deliver, as potentially available, a double quantity of the active substance than that contained in their thiazolidinic equivalents. Some toxicological and radiopharmacological results are presented in **Table 2**.

Table 2 : *Toxicity and radioprotective activity of metalladithioacetals, germaselenoacetals and their parental organic derivatives [NEI= 2-(1-naphthylethyl)-2-imidazoline].*

Metalladithioacetals and germaselenoacetals									
	M	X	R_1	R_2	R_3	R_4	$LD_{50tox/48h}$ (mg/kg)[a]	Irr. dose (Gy)	Survival at 30 days (%) [t, min][b]
11	Ge	S	p-$CH_3C_6H_4$	CH_3	CH_3	$-NHCH_2CH_2NEI$	200	8.1	40 [15]
12	Ge	S	p-$CH_3C_6H_4$	CH_3	H	$-NHCH_2CH_2NEI$	184	8.1	50 [15]
13	Ge	S	n-C_6H_{13}	n-C_6H_{13}	CH_3	$-NHCH_2CH_2NEI$	300	8.1	40 [15]
14	Ge	S	i-C_5H_{11}	n-C_6H_{13}	CH_3	$-NHC_6H_4(C{=}O)C_6H_5$	600	8.1	60 [15]
15	Ge	S	n-C_6H_{13}	n-C_6H_{13}	H	$-NHC_6H_4(C{=}O)C_6H_5$	600	8.1	40 [15]
16	Ge	Se	n-C_6H_{13}	n-C_6H_{13}	H	NH_2	260	8.1	20 [15]

Parental organic derivatives				
	Compound	$LD_{50tox/48h}$ (mg/kg)[a]	Irr. dose (Gy)	Survival at 30 days (%) [t, min][b]
17	$NEICH_2CH_2NHCH_2CH_2SH$	300	8.1	0 [15]
18	$NEICH_2CH_2NHCH_2CH(CH_3)SH$	212	8.1	0 [15]
19	$H_2NCH_2CH_2SeH$, HCl	17	8.1	0 [15]
20	Ph-$(C{=}O)C_6H_4NHCH_2CH_2SH$	> 300	8.1	30 [15]
21	Ph-$(C{=}O)C_6H_4NHCH_2CH(CH_3)SH$	> 300	8.1	0 [15]

a) *The administered dose is ½ $LD_{50tox/48h}$* b) [t] = *time between administration of the compound and irradiation.*

This study highlighted an overall behaviour equivalent between the sila- and germadithioacetals in the field of toxicity and protection against radiation, even if some germadithioacetals are characterized by a radioprotective activity, definitely more marked than their silicon isologs. For the majority of the metalladithioacetals which we studied, this structure gets a reduction in the toxicity of the purely basic organic derivatives as well as an exaltation, sometimes by a large factor, of their radioprotective properties. This observation is in agreement with the idea of a potentiation of the basic organic substances in mice, owing to the metalladithioacetal structure. We showed that *in vivo* vectorization of metalladithioacetals was taking place in lipophilic medium. Thus, the greater the lipophilic character of the various vectors related to the metal derivative is, the more important the radioprotective activity.

Phosphorous radioprotectors

The introduction of an hydrosoluble phosphorous group increases the hydrosolubility and the activity of these molecules by supporting their transport *in vivo* in comparison with their parental aminothiols. Moreover, these kinds of phosphorous derivatives are often more active because they have an intrinsic radioprotective activity and their metabolites have radioprotective activity too. Phosphorothioates offer a slow and prolonged release of radioprotective substances.

WR-2721 ($H_2NCH_2CH_2CH_2HNCH_2CH_2SPO_3H_2$) is one of the best currently known radioprotectors. It was the subject of many studies because of its very high radioprotective effect [11-12]. In spite of some side effects (vomiting in 25% of the cases, hypotension, etc.), WR-2721 is currently used for a clinical use, because it preferentially protects normal cells. The structure of these phosphorated derivatives is presented in **Figure 5**:

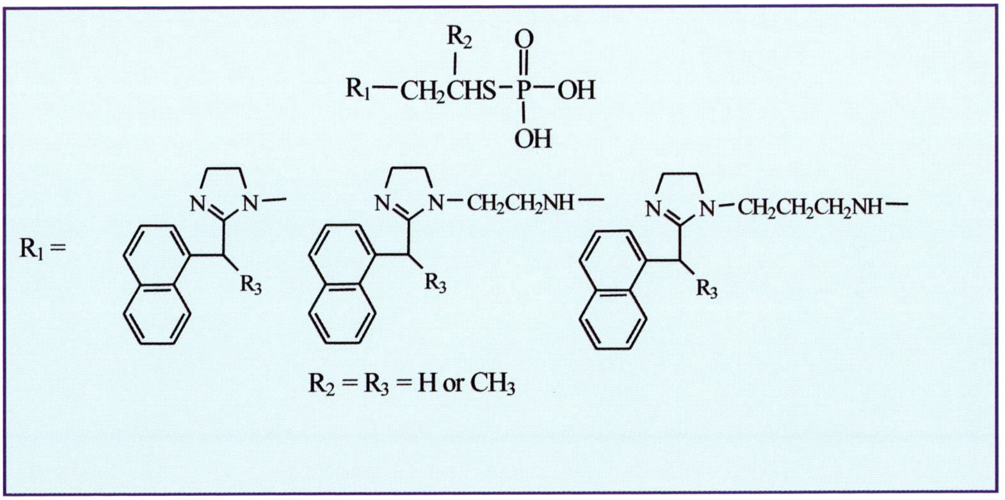

Figure 5 : *Structures of phosphorothioates.*

Toxicity and radioprotective activity of phosphorothioates and their parental organic derivatives are presented in **Table 3**. Addition of a phosphorothioate moiety to the organic structures makes it possible to obtain, for these molecules, similar properties to those observed with the incorporation of silicon or germanium atom in the case of organometallic compounds (metallathiazolidines or metalladithioacetals) developed by our laboratory [13-19], for example : delayed or prolonged *in vivo* release of one or more organic substances having radioprotective properties (aminothiols), fundamental modification of the chemical or biological properties.

Table 3. *Toxicity and radioprotective activity of phosphorothioates and their parental organic derivatives [NEI= 2-(1-naphthylethyl)-2-imidazoline; NMI=2-(1-naphtylmethyl)-2-imidazoline].*

	R	$LD_{50tox/48h}$ (mg/kg)[a]	Irr. dose (Gy)	Survival at 30 days (%) [t, min][b]
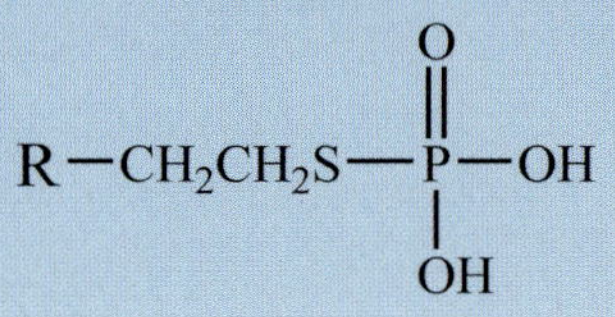				
22	-NMI	168	8.1 10.1	100 [15] 70 [15]
23	-NEI	184	8.1 10.1	100 [15] 80 [15]
24	$-HNCH_2CH_2NMI$	280	8.1 10.1	10 [15] 0 [15]
25	$-HNCH_2CH_2NEI$	400	8.1 10.1	90 [15] 60 [15]
26	$-HNCH_2CH_2CH_2NMI$	150	8.1 10.1	100 [15] 50 [15]
27	$-HNCH_2CH_2CH_2NEI$	260	8.1 10.1	100 [15] 70 [15]
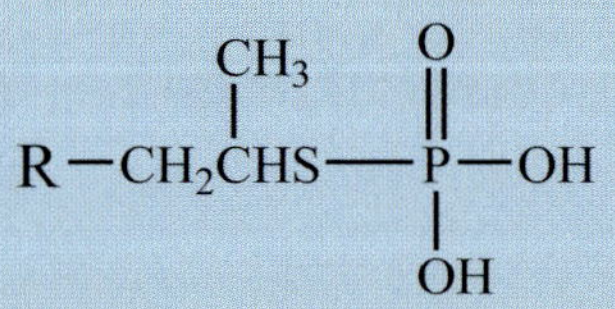				
28	-NMI	184	8.1 10.1	100 [15] 70 [15]
29	-NEI	260	8,1 10,1	40 [15] 0 [15]
30	$-HNCH_2CH_2NMI$	150	8.1 10.1	80 [15] 70 [15]
31	$-HNCH_2CH_2NEI$	360	8.1 10.1	100 [15] 70 [15]

a) *the administered dose is ½ $LD_{50tox/48h}$.* b) [t] = *time between administration of the compound and irradiation.*

The presence of phosphorus atom does not modify the toxicity. Indeed, phosphorothioate and corresponding aminothiols have similar $LD_{50tox/48h}$. On the other hand, phosphorated pro-drugs, in comparison of aminothiols, have an excellent radioprotective activity. These molecules are characterized by properties intermediate between lipophilic and hydrophilic.

Most of these phosphorated derivatives present still an interesting radioprotective activity even when mice have been exposed to $LD_{100irr/30d}$ + 2 Gy. Compounds derived from naphthylethylimidazoline are less toxic and sometimes more active. Compounds (**30**, **31**) are less toxic than compounds **28** and **29**. However, in derivative **23**, the most active of the series, the substitution of an hydrogen atom by a methyl group, giving compound **29**, induces a great decrease in the activity, and even a total decrease of radioprotective activity at ≈ 10 Gy. In contrast, in the case of phosphorothioate **30**, the presence of a methyl group greatly improves radioprotective properties in comparison with compound **24** which is the less active derivative. The Dose Reduction Factor has been determined as previously described for compounds **27, 28** and **31**. Their values are ≈ 1.6, 1.8 and 1.6, respectively (**Tab. 4**).

Table 4. *Dose Reduction Factor (DRF) of phosphorothioates.*

Compound	$LD_{50irr/30days}$ (Gy)	r^2	DRF
No compound	6.2	0.970	-
27	10.1	0.982	1.6
28	11.4	0.999	1.8
31	9.9	0.999	1.6

Conclusion

In this work, several silicon, germanium, selenious or phosphorated derivatives have been synthesized : a great majority presents an interesting radioprotective effect. In many cases, when the structure was cyclic, we observed a delayed effect. Let us underline the more important radioprotective activity and the generally lower toxicity of the silicon or germanium derivatives compared to basic organic materials. This corresponds to a clear potentiation of these molecules by the organometallic substituents. The results show clearly the important contribution of silicon, germanium and phosphorus in the origin of the radioprotective properties of these structures. Phosphorated derivatives from NEI and NMI are the most efficient. However, they also are more toxic than silicon, germanium or selenious derivatives.

The development of efficient and non-toxic radioprotective compounds should present an interest not only for military applications or nuclear energy accidents as Chernobyl, but also in the chemo- and radiotherapy treatments.

Acknowledgements

The authors want to thank the *Délégation Générale pour l'Armement (DGA/STTC/DT/SH), Ministère de la Défense Nationale, France,* for its financial support and interest in this research.

References :

[1] Dale W.M., Gray L.H., Meredith W.J., The inactivation of an enzyme (carboxypeptidase) by X- and α-radiation, Phil. Trans. Roy. Soc., 1949, **242A**, 33-62.

[2] Patt H.M., Tyree E.B., Straube R.L., Smith D.E., Cysteine protection against X Irradiation, Science, 1949, **110**, 213.

[3] Bacq Z.M., Hervé A., Lecomte J., Fischer P., Blavier J., Dechamps G., Le Bihan H., Rayet P., Protection contre le rayonnement X par la β-mercaptoethylamine, Arch. Int. Physiol., 1951, **59**, 442-446.

[4] Kligerman M.M., Glover D.J., Turrisi A.T., Norfleet A.L., Yuhas J.M., Coia L.R., Simone C., Glick J.H., Goodman R.L., Toxicity of WR-2721 administered in single and multiple doses, Int. J. Radiat. Oncol. Phys., 1984, **10**, 1773-1776.

[5] Piper J.R., Stringfellow C.R., Elliot R.D., Johnston T.P., S-2-(omega-aminoalkylamino)ethyl dihydrogen phosphorothioates and related compounds as potential antiradiation agents, J. Med. Chem., 1969, **12**, 236-243.

[6] Terol A., Fernandez J.P., Robbe Y., Chapat J.P., Granger R., Fatôme M., Study on 2-phenylthiazolidine derivatives as radioprotectant agents, Eur. J. Med. Chem., 1978, **13**, 149-151.

[7] Fatôme M., Poutrain P., Granger R., Orzalesi H., Robbe Y., Randon M., Fernandez J.P., Radioprotective effects of thiazolidines, Chim. Ther., 1970, **5**, 312-317.

[8] Voronkov M.G., Biologically active silicon compounds, Pure Appl. Chem., 1969, **19**, 399-416.

[9] Voronkov M.G., Zelchan G., Lukevics E., Silicon and Life, Akademii Verlag, Berlin, 1975, p. 370.

[10] Fatôme M., Poutrain P., Granger R., Orzalesi H., Robbe Y., Randon M., Valentin M., Chevallet P., Femandez J.P., Radiation protective agents. Dithiolanes, Chim. Ther., 1970, **5**, 327.

[11] Piper J.R., Stringfellow C.R., Johnston T.P., Terminally substituted S-2-(omega-aminoalkylamino) ethyl dihydrogen phosphorothioates and related compounds as potential antiradiation agents, J. Med. Chem., 1969, **12**, 244-253.

[12] Davidson D.E., Grenan M.M., Sweeney T.R., Biological characteristics of some improved radioprotectors *in* "Radiation Sensitizers. Their use in the clinical management of cancer", Brady L.W. (ed), Masson, NY, 1980, 309-320.

[13] Célariès B., Amourette C., Lion C., Rima G., Synthesis and radioprotective study of new siladithioacetals and germadithioacetals, Appl. Organomet. Chem., 2003, **17**, 561-569.

[14] Rima G., Satgé J., Dagiral R., Lion C., Fatôme M., Roman V., Laval J.D., Synthesis and radioprotective activity of new organosilicon and germanium compounds, Metal-Based Drugs, 1998, **5**, 139-146.

[15] Rima G., Satgé J., Dagiral R., Lion C., Fatôme M., Roman V., Laval J.D., Synthesis and application of new organometallic compounds of silicon and germanium in chemical radioprotection, Appl. Organomet. Chem., 1999, **13**, 583-594.

[16] Satgé J., Rima G., Fatôme M., Sentenac-Roumanou H., Lion C., Synthesis and radioprotective activities of germatranes, silatranes, germylated oxides and sulfides with cysteamine, methylcysteamine and N(2-thioethyl)-1,3-diaminopropane hydrochloride ligands, Eur. J. Med. Chem., 1989, **24**, 48-54.

[17] Yuhas J.M., Storer J.B., Differential chemoprotection of normal and malignant tissues, J. Natl-Cancer Inst., 1969, **42**, 331-335.

[18] Célariès B., Amourette C., Lion C., Rima G., New phosphorothioates derived from naphthyl-methylimidazoline and naphthylethylimidazoline: application in chemical radioprotection, Radioprotection, 2005, **40**, 57-71.

[19] Célariès B., "Étude et applications de nouveaux dérivés organosiliciés, -germaniés, -phosphorés et -séléniés en radioprotection chimique", Ph. D. Thesis, Université de Toulouse, 2003, p. 292.

Chapter 20

Advances in radiotherapy : new principles

Nicolas FORAY and Jacques BALOSSO

Introduction

Since their discovery, ionizing radiations have been recognized to induce deleterous effects on living cells and locally used to cure tumours. The powerful penetration of high-energy photon beams has made easier the treatment of a number of deep-seated tumours, while kilovoltage X-rays are more appropriate for radiotherapy of superficial cancers. Nevertheless, high-energy photon radiotherapy does not produce optimal radiobiological effects within targeted tissue region and its development has not fully eliminated the central problem of irradiated surrounding normal tissues. Moreover, lesions induced in DNA of tumours (Chapters 12-14, 18) may be repaired (Chapter 15), a process which decreases the treatment efficiency. Particularly, it must be reminded that proliferating cells (S phase) (Chapter 15) are more radioresistant than quiescent cells (G_0/G_1 phase), whatever the type of radiation used, high-energy photons or X-rays. Hence, two major innovating anti-cancer strategies involving radiotherapy are proposed to date:

1) to apply powerful radiation to kill a larger yield of tumour cells (notably proliferating ones) and to use multiple radiation beams whose geometry fits to the tumour shape by using computerised data obtained from imaging. These requirements are notably reached by *3D conformal radiotherapy* and the *intensity modulated* approach.

2) to use drugs, the activities of which are exacerbated by radiation. Radio-chemotherapy does not systematically lead to strong synergetic effects and a low number of chemotherapy drugs are known to be overactivated by radiation. The use of radiolabelled drugs and antibodies is an interesting strategy and has provided encouraging results. However, again, the success depends on the specific penetrability of agents into the tumour cells and the cell cycle dependence of their action. By contrast, the X-rays-induced photoactivation of high Z elements that are contained in some specific drugs used routinely in diagnostics or in anti-cancer therapy is an alternative to overcome the difficulties exposed above (as far as photoactivation does not destroy the drug itself).

New advances in radiotherapy

So far, despite of the considerable progress in our understanding of the biology of tumours, the most efficient way to cure cancers still consists in destroying as much malignant cells as possible. To this aim, a plethora of physical agents have been proposed. This was notably the case of powerful ultra-sounds, micro-waves, lasers, etc. However, none of them is more efficient than ionising radiation. Nowadays, most of cancers are treated with radio-therapy and about half of them with chemo-radiotherapy.

Anatomical problems

As with surgery, the first difficulty encountered by radiotherapists is *anatomy*. Radiotherapy and surgery show the same limits: the infiltrating activity of tumours and their intimate interplay with normal tissues frequently involve vital organs and render difficult a precise targeting of the tumour. Surgery as well as any means leading to the physical destruction of the tumour are applied only in limited cases, notably when tumours are situated in some anatomical scaffolds that can be crossed without danger. In any other case (*i.e.* the great majority), radiotherapy is applied but raises the problem of damage formation into surrounding tissues: a significant differential effect between healthy tissues and tumour is therefore required. To date, technological advances in radiation production have progressively solved the geometrical problem raised by the tumour shapes. However, the differential effect is mainly obtained by the fractionation of the treatment in multiple sessions and/or by the use of concomitant chemotherapy to impede the tumour re-growth during the therapeutic course. Four major approaches are now possible to enhance the success of the treatment:

– to better use computer data generated by tumour imaging;

– to perform the most precise dosimetry;

- to take into account the natural movements of patients, notably breathing,

- and, if possible, to use charged particle beams, such as protons, when therapy with extreme accuracy in depth is needed.

3D conformal radiotherapy

The radiation dose to be delivered and actually deposited in the different volumes to be treated unavoidably differs for several reasons, notably:

- because tumours are very heterogeneous tissues and they may elicit very proliferating and/or very dense regions in which a higher dose is required whereas some quiescent and/or weakly invading regions require lower doses;

- some radiosensitive and/or vital organs situated in the vicinity of the tumours must not be submitted to radiation. This is notably the case with eyes and optical nerves, brain stem, lungs and kidneys, ...

The tumour shape can be therefore complex, rendering quite difficult a selective exposure to radiation. One of the major recent advances in external radiotherapy consists in using *numerised tumour imaging* from CTscan and/or from nuclear magnetic resonance (NMR) technology in order to compute the anatomical extensions of the tumour and to modelise it in 3 dimensions at the scale of the radiation treatment conditions: this is the *3D conformal radiotherapy*. Such an approach allows, as has never been done before, both the tumour volume to be targeted and sensitive organs to be materialised and protected (**Fig. 1**) [1].

Figure 1 : *A representative example of multiple isocentric radiation beams for the 3D conformal treatment of prostate cancer.*

Intensity-modulated radiotherapy

According to the tumour shape, data acquired from imaging in the frame of the 3D conformal radiotherapy, the geometry and multiple radiation beams must be set up. To date, the standard approach has consisted in using interactive softwares that provide the calculation of the dose distribution to virtual model of the patient reconstructed from computed imaging. These data are finely adapted to each individual case by iterative manual correction. Unfortunately, such a standard approach cannot manage a large number of beams, differing by their size, intensity and time of exposure. In order to avoid an increasing amount of data to be acquired, an innovating approach is to generate a system leading by automatic corrections. This is the so-called principle of *inversed treatment planning system*: the radiation doses prescribed by the radiotherapist determine the radiation beam parameters. Obviously, powerful computers are necessary to reach such requirements, notably for managing complex geometries. In fact, the real technological advance lies in the development of sophisticated computed operated linear accelerators that deliver X-rays of high energy through a multi-leaf collimator. By automatically generating a full set of complementary beams with differential intensity according to the different part of their sections, the complex requirements of highly conformal radiotherapy can be reached: this is the *intensity modulated radiotherapy* (IMRT). This new technology represents the ultimate physical frontier of X-rays radiotherapy (**Fig. 2**) [1].

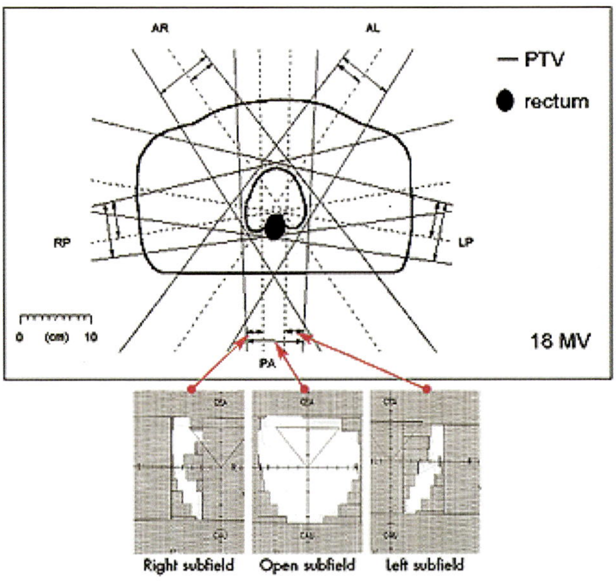

Figure 2 : *A representative example of radiation beams modulated by collimators in an IMRT rectum treatment. PTV : planning target volume.*

Pharmaco-modulation

The chemo-radiotherapy concept gained the status of a global standard for curative radiotherapy across the 1990s. Basically, it consists in impeding the tumour regrowth during the therapeutic course. This effect is tremendously important and can explain most of the effects of regular chemo-radiotherapy. Together with this largely applied concept, we propose to present innovative approaches that are still in their preclinical state of development.

First clinical trials with photoactivated drugs

Mainly based on the enhancement of photoelectric effect, the concept of photoactivation of high Z elements has long been developed by physicists but its application to anti-cancer therapy is recent and can legitimately be attributed to the researchers of the Norman's group [2, 3]. In the 1970s, Norman observed chromosomal aberrations in circulating lymphocytes of patients submitted to urography. Since such a radiodiagnostic treatment involves iodinated contrast agents, he hypothesised that such aberrations resulted from a local excess of radiation dose, due to a maximal energy absorption in iodine atoms contained in contrast agents. This physical effect is notably obtained after irradiating iodine atoms in water at 33 keV [2-4]. Norman proposed to "exploit" these chromosome damaging effects by applying it to tumours in an innovating therapy approach consisting in:

- loading the tumour with iodinated contrast agent;
- imaging the tumour with a modified X-rays scanner system and computing the tumour position;
- treating the tumour with the same irradiation X-ray set up by using the computing data obtained during imaging and by producing photoactivation of iodine contained in contrast agents.

This technique had the considerable advantage of minimizing patient displacement since the beam is mobile. Although such a strategy did not overcome the problem of chromosomal aberrations in normal tissues (observations on which this technique was based), it opened the wide field of photoactivation of pharmacological compounds containing high Z elements. This technique has been applied to animals with a limited success and to humans in a single clinical trial combined with a standard treatment [2,3]. To our knowledge, since 1990, no other clinical trial has been performed with this technique, maybe due to its relative toxicity for normal tissues [4].

Synchrotron radiation rescues the medical application of photoactivation

In addition to the crucial choice of the photoactivable drugs, the main problem for triggering photoactivation is to produce stable and sufficiently intense monochromatic X-rays, tunable to the appropriate energy. As mentioned above, Norman's group had modified a standard scanner machine to perform experiments with iodinated contrast agents but the energy was not tuned [3]. In the 1990s, the development of synchrotron in the medical research area solved this problem. Briefly, synchrotrons are electrons accelerators that provide extremely high-fluenced X-rays resulting from changing direction the GeV electrons by bending magnets. The synchrotron X-rays fluence is about 10^6 times higher than that of standard medical irradiators and renders possible the production of monochromatic X-rays by refraction on appropriate crystals. According to the type of inserted devices, the produced synchrotron X-rays range from a few to hundred keV permitting a theoretical photoactivation of a number of high-Z elements (at least heavier than iron). However, once solved the problem of production of high-fluence monochromatic X-rays, physicists and physicians faced another difficulty: while an X-ray scanner source turns around patients during diagnostic exams, the synchrotron X-ray beam is necessarily horizontal and fixed. Consequently, in order to ensure a homogeneous distribution of radiation dose, engineers developed a turning set-up allowing circular irradiation of samples and patients [5,6].

With regard to the choice of photoactivable drugs, platinum-containing drugs such as cisplatinum, carboplatinum and oxaliplatinum appeared early to be the best candidates for photoactivation of chemotherapeutic agents since:

- they contain platinum atoms making possible their photoactivation at 78.4 keV, corresponding to the K-edge of platinum;
- they are extensively used with success in various chemotherapy and chemo-radiotherapy treatment;
- they target more specifically the proliferating cells with a powerful penetration capacity into cells;
- they induce severe DNA adducts preventing successful repair in tumour cells [5,6].

Recently, photoactivation of cis-platinum (called PAT-Plat) provided by synchrotron X-rays was applied to rats bearing radioresistant gliomas [6]. After a cisplatinum intratumoral injection, a dose of 15 Gy X-rays was delivered by synchrotron radiation into tumour at a tuned energy corresponding to the platinum K-edge (78.4 keV). This treatment resulted in the cure of 33% rats and provides still to date the most protracted survival of rats bearing gliomas [6]. New developments of photoactivation of heavy elements are in progress and clinical trials in humans may be seriously envisaged and scheduled in the coming years. In parallel, intensive *in vitro* experiments still are necessary to propose a molecular model of the mechanisms involved in the PAT-Plat approach in order to secure its clinical transfer.

Why PAT-plat approach is more efficient to kill tumours?

DNA double-strand break (DSB) is a significant lesion that, if unrepaired, can result in loss of genetic material (cell lethality), or if misrepaired, can cause genomic rearrangements, (potential cancer onset) [7]. Mammalian cells possess two major mechanisms for repairing DSB (Chapter 15): homologous recombination (HR), mediated by the RAD51 protein and non-homologous end-joining (NHEJ), dependent upon the DNA-PK protein trimeric complex [7]. NHEJ acts predominantly during G1 phase and its impairment is likely to be associated to marked radiosensitivity and immunodeficiency since it is essential for the production of immunoglobulins [7]. In parallel, HR is a minor DSB repair pathway, alternative to NHEJ. HR functions to repair breaks that arise during the S/G2 phase (Chapter 15). Defects in HR have been identified to be more likely in cancer-proned diseases and in cells sensitive to alkylating agents [7].

Briefly, NHEJ consists in the translocation of the Ku heterodimer all along DNA (Chapter 15). At a break site, the Ku dimeric protein, component of the DNA-PK complex, clamps and recruits the catalytic subunit of DNA-PK (DNA-PKcs). Altogether, Ku and DNA-PKcs forms the serine-threonine kinase DNA-PK that phosphorylates a number of substrates, leading to the recruitment of ligases that are able to join DNA-ends. Interestingly, whereas NHEJ is the major DSB repair pathway in mammalians, the translocation of Ku all along DNA makes it sensitive to any sterical blockage and to large gaps in DNA [7]:

- the first situation (DNA-binding molecules) is notably encountered with cisplatinum-induced DNA adducts that are known to inhibit the NHEJ process [8]. Consequently, association between ionising radiation producing DSB and cisplatinum may render the radiation-induced breaks irreparable (**Fig. 3A**) as far as:
    - the concentration of DNA adducts is sufficient but not toxic for normal tissues ;

- the succession of the radiotherapy session and the cisplatinum treatment are kinetically close [5,6];

– the second situation (large gaps) can be encountered with hadrontherapy that allows very dense energy clusters responsible for large holes in DNA.

In the case of PAT-Plat conditions (**Fig. 3B**), the presence of cisplatinum DNA-adducts in tumours prevents the repair of the radiation-induced DSBs. Furthermore, PAT-Plat results in the emission of low-energy electrons (LEE; Chapter 13) from cisplatinum molecules (for 3% of the atoms), and the range of such electrons is limited to DNA surrounding cisplatinum molecules. Consequently, additional DSBs directly due to the photoactivation are induced at the close vicinity of DNA adducts, making them irreparable by NHEJ. These specific PAT-Plat-induced DSBs are likely to enhance the therapeutic index of the PAT-Plat approach [5,6]. However, as mentioned above, although playing a minor role in the DSB repair of mammalians, HR may serve as an alternative to NHEJ and partly reduce the efficiency of the PAT-Plat strategy. The relevance of this hypothesis is in progress by testing different models of tumours. One can however consider to date that HR-deficient tumours are preferentially targeted by PAT-Plat approach whereas therapeutic index is lower for HR-proficient ones (**Fig. 3**).

Toward anti-cancer treatments specific to each individual case

The PAT-Plat strategy is a representative example of innovating anti-cancer strategy that endeavours to associate benefit of both radiation and chemotherapy. However, such association raises the problem of co-toxicities with concomitant induction of different DNA damage types that are repaired by different and sometimes interplaying repair pathways. Hence, pharmaco-modulation of DNA repair necessarily implies a differential tumour targeting according to their genetic status. In parallel, the tumour shape may be also strongly dependent upon each individual case and these two physical and chemical approaches raise an actual strategic problem: the systematic repair gene mutations screening in tumours is required in order to ensure that each individual case receives the most specific and appropriate anti-cancer treatment. Personalised anti-cancer treatments will undoubtedly be one of the major challenges of the radiotherapy of the future.

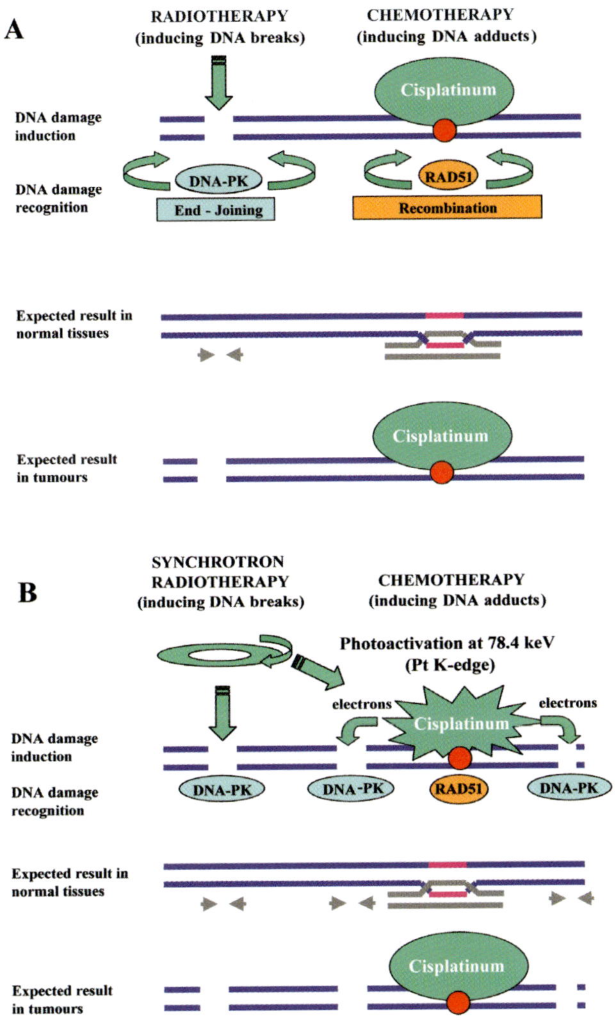

Figure 3 : ***DNA damage in radio-chemotherapy.*** **A)** ***A molecular explanation for the synergetic effects of a radio-chemotherapy with cisplatinum.*** *The expected result of any radio-chemotherapy is to induce unrepairable DNA damage into tumours while sparing normal tissues. Radiotherapy induces DNA double-strand breaks (DSB) that are mainly managed via the NHEJ repair pathway which is active in G_0 / G_1 cells (Chapter 15). Chemotherapy with cisplatinum induces large volume DNA adducts that preferentially target proliferative cells. Interestingly, when the DNA adduct is sufficiently close to a DNA break, the DNA-PK protein, that is essential for repairing DSBs by NHEJ, is sterically blocked. Consequently, the radiation-induced DSBs become unrepairable.* **B)** ***The X-ray photoactivation of cisplatinum additionally enhances the synergy of radio-chemotherapy. Synchrotron X-rays induce also DSBs at the same rate as standard radiotherapy.*** *However, by using X-rays synchrotron at an energy tuned at the K-edge of platinum (78.4 keV), low energy electrons (LEE) are emitted locally from platinum atoms, which play the role of a radiosensitizer, and induce additional DSBs in the close vicinity of DNA adduct sites. Consequently, such unrepairable DSB are more numerous and localized than in a standard radio-chemotherapy. An extra-synergy is therefore expected into tumour treatment.*

Acknowledgements

We would like to thank all our Ph.D. students whose contributions to the development of the PAT-Plat approach was crucial: S. Corde, A. Joubert, M.-C. Biston, Z. Bencokova and J. Gastaldo. This work was also supported by the *Association Pour la Recherche sur l'Ataxie-Telangiectasie* (APRAT), the *Association pour la Recherche contre le Cancer* (ARC), the *Électricité de France* (*Comité de Radioprotection*), the ETOILE Project (*Région Rhône-Alpes* and Lyon University), the *Région Rhône-Alpes* and ROCHE-France.

References

[1] Mazeron J.-J., Maugis A., Barret C., Mornex F., Technique d'irradiation des cancers à La radiothérapie conventionnelle. Maloine, Paris, 2005.

[2] Norman A., Adams F., Riley R., Cytogenetic effects of contrast media and triiodobenzoic acid derivatives in human lymphocytes. Radiol., 1978, **129**, 199-203.

[3] Rose J.H., Norman A., Ingram M., Aoki C., Solberg T., Mesa A., First radiotherapy of human metastatic brain tumors delivered by a computerized tomography scanner (CTRx), Int. J. Radiat. Oncol. Biol. Phys., 1999, **45**, 1127-1132.

[4] Joubert A., Biston M.C., Boudou C., Ravanat J.L., Brochard T., Charvet A.M., Esteve F., Balosso J., Foray N., The concomitant use of iodinated contrast media and ionizing radiation may result in severe DNA breaks and radiosensitization of endothelial cells, Int. J. Radiat. Oncol. Biol. Phys., 2005, **62**, 1486-1496.

[5] Corde S., Balosso J., Elleaume H., Renier M., Joubert A., Biston M.C., Adam J.F., Charvet A.M., Brochard T., Le Bas J.F., Estève F., Foray N., Synchrotron photoactivation of cis-platin (PAT-Plat) elicits an extra-number of DNA breaks that stimulate RAD51-mediated repair pathways., Cancer Res., 2003, **63**, 3221-3227.

[6] Biston M.C., Joubert A., Adam J.F., Elleaume H., Bohic S., Charvet A.M., Estève F., Foray N., Balosso J., Cure of Fisher rats bearing radioresistant F98 glioma treated with cis-platinum and irradiated with monochromatic synchrotron X-rays, Cancer Res., 2004, **64**, 2317-2323.

[7] Chu G., Double-strand break repair, J. Biol. Chem., 1997, **272**, 24097-24100.

[8] Turchi J.J., Henkels K.M., and Zhou Y., Cisplatin-DNA adducts inhibit translocation of the Ku subunits of DNA-PK, Nucleic Acids Res., 2000, **28**, 4634-4641.

Index

O

P

R

W

X

Y

Z

Achevé d'imprimer sur les presses de l'Imprimerie BARNÉOUD
B.P. 44 - 53960 BONCHAMP-LÈS-LAVAL
Dépôt légal : mai 2008 - N° d'imprimeur : 804088
Imprimé en France